Biophysical Analysis of Membrane Proteins

Edited by
Eva Pebay-Peyroula

Related Titles

Tamm, L. K. (ed.)

Protein-Lipid Interactions

From Membrane Domains to Cellular Networks

2005

ISBN: 978-3-527-31151-4

Nierhaus, K. H., Wilson, D. N. (eds.)

Protein Synthesis and Ribosome Structure

Translating the Genome

2004

ISBN: 978-3-527-30638-1

Schliwa, M. (ed.)

Molecular Motors

2003

ISBN: 978-3-527-30594-0

Biophysical Analysis of Membrane Proteins

Investigating Structure and Function

Edited by
Eva Pebay-Peyroula

WILEY-VCH Verlag GmbH & Co. KGaA

The Editor

Prof. Eva Pebay-Peyroula
Institut de Biologie Structurale
CEA-CNRS-Université J. Fourier
41, rue Jules Horowitz
38027 Grenoble Cedex 1
France

Library of Congress Card No.:
applied for

British Library Cataloguing-in-Publication Data
A catalogue record for this book is available from the British Library.

Bibliographic information published by the Deutsche Nationalbibliothek
Die Deutsche Nationalbibliothek lists this publication in the Deutsche Nationalbibliografie; detailed bibliographic data are available in the Internet at <http://dnb.d-nb.de>.

Composition SNP Best-set Typesetter Ltd., Hong Kong

Printing Betz-Druck GmbH, Darmstadt

Bookbinding Litges & Dopf GmbH, Heppenheim

Cover Design Adam Design, Weinheim

Printed in the Federal Republic of Germany
Printed on acid-free paper

ISBN: 978-3-527-31677-9

Contents

Biophysical Analysis of Membrane Proteins. Investigating Structure and Function. Edited by Eva Pebay-Peyroula

ISBN: 978-3-527-31677-9

Preface

Membrane proteins are known to be key molecules in cellular communications, from signal transduction to ion exchanges or transport of metabolites and other molecules. They also participate in the synthesis of ATP, by generating the proton gradient necessary for the rotatory motor of ATP-synthetase to function and to catalyze ATP formation from ADP and inorganic phosphate. Membrane proteins are necessary for the import of soluble or membrane proteins from the cytosol, where they are synthesized into various compartments such as the mitochondrial matrix or outer and inner mitochondrial membranes. Living organisms have also designed efficient machineries that protect cells from toxic elements. Bacteria or eukaryotic cells have, in their membranes, efflux pumps that will clean the cell. The efflux of toxic elements also has drastic consequences for the efficiency of drugs that may find difficulties in penetrating the cell in order to be active. In contrast to soluble proteins, membrane proteins are embedded in a medium which is organized continuously from the atomic level (at the nanoscale) to the micron range. However, the mesoscopic organization of membranes influences, through long-range effects, the properties of the molecules that are embedded in the membranes. Therefore, an understanding of the function of membrane-integrated molecular machineries necessitates a description of the proteins on the atomic level, their various conformations, their specialized organization, as well as their dynamics within the membrane.

Despite attracting great interest, membrane proteins are still difficult to study at the molecular level. Indeed, they are difficult to produce, to extract from their natural environment, and to purify in a native conformation. However, during the past decade efforts have been stepped up worldwide such that several new structures have been resolved at high resolution and their details published within the past two to three years. All of these structures have opened a wide field of discussion about the function and the topology of membrane proteins, their interactions with lipids, the need for such interactions, interactions with ligands or cofactors, and a large number of functional mechanisms could be postulated. At the same time, it has also become clear from the results of many studies that, even with very high-resolution structures, the atomic details were insufficient to understand the function. Further information was needed on the identification and characterization of different conformations, on the dynamics that are necessary for

Biophysical Analysis of Membrane Proteins. Investigating Structure and Function. Edited by Eva Pebay-Peyroula

ISBN: 978-3-527-31677-9

conformational changes, on how membrane proteins are inserted in their natural environment, and on how they are organized within the membrane. Although, crystallography represents an extremely powerful method by which to describe the atomic structures of proteins, an ensemble of complementary biophysical approaches is essential in order to fully describe the structure–function relationships of proteins in general, and of membrane proteins in particular.

This book will serve as a cutting-edge resource for the biophysical methods that are – or soon will be – the major techniques used in the field. Each chapter is dedicated to a specific approach, describing the method involved, highlighting the experimental procedure and/or the basic principles, and offering an up-to-date understanding of what is measured, what can be deduced from the measurements, as well as the limitations of each procedure. This comprehensive reference book will be helpful to junior scientists whose target is to solve structure–function problems associated with membrane proteins, an will surely guide them in their experimental choices. Indeed, this book will also serve as a resource for anybody who is interested in membranes.

Following a general introduction to membrane protein structures and X-ray crystallography, the book is divided in five sections. Part I (the Introduction) is dedicated to structural approaches, while in Part II, Chapter 2 describes several aspects of electron microscopy either on single particles or on two-dimensional and tubular crystals, and Chapter 3 illustrates the current possibilities of NMR, and their future. Part III is centered on molecular interactions and the study of large molecular assemblies, with Chapter 4 illustrating how analytical ultracentrifugation can be used to address the study of membrane proteins solubilized in detergent micelles. Chapter 5 discusses how surface plasmon resonance – a well-known method used to study molecular interaction with soluble proteins – can also be adapted to membrane proteins. Molecular interactions and the topology of large assemblies of membrane proteins, either in reconstituted systems or in natural membranes, can also be studied by using atomic force microscopy, as shown in Chapter 6. Part IV is focused on dynamics, either by computational or experimental approaches. Here, Chapter 7 illustrates the possibilities of molecular dynamic calculations, while Chapter 8 describes how transport pathways can be followed by free energy calculations and Chapter 9 highlights the power of neutron scattering for studying membrane protein in their natural environment. Part V focuses on spectroscopies of various types. For example, circular dichroism can be extended to membrane proteins, as shown in Chapter 10, whilst infrared or Raman spectroscopy is able to probe either global folding properties or very fine local information, as demonstrated in Chapters 11 and 12, respectively. Finally, Part VI is devoted to functional approaches in whole cells, wherein Chapter 13 explains the possibilities offered by FRET or BRET experiments.

The Editor

Eva Pebay-Peyroula is a professor in the Physics Department at the University of Grenoble. Having gained her PhD in molecular physics in 1986, Prof. Pebay-Peyroula began working the Laue-Langevin Institut, where her interests shifted from physics to biology. Subsequently, after studying the structural properties of lipidic membranes, mainly by neutron diffraction, she moved into the field of protein crystallography, which in turn aroused an interest in membrane proteins. During the past years, Prof. Pebay-Peyroula's main area of study has included light-driven mechanisms achieved by bacterial rhodopsins, membrane proteins from archaeal bacteria and, more recently, the ADP/ATP carrier, a mitochondrial membrane protein. Currently, Prof. Pebay-Peyroula heads the *Institut de Biologie Structurale* in Grenoble and, since 2005, has belonged to the French Academy of Science.

Biophysical Analysis of Membrane Proteins. Investigating Structure and Function. Edited by Eva Pebay-Peyroula

ISBN: 978-3-527-31677-9

List of Contributors

Mohammed-Akli Ayoub
Institut de Génomique Fonctionnelle
CNRS UMR5203
Universités de Montpellier 1 & 2
34000, Montpellier
France

John E. Baenziger
Department of Biochemistry, Microbiology, and Immunology
University of Ottawa
451 Smyth Road
Ottawa
ON K1H 8M5
Canada

Nikolay Buzhynskyy
Institut Curie
UMR168-CNRS
26 Rue d'Ulm
75248 Paris
France

Christophe Chipot
Equipe de Dynamique des Assemblages Membranaires
UMR CNRS/UHP 7565
Université Henri Poincaré
BP 239
54506 Vandoeuvre-lès-Nancy cedex
France

Corrie J. B. daCosta
Department of Biochemistry, Microbiology, and Immunology
University of Ottawa
451 Smyth Road
Ottawa
ON K1H 8M5
Canada

Christine Ebel
CNRS, IBS
Laboratoire de Biophysique Moléculaire
41 rue Jules Horowitz
38027 Grenoble Cedex 1
France

Rui Pedro Gonçalves
Institut Curie
UMR168-CNRS
26 Rue d'Ulm
75248 Paris
France

John Holyoake
Department of Biochemistry
University of Oxford
South Parks Road
Oxford
OX1 3QU
United Kingdom

Biophysical Analysis of Membrane Proteins. Investigating Structure and Function. Edited by Eva Pebay-Peyroula

ISBN: 978-3-527-31677-9

Nadège Jamin
CEA/iBiTecS/SB^{2}SM et URA CNRS 2096
Laboratoire des protéines membranaires
CE Saclay Bat 532
91191 Gif sur Yvette Cedex
France

Szymon Jaroslawski
Institut Curie
UMR168-CNRS
26 Rue d'Ulm
75248 Paris
France

Syma Khalid
Department of Biochemistry
University of Oxford
South Parks Road
Oxford
OX1 3QU
United Kingdom

Jean-Jacques Lacapère
INSERM U773
Centre de Recherche Biomédicale Bichat-Beaujeon (CRB3)
Faculté de Médecine Xavier Bichat
16 rue Henri Huchard
BP 416
75018 Paris
France

Marc le Maire
CEA, iBiTecS
Service de Bioénergétique Biologie
Structurale et Mécanismes
Laboratoire Protéines membranaires
91191 Gif sur Yvette Cedex
France

Damien Maurel
Institut de Génomique Fonctionnelle
CNRS UMR5203
Universités de Montpellier 1 & 2
34000, Montpellier
France

Jesper V. Møller
Institute of Physiology and Biophysics
University of Aarhus
Ole Worms Allé 1185
8000 C, Aarhus C
Denmark

David G. Myszka
Center for Biomolecular Interaction Analysis
School of Medicine 4A417
University of Utah
Salt Lake City
Utah 84132
USA

Iva Navratilova
Center for Biomolecular Interaction Analysis
School of Medicine 4A417
University of Utah
Salt Lake City
Utah 84132
USA

Andrew Aaron Pascal
Institut de Biologie et de Technologie de Saclay (iBiTec-S)
CEA Saclay
91191 Gif sur Yvette Cedex
France

Eva Pebay-Peyroula
Institut de Biologie Structurale Jean-Pierre Ebel
Université Joseph Fourier-CEA-CNRS
41 rue Jules Horowitz
38027 Grenoble cedex 1
France

Julie Perroy
Institut de Génomique Fonctionnelle
CNRS UMR5203
Universités de Montpellier 1 & 2
34000, Montpellier
France

Jean-Philippe Pin
Institut de Génomique Fonctionnelle
CNRS UMR5203
Universités de Montpellier 1 & 2
34000, Montpellier
France

Rebecca L. Rich
Center for Biomolecular Interaction Analysis
School of Medicine 4A417
University of Utah
Salt Lake City
Utah 84132
USA

Bruno Robert
Institut de Biologie et de Technologie de Saclay (iBiTec-S)
CEA Saclay
91191 Gif sur Yvette Cedex
France

John L. Rubinstein
Research Institute
The Hospital for Sick Children
555 University Avenue
Toronto M5G 1X8
Canada

Mark S. P. Sansom
Department of Biochemistry
University of Oxford
South Parks Road
Oxford
OX1 3QU
United Kingdom

Simon Scheuring
Institut Curie
UMR168-CNRS
26 Rue d'Ulm
75248 Paris
France

Klaus Schulten
Theoretical and Computational Biophysics Group
Beckman Institute
University of Illinois at Urbana-Champaign
Urbana
Illinois 61801
USA

Christopher G. Tate
MRC Laboratory of Molecular Biology
Hills Road
Cambridge CB2 2QH
United Kingdom

Eric Trinquet
CisBio International
BP 84175
30204 Bagnols sur Cèze cedex
France

Krisztina Varga
University of Oxford, Department of Biochemistry
South Parks Road
Oxford OX1 3QU
United Kingdom

Anthony Watts
University of Oxford,
Department of Biochemistry
South Parks Road
Oxford OX1 3QU
United Kingdom

Katy Wood
Institut Laue Langevin
6 rue Jules Horowitz
BP 156
38042 Grenoble cedex
France

Giuseppe Zaccai
Institut Laue Langevin
6 rue Jules Horowitz
BP 156
38042 Grenoble cedex 9
France

Part I
Introduction

1
High-Resolution Structures of Membrane Proteins: From X-Ray Crystallography to an Integrated Approach of Membranes

Eva Pebay-Peyroula

1.1
Membranes: A Soft Medium?

Membranes delineate cells and cellular compartments, and are efficient barriers that allow the compartmentalization necessary for the functional specificity of each cell or organelle. Membranes are mainly composed of lipids and proteins. As a first approximation, lipids – which spontaneously form bilayers in water – ensure the mechanical properties of the membranes, such as shape, watertightness, robustness and plasticity, whereas proteins are responsible for the communications between compartments or cells, and ensure signaling, channel or transport activities. In fact, membranes are much more complex, and proteins also participate in mechanical properties whereas lipids play a role in the function. Some integral membrane proteins such as the ATP-synthetase located in the inner mitochondrial membrane are described to induce a local curvature of the membrane by dimerization, and could therefore be responsible for the topology of this membrane [1]. Membrane-associated proteins such as clathrin, and associated proteins, by coating the membrane of vesicles formed during endocytosis, may also strongly influence the mechanical properties of the membrane [2]. Likewise, lipids are described now as important players in the function. For example, phosphatidylserine is known to be exposed at the surface of apoptotic cells and used as a signal for the immune system to eliminate the cell [3]. Various sugars participate also both in the mechanical properties and functional aspects of membranes. These play major roles in molecular recognition as illustrated by the role of heparan sulfate molecules [4]. Among all the molecular components of biological membranes, proteins are the only ones to be structured at an atomic level. With the exception of a few individual lipids that are tightly bound to proteins, most of the lipids are organized within a bilayer that can be described at a so-called "mesoscopic" scale by a mean bilayer thickness, a surface area per lipid, lipid order parameters describing chain dynamics and possibly local domain structures [5]. Strong thermal fluctuations of each individual molecule within the membrane make an atomic description irrelevant. Therefore, membranes must be described

Biophysical Analysis of Membrane Proteins. Investigating Structure and Function. Edited by Eva Pebay-Peyroula

ISBN: 978-3-527-31677-9

at various scales in order to take into account all the molecular components and the high protein concentration of some membranes where proteins account for 50% or more of the membrane [6]. Structure–function analyses of membrane proteins at an atomic level are thus of major importance, and shed light on major cellular processes, signaling pathways, bioenergetics, the control of synaptic junctions, and many others. Indeed, membrane proteins in general – and G-protein-coupled receptors (GPCRs) and ion-channels in particular – are known to be the target for many drugs (60% of drug targets are estimated to be membrane proteins). However, despite high potential interest – both for fundamental understanding and also for pharmaceutical applications such as drug design – very little is still known regarding the structure of membrane proteins compared to soluble proteins. This lack of information reflects the difficulty of producing large quantities of stable membrane proteins and crystallizing them in order to solve their structure by using X-ray diffraction (XRD). In addition, the functional state of membrane proteins is often tightly linked to their natural environment, a lipid bilayer, with some individual lipids bound specifically to the proteins. In order to proceed to structural studies it is first essential to mimic, as best as possible, the natural environment.

1.2 Current Knowledge on Membrane Protein Structures

1.2.1 An Overview of the Protein Data Bank

Currently, amongst more than 40 000 entries, the "Protein Data Bank" (PDB) contains about 250 membrane protein structures, representing at least 120 unique proteins (http://blanco.biomol.uci.edu/Membrane_Proteins_xtal.html). Since 1985, when the first membrane protein structure – a photoreaction center from *Rhodopseudomonas viridis* – was resolved [7], the number of structures solved per year has increased almost exponentially, with the progression resembling that of soluble proteins with a slight shift toward lower values [8]. This progression is rather encouraging, and has resulted from the large-scale efforts undertaken recently in several countries. Several programs dedicated to the structural genomics of membrane proteins were started. Some of these are based on large networks and focus on the exploration of various expression systems and on the set up of automated procedures that facilitate these explorations. Smaller networks help to share expertise on membrane protein biochemistry and the physical chemistry of amphiphiles and lipids, and favor interdisciplinary developments that are valuable for structural and functional studies of the proteins in a natural environment.

Most of the structures deposited in the PDB were solved by X-ray crystallography to typical resolutions ranging from 3.5 to 1.5 Å. A few structures were solved by using electron diffraction with two-dimensional crystals. Among these, bacteriorhodopsin – a light-activated proton pump, which is well ordered in two dimensions in the native membrane – was the first membrane protein structure to be determined [9], and was later solved at a resolution of 3.5 Å, or better [10, 11]. Although the nicotinic acetylcholine receptor could never be crystallized in three dimensions, two-dimensional (2-D) tubular arrangements allowed the structure to be solved at 4 Å resolution, revealing the overall topology [12]. Electron diffraction was also used successfully for aquaporins, with AQP4 – a water channel from rat glial cells – being solved to 1.8 Å resolution [13]. As described in Chapter 2, electron microscopy (EM) provides an alternative structural method for membrane proteins, in some cases with a lipidic environment that is close to the native one. This is of particular interest when proteins are present as oligomers in the membrane, possibly in a lipid-dependent manner. EM is also relevant for the characterization of structural modifications that are more likely to be induced in 2-D crystals than in 3-D crystals where crystal contacts might hinder larger movements, as demonstrated for bacteriorhodopsin. More recently, a few structures were reported that had been solved with nuclear magnetic resonance (NMR), including three β-barrel proteins from the *Escherichia coli* outer membrane in dodecylphosphocholine (DPC) or octyl-glucoside micelles [14–16], and one human helical protein, phospholamban, from the sarcoplasmic reticulum [17]. NMR also represents a very useful approach for probing ligand pockets and detecting structural modifications induced by ligand binding [18].

1.2.2
Protein Sources for Structural Studies

The majority of membrane proteins for which structures were solved are derived from bacterial sources, and less than 20% of these are eukaryotic. Indeed, some are specific to bacteria, and solving their structures might create new openings for antibiotics or bioremediations. Others can be used as models for eukaryotic homologues (ion channels, ABC transporters). Unfortunately, even if these models are able to provide the first insights into important structural features, they are certainly not informative enough to provide a full understanding of the functional mechanisms and specifically of functional mechanisms that might achieve an efficient drug design. There remains a broad range of membrane proteins of new classes or different functions and/or different species for which structures are needed. Despite many efforts, the expression of membrane proteins remains a hazardous task, with success relying on the outcome of many different investigations [19, 20]. Obvious restrictions to overexpression result from the limited volume of membranes compared to the cytosol when expressing soluble proteins. Insertion and correct folding in the membranes are also non-trivial issues that must be addressed with appropriate signals within the amino-acid sequence of the protein. Investigations into insertion mechanisms are still under way (e.g., Ref. [21]). Finally, expressing a protein at high level in the membrane causes significant perturbation to the cell, and this often causes a highly toxic effect. However, among the success stories, many bacterial proteins have been expressed in large quantities in *E. coli*; indeed, only recently several eukaryotic proteins were expressed in sufficient quantity and quality in heterologous systems to allow crystallization and structure determination. For example, a voltage-gated potassium channel, Kv1.2, from *Rattus norvegicus* was expressed in *Pichia pastoris* and its structure solved to 2.9 Å resolution [22]. Elsewhere, a plant aquaporin, soPIP2;1, from spinach was also expressed in *Pichia pastoris* and solved to 2.1 Å resolution in its closed state and 3.9 Å in its open state [23]. A recent breakthrough of heterologous expression was achieved for the sarcoplasmic Ca-ATPase. This protein, which is highly abundant in the rabbit sarcoplasmic reticulum, was purified from the native membrane and extensively studied, leading to the structures of different conformations from which a functional mechanism was postulated. Recently, the protein was overexpressed in yeast, whereupon it could be purified, crystallized and the structure solved, thus opening the way to functional and structural studies of mutants, which serve as an essential link in a complete structure–function analysis [24]. These examples of recent successes in heterologous expression demonstrate that such as approach is possible, and that the rate of success depends not only on the exploration of various expression systems but also on the knowledge of the biochemical behavior of the protein itself.

1.2.3
The Diversity of Membrane Protein Topologies

The only structural motifs of membrane-inserted peptides are α-helices and β-barrels Transmembrane helices (TMH) are identified by hydrophobic scoring from the protein sequence. β-barrels are found in bacterial outer membranes and are more difficult to predict from their amino-acid composition, although recent progress in β-barrel prediction has emerged. Extensive internal hydrogen bondings in α-helices and β-barrels ameliorate the high energetic cost of dehydrating the peptide bonds, which is necessary for the insertion of peptides into membranes [25]. Although very few membrane proteins are known to be structurally organized in multi-domains, the structures currently available in the PDB highlight the diversity of transmembrane arrangements. TMH bundles create various topologies, depending on the tilt and the kinks that are possible for each individual helix. Some examples of overall membrane protein structures are illustrated in Fig. 1.1. Heteromeric or homomeric associations of TMHs also contribute to the variety of membrane protein topologies. Setting apart proteins with a single TMH (for which the TMH is mainly a membrane anchor and in some cases is responsible for signal transduction through protein dimerization), most membrane proteins that have a function in the membrane have more than six TMHs. Channels are constituted by more than eight TMHs (an octamer of one TMH for WZA, tetramer of two TMHs for various potassium channels, pentamer of two TMHs for the nicotinic receptor). In these examples, channels are formed by several TMHs, each of which is derived from one of the monomers, whereas transport pathways can also be formed within a single monomer of several TMHs (seven TMHs for aquaporins, six for the mitochondrial ADP/ATP carrier, and 12 for lactose permease), which in turn form multimers in the native membrane (tetramer for aquaporins, dimer for lactose permease). Currently, DsbB (a component of a periplasmic

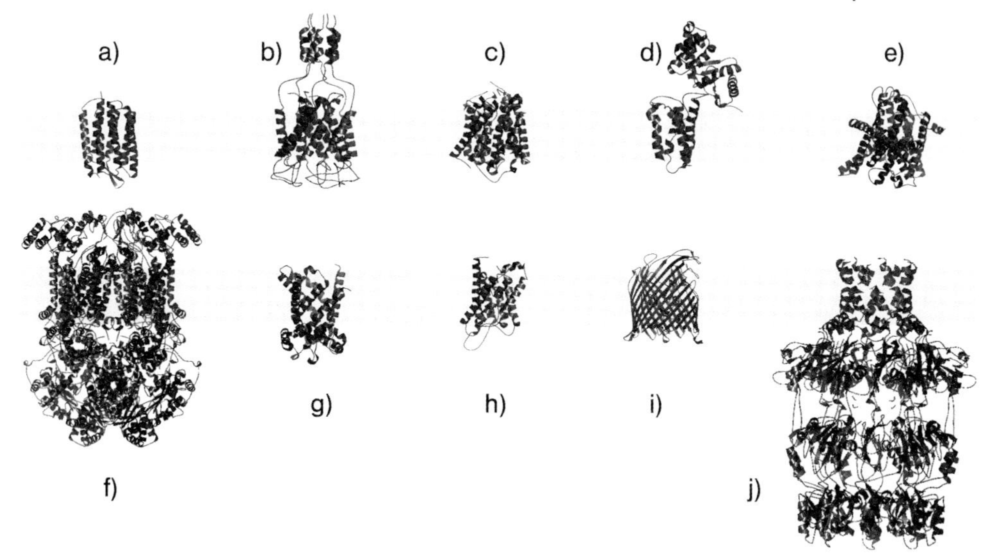

Fig. 1.1 Various topologies of membrane proteins. The Fig. depicts several α-helical proteins showing the diversity of transmembrane helices, and one β-barrel protein. (A) Monomer of bacteriorhodopsin (BR), BR forms a trimer (1qhj). (B) Mechanosensitive channel, a homopentamer with 10 TMHs (1msl). (C) Monomer of the Ammonium transporter AmtB, 11 TMHs per monomer, forms a dimer (1u77). (D) DsbB, four TMHs (2hi7). (E) The protein-conducting channel SecY, heterotrimer with 12 TMHs in total (1rhz). (F) The cytochrome bc1 complex from bovine heart mitochondria, 11 subunits and 12 TMHs per monomer, forms a dimer (1bgy). (G) The ADP/ATP carrier from bovine heart mitochondria, six TMHs (1okc). (H) Monomer of the AQP1 water channel from bovine blood, a homotetramer with seven TMHs per monomer (1j4n). (I) FptA, a pyocheline receptor from the *Pseudomonas aeruginosa* outer membrane, representative for β-barrel structures (1xkw). (J) WZA, the first α-helical protein characterized from the *E. coli* outer membrane (2j58).

oxidase complex with four TMHs) is a membrane protein of known structure, which has the smallest number of TMHs. However, this protein is known to interact with another membrane protein, DsbC, and therefore in the native membrane the total number of TMHs present in the functional complex might be higher. The structure of membrane proteins in a lipidic environment might be energetically more favorable to a larger number of TMH helices. Indeed, it was proposed that helix associations are probably driven by van der Waals interactions through helix–helix interactions rather than hydrophic effects such as those which lead to the folding of soluble proteins [26]. Such stabilizing van der Waals interactions could thus be favored by a larger number of TMHs.

The functional properties of membrane proteins, when driven by dynamic properties, will also constrain the topology. The main role of α-helices in the transmembrane domains of photosynthetic complexes is to locate precisely all of the pigments necessary for the efficiency of photon absorption and their conversion into an electron transfer. The dynamics of such helices must therefore be limited. In contrast, transporters which have to shuttle large metabolites in a very specific manner over the membrane, must undergo large conformational changes that necessitate the molecule to be highly dynamic. Based on these extreme examples, it is easy to imagine that the number of TMHs of the functional entity within the membrane will play a crucial role.

1.2.4
Genome Analyses

What can be learned from the genome data available so far? The analyses of the genomes were performed in order to identify membrane proteins and to classify them into families. A recent analysis showed that membrane proteins cluster in fewer structural families than do their soluble counterparts [27]. However, because of the physical constraints of the lipidic environment, this smaller number of families is rather logical; indeed, some authors have even proposed that membrane proteins have 10-fold fewer families [28]. For example, Oberai et al. estimate that 90% of the membrane proteins can be classified into 1700 families and are structured with 550 folds, while 700 families structured in 300 folds cover 80% of the membrane proteins. This study is based on the search of the TM segment defined by hydrophobic sequences, and is therefore appropriate to helical rather than to β-barrel proteins. Furthermore, these authors also noted that their estimate was based on a limited number of known structures, and may have been biased by present knowledge. Today, new features continue to emerge from recent experiments. For example, TMHs were characterized in a bacterial outer membrane protein, WZA, the translocon for capsular polysaccharides in *E. coli* [29]. The eightfold repeat of a single TM of the octameric protein forms a 17 Å pore in the outer membrane, showing the first α-helical-barrel in the outer membrane of *E. coli*. Unfortunately, this example clearly illustrates that our current structural knowledge is still limited, and that further experimentally determined structures will provide new data for global genome analyses.

Further interesting information has emerged from the comparison between the size of the families and current structural knowledge (see Table 1 in Ref. [27]). For the most important families – rhodopsin-like GPCRs (5520 members) and major facilitators (3680 members) – only one and three structures, respectively, have yet been determined. Moreover, the situation is no better for other families– some are completely absent from the PDB, and even if a few representatives of a family are structurally known, the overall fold might not be sufficient to provide an understanding of the functional mechanism and to help derive structure-based drug designs.

1.3
X-Ray Crystallography

This section will briefly describe some general aspects of crystallization and crystallography, after which attention will be focused on those features more specific

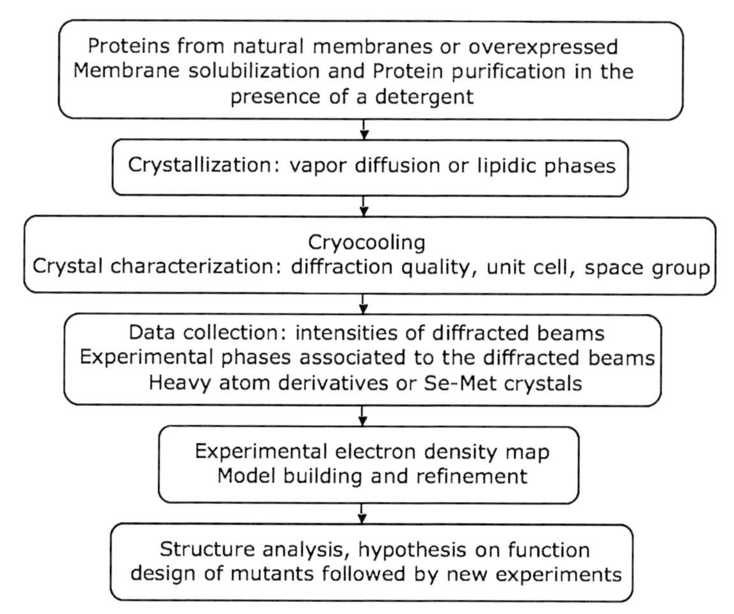

Fig. 1.2 A flowchart for membrane protein crystallography.

to membrane proteins. The basic principles of protein crystallography have been addressed in several handbooks (e.g., see Ref. [30]); the various stages are depicted schematically in Fig. 1.2.

1.3.1
Crystallization of Membrane Proteins

In some membranes large quantities of proteins are present naturally. For example, whilst various proteins from the respiratory chain either in mitochondria or in bacteria can be purified from natural membranes, others must be overexpressed. In both cases, the protein must be extracted – by using detergents – from its natural medium where it is properly folded in the lipidic bilayer – and purified to sufficient quantity and quality.

Detergents are amphiphilic molecules which have a hydrophobic head group and a hydrophobic tail. They are soluble below the critical micellar concentration (cmc), but form aggregates called "micelles" above the cmc. Detergents have the ability to penetrate the membrane, to disrupt it, and then to extract the protein from the membrane by surrounding the hydrophobic part of the protein with their hydrophobic chains, thus creating a soluble protein–detergent complex (PDC). Detergents with different chemical compositions have different properties. For example, their head groups may be either charged, polar or neutral and are more or less bulky, whilst their aliphatic chains can vary in length. As a result of these differences, the micelles will vary in both form and size, though their overall ability to solubilize membranes will remain strong. The choice of detergent used for membrane solubilization and protein purification can have drastic consequences on the protein product, as the amount of lipids co-solubilized with the PDC will depend on the detergent chosen. In addition, the hydrophobic match between protein and detergent micelle is far from being sufficient to mimic the native protein environment. In particular, long-range effects induced by the membrane (lipids or other supramolecular arrangements) are not reproduced. Proteins in micelles are therefore less stable, and very labile proteins will lose their 3-D structure. Thus, it is essential that, prior to any crystallization attempt, the protein solution is characterized as much as possible by using techniques such as negative-stain electron microscopy (Chapter 2), analytical ultracentrifugation (Chapter 4), circular dichroism (Chapter 10), and/or infrared (IR) spectroscopy (Chapter 11). Each of these methods represents an appropriate method by which to explore the protein solutions as a function of the different detergents. Considering the extended time required to produce and crystallize membrane proteins, it is worthwhile first to check for correct folding, stability, monodispersity, conformational homogeneity, and lipid composition. These factors are important because if they can be controlled the amount of protein used for crystallization screening will be reduced and the success rate dramatically increased.

Following the production of pure and stable proteins, the next bottleneck of structure determination is that of crystallization. Previously, most X-ray structures have been obtained using a classical approach which consists of exploring various precipitants by using vapor diffusion methods in the presence of different detergents. The detergent can be exchanged during the purification protocol, although charged detergents should be avoided as they may induce repellent effects between the PDCs and thus discourage crystal nucleation. The aliphatic chains of the detergents must be sufficiently long so as to cover the hydrophobic surface of the protein, while the volume of detergent surrounding the protein should not prevent protein–protein interactions. It should be noted that the crystallization process is somewhat more complex for soluble proteins (Fig. 1.3A); this is because the detergent solution also undergoes phase separation at high detergent and precipitant concentrations, from a micellar phase towards two non-miscible phases (Fig. 1.3B). Crystallization is often benefited by the consolution boundary defining the phase separation, since when approaching this boundary the probability of protein–protein interaction is increased as a consequence of micelle–micelle interactions [31]. This type of mechanism is very sensitive, however, and uncontrolled deviations caused by different detergent concentrations, by impurities in the detergent, contamination due to another conformer of the detergent, or by endogenous lipids present in the protein solution, may lead to non-reproducible results between protein batches.

In recent years a number of alternative crystallization methods based on lipid phases have emerged. Lipidic cubic phases have been particularly successful for bacterial rhodopsins, which have seven TMHs and short extra-membrane loops [32]. Although these phases are difficult to use in the non-specialized laboratory, their automation and miniaturization allows the use of nano-volumes per trial,

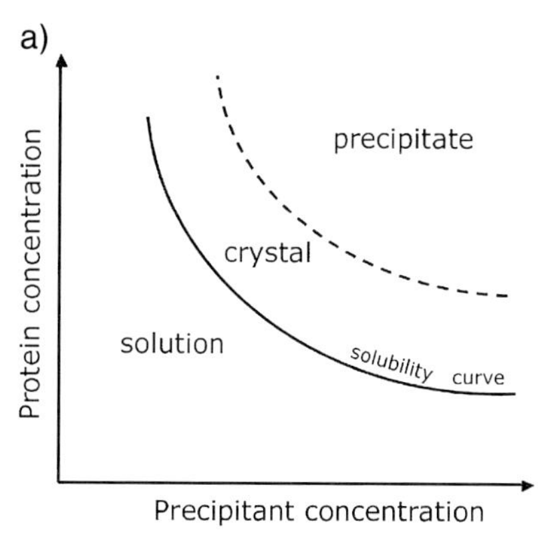

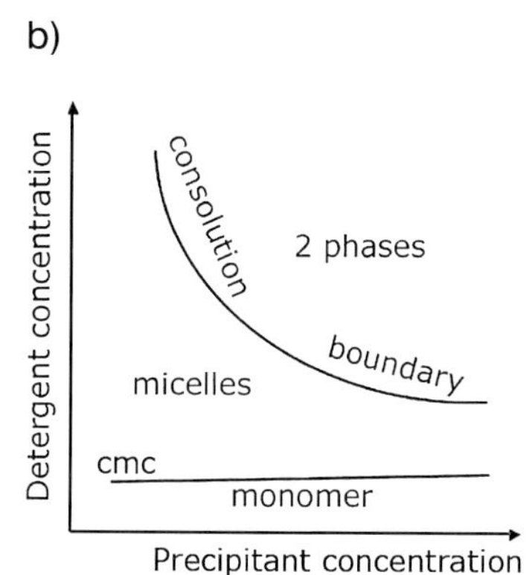

Fig. 1.3 Schematic phase diagrams. (A) Protein concentration as a function of precipitant concentration. Above the solubility curve and below the dashed line, proteins remain soluble in a metastable way. In this region of the phase diagram crystals can be obtained. This scheme is general for soluble and membrane proteins. (B) Detergent concentration as a function of precipitant concentration. Below the critical micellar concentration (cmc), the detergent is soluble as monomers. Above the cmc, it forms micelles. At higher concentrations of detergent and precipitant, the solution undergoes a phase transition and forms two non-miscible phases. For membrane proteins, the phase diagram in (B) will interfere with that represented in (A), adding to the complexity.

and this will undoubtedly contribute to their rapid development [33]. Cubic phases are certainly not the only lipid phases usable for the crystallization, and there are no doubts that the current explorations will lead to new approaches.

The massive efforts in automation that have been undertaken in structural genomics have led to drastic improvements in strategies for membrane protein production and crystallization. These advances have been made not only because the methods allow several conditions to be followed in parallel for the same protein (from cloning to the soluble, stable protein), but also because the crystallization robots in the laboratory are capable of pipetting, reproducibly, aliquots of 50 nL, thus reducing by a factor of 20 the total volume of protein solution required. Hence, currently a typical membrane protein study requires only a few hundred milligrams (typically 100–300 mg) of a pure protein. Moreover, with improved miniaturization and better control of the protein solution in the final purification stages, even this quantity may be reduced significantly, thus opening the field of crystallography to proteins expressed in mammalian cells.

1.3.2 General Aspects of Crystallography

X-ray crystallography for proteins, which was developed during the 20th century, led to the first structure of a soluble protein during the 1960s and to the first membrane protein structure in 1985. Waves interacting with a protein are scattered, and are described by an amplitude and a phase that can be represented by a complex number, the structure factor F(**s**). This factor is related to the molecular structure by:

$$F(\mathbf{s}) = \int_{\text{molecule}} \rho(\mathbf{x}) \exp(-2\pi i\, \mathbf{s}.\mathbf{x}) d\mathbf{x}$$

where $\rho(\mathbf{x})$ is the electron density and **s** the scattering vector. Knowledge of the amplitude and phase of the scattered waves all around the molecule is, in theory, sufficient to calculate the electron density and to characterize atomically the molecule responsible for the scattering. In practice, the signal from one molecule is far too low to be measured and must be increased by taking into account a large number of identical molecules. The principle of crystallography is to amplify drastically the signal by using a 3-D crystal in which a large number of molecules are ordered. The wave scattered by the crystal is the sum of the waves scattered by each molecule in the crystal. All these waves are in phase along well-defined directions, and sum up coherently resulting in strong spots in these directions in which the intensity is increased by a factor N^2, where N is the number of molecules in the crystal. Elsewhere, the resulting wave has a zero amplitude. As a consequence, the so-called "diffraction pattern" is a series of strong spots in directions given by Bragg's law. A diffraction experiment consists of measuring the position and intensity of all the spots. A more complete description of the basic principle of diffraction can be found elsewhere [30].

Protein crystals are analyzed by X-rays produced by a home-source, by a rotating anode usually operating at a wavelength of 1.54 Å, or by synchrotron radiation. Preliminary experiments prior to the final data collection that will be used for the structure determination, are carried out with home sources and used to characterize the unit cell, space group and diffraction quality, to identify the appropriate conditions for freezing the crystals, and to screen for good quality crystals. Synchrotron sources produce a white beam that is much brighter than rotating anodes, with a very low angular divergence. Normal diffraction experiments are performed using a monochromatic beam, with the wavelength being selected by a monochromator. The quality of data collected with synchrotron radiation sources is far better in terms of resolution and signal-to-noise ratios of intensities. Even for routine crystals the final data sets are preferentially collected on such sources. In addition, they allow more difficult cases to be tackled, such as micro-crystals (which typically are up to 5 or 10 μm in size) or crystals with very large unit cells (typically 1000 Å). Very small crystals can be analyzed with an optimized signal-to-noise ratio by collimating the beam to a size that is comparable to the crystal size. Microdiffractometers produce a narrow focused beam, and the appropriate visualization devices allow precise centering of the crystal in the beam. Several membrane protein structures have been solved using microcrystals [34, 35]. Diffraction from crystals with large unit cells can be collected by narrowing the beam with slits and increasing the crystal-to-detector distance. In this way the low angular divergence of the beam will allow the spots to be separated, while large charge-coupled device (CCD) cameras allow data collection in the appropriate resolution range. The success achieved in the highly challenging field of ribosome structural studies demonstrates the efficiency of synchrotron radiation [36].

1.3.3
Determining the Phases Associated with Diffracted Waves

Structures are solved with X-ray crystallography according to two main methods. The first method is based on heavy-atom derivatives (atoms which have a large number of electrons and therefore interact strongly with the incident X-ray beam), and consists of the experimental determination of phases associated with the diffracted beams. The second method – molecular replacement – is based on a known similar structure (or partially similar structure) that is used as an initial model to provide a first set of estimated phases. Since the widespread introduction of synchrotron radiation, the "heavy-atom" method has evolved and currently utilizes not only a large number of electrons of heavy atoms bound specifically to the protein but also the absorption property of X-rays by electrons, with the anomalous signal occurring at a well-defined wavelength for each atom. The use of this method is possible with synchrotron radiation because an adequate wavelength can be selected from the white beam – a procedure which is not possible when using a home X-ray source that is limited to one wavelength. When combined with developments in molecular biology which allow the production of large quantities of engineered proteins, the method of choice for soluble proteins is based on the anomalous signal of selenium incorporated into the protein by replacing methionines by selenomethionines. By tuning the wavelength of the X-ray beam, it is then possible to use the anomalous signal of selenium and to collect diffraction data for one or several wavelengths using methods of either single anomalous dispersion (SAD) or multiple anomalous dispersion (MAD) near the absorption edge of selenium. The structure of many soluble proteins overexpressed in *E. coli*, were solved by using this very elegant approach.

Several user-friendly programs allow the data collection, data treatment and integration of diffraction spots. Experimental improvements – notably, broader and easier access to synchrotron beams, and improvements in data treatment, and in particular a better weighting scheme in phase calculations – now lead to electron density maps of much higher quality from which, in some cases, the protein model is built automatically. As a consequence of the general principles of crystallography, atoms that are sufficiently well ordered in the crystal will be located precisely, whereas flexible loops or flexible N- and C-termini will not be visualized experimentally. Compared to carbon, nitrogen or oxygen atoms, hydrogen atoms have only a single electron and interact weakly with X-rays, and hence are generally not detectable when using XRD.

1.3.4 Structure Determination of Membrane Proteins

1.3.4.1 Crystal Quality

Interestingly, crystals obtained from lipidic phases often grow as small plates, whereas no general tendency is observed in the classical approach. This can be explained by the crystalline arrangement. In lipidic cubic phases, the crystals grow as stacks of 2-D crystal layers that are ordered in the third dimension [37]. The growth rate perpendicular to the layers is often slower because crystal contacts are weaker along this direction, which explains the thinness along this axis. Despite their very small sizes, crystals of bacterial rhodopsins diffract to high resolution, thus demonstrating the high diffraction power of these crystals. In contrast, crystals grown by vapor diffusion from a protein–detergent solution, have no particular crystal packing. However, even with large sizes, their diffraction is in general more limited and the crystal quality is usually poorer than for soluble proteins. This can be explained by the fact that the detergent occupies a non-negligible volume in the crystal (ca. 20%), and might interfere with crystal contacts. Whilst several structures were solved at 3 Å resolution or less, many crystals which are of good appearance diffract to less than 10 Å and consequently are completely useless for structure determination.

1.3.4.2 Phase Determination

In contrast to soluble proteins, molecular replacement is not a method of choice for solving the phase problem because of the low number of known membrane protein structures. MAD or SAD methods based on selenomethionine-containing proteins are not yet used extensively. Except, for some bacterial proteins, few membrane proteins are currently expressed in *E. coli*, and most of the phases have been obtained using the classical "heavy-atom" method which involves soaking the crystals in heavy-atom solutions. These chemical compounds (whether ions or small molecules) are able to diffuse into the crystals and eventually bind to specific sites. This method tends to be destructive towards crystals because the soaking solution, by definition, is different from the solution in which the crystals were grown; thus, the diffraction quality is decreased compared to that obtained from Se-Met-containing crystals. In addition, binding sites for heavy atoms are more restricted for a soluble protein of equal size or weight, because only hydrophilic surfaces are accessible to heavy atoms, the hydrophobic surface of the protein being surrounded by amphiphilic molecules such as detergents or lipids. If hydrophobic compounds are used they may easily diffuse within the detergent phase and become trapped by the many unspecific interactions that compete with specific binding sites.

1.3.4.3 Crystal Freezing

Finally, in order to limit the radiation damage that may occur during XRD experiments (this is especially important when using synchrotron radiation), the crystals are maintained at 100 K under a cold nitrogen stream. This low temperature reduces the diffusion of free radicals created by the intense beam, and therefore limits their destructive power. For all types of protein, cryocooling of the crystals is achieved under conditions where water is transformed to vitreous ice rather than crystalline ice, in order to avoid changes in its specific volume and to maintain crystal integrity. For membrane proteins, the amphiphilic molecules as well as water will undergo phase transitions that will modify their specific volumes and perturb crystal contacts. In general, the freezing of membrane protein crystals is more difficult and, indeed, for some crystals the appropriate conditions could not be identified. In such cases, the diffraction data are collected by reducing the crystal temperature to 4 °C, which is a compromise but does limit the radiation damage to some extent. It is also worthwhile investigating smaller crystals (and even microcrystals), as these are often easier to freeze but can still be used on synchrotron beams.

Complementary approaches, such as neutron scattering (see Chapter 9), provide in some cases very powerful means of locating hydrogen atoms at high resolution by standard neutron crystallography [38]. In addition to the spatial position, XRD experiments also provide a thermal parameter (the B factor or temperature factor) that sums up both statistical disorder (different atomic positions from one unit cell in the crystal to the other) and dynamic fluctuations. Water, detergent or lipid molecules will only be visible if their interactions with the protein are sufficient such that the molecules are ordered at the atomic level in the crystal. These interactions are possible for individual molecules [39], and even if more detergent or lipids are present in the crystal, X-ray crystallography will not show the bulk organization of such molecules. Low-resolution neutron diffraction is, however, able to locate the detergent organized around the protein in the crystal, as shown in Fig. 1.4 [40]. In addition, as illustrated in the recent examples below, crystallography will provide high-resolution structures of one possible conformation of the protein that was crystallized. A key point here is that if the conformation is correlated to a functional state it will serve as a strong basis for postulating functional mechanisms that must be confirmed using other experimental approaches.

In conclusion, membrane protein crystallography is more difficult to perform than soluble protein crystallography, as the crystals are smaller (microcrystals), they have a lower diffraction power, and are more difficult to freeze. Each of these reasons generally entails lower resolution, and electron density maps are therefore more difficult to interpret and models can rarely be constructed automatically. In general, when considering the stages through crystallization, diffraction experiments, phase determination, and model building and refinement, the X-ray crystallography of membrane proteins requires greater expertise than for soluble proteins, where automatic procedures may be highly efficient and serve as "black boxes". The following section illustrates how high-resolution structures are essential when characterizing the precise conformational states of a protein, for obtaining information from such structures, and for postulating the functional mechanisms involved. The limitations and development of these methods are also highlighted.

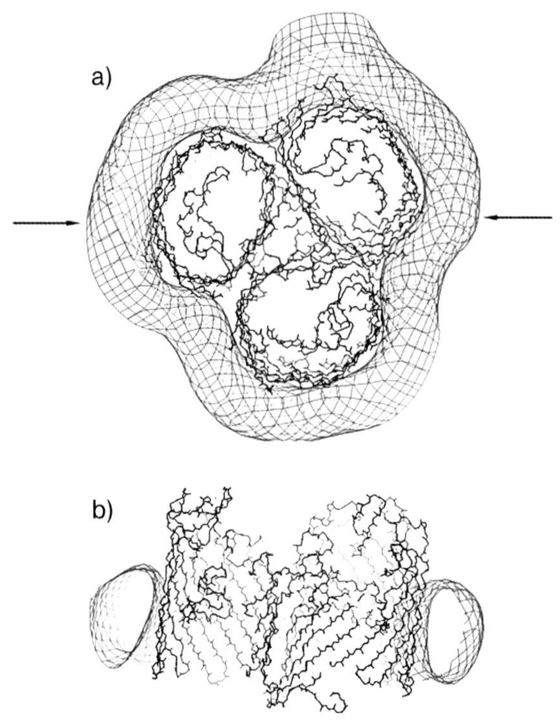

Fig. 1.4 The detergent belt around OmpF trimers. The detergent belt around the porin trimer in the crystals was determined experimentally by neutron diffraction, and is viewed from the top (a), and a section through the detergent–protein complex viewed from the side (b).

1.4 Recent Examples

In this section, two examples of structure–function studies of membrane proteins are described in order to highlight the power of high-resolution X-ray crystallography. More generally, in addition to high-resolution structures, the oligomerization state and/or the formation or larger assemblies of membrane proteins in the membrane are important issues for understanding functional mechanisms. Such studies raise questions that must be further investigated with *in-situ* integrative studies based on methods which are complementary to X-ray crystallography.

1.4.1 Bacterial Rhodopsins

Bacteriorhodopsin was discovered during the 1970s and remains the most studied membrane protein by several means. This bacterial membrane protein, which is found in halophilic organisms, is responsible for the first event in the bioenergetic pathway as it absorbs visible light and in turn expels a proton outside the cytosol. The proton gradient thus created is then used by the rotatory motor of the ATP synthase, which provides the catalytic energy to synthesize ATP from ADP and inorganic phosphate. Because the protein contains a chromophore which absorbs visible light – namely, a retinal that is covalently bound in the center of the protein to a lysine via a Schiff base – various spectroscopies serve as the ideal tools to follow the consequences of photon absorption on the chromophore and its environment. Other similar membrane proteins also utilize light energy via the same chromophore, but achieve completely different functions. For example, *halorhodopsin* is a light-activated chloride pump, while *sensory rhodopsins* are implicated in light sensing. These materials are connected to the machinery of flagellar motors and drive the bacteria in or out the vicinity of the light source as a function of light intensity. Other bacterial rhodopsins with similar structures have been identified and characterized more recently, though their function is not yet clear.

The structure of bacteriorhodopsin was solved from crystals grown in lipidic cubic phases [41, 42]. These highly diffracting crystals served as the "proof of concept" of the method [32, 34]. The structure consists of seven tightly packed TMHs linked by very short loops, and forming a central pocket in which the retinal in its all-*trans* configuration is located (Fig. 1.5A–C). Before the structure was known, extensive absorption spectroscopic studies identified the photocycle of bacteriorhodopsin with different intermediate states in light with the proton translocation. Globally, the photocycle can be divided in two halves. During the first half, the retinal absorbs a photon and changes its conformation to 13-*cis*; this induces local structural rearrangements towards the extracellular side, with proton movements from the Schiff base linking the retinal to the protein, and ultimately a proton being expelled outside the cytosol. All these events occur with small structural modifications of the protein. During the second half of the photocycle, the Schiff base must be reprotonated from the cytoplasm, and this requires much larger structural movements. X-ray crystallography is well suited to determine the high-resolution structures of bacterial rhodopsins in their ground states [41–44]. The comparison of the structures and, in particular, the retinal structures (Fig. 1.5D) and their binding pockets linked to spectroscopic properties or functional consequences are of special interest. X-ray crystallography has also been shown as an appropriate tool to follow fine structural modifications, the movements of water molecules, rearrangements of side chains, and small backbone movements that allow proton translocation to occur. However, the larger structural modifications which occur in the second half of the photocycle are better characterized by electron microscopy [45] or by neutron scattering (see Chapter 9).

1.4.2 ADP/ATP Carrier

The mitochondrial ADP/ATP carrier (AAC) from bovine heart belongs to a family of membrane proteins located in the inner mitochondrial membrane, which are

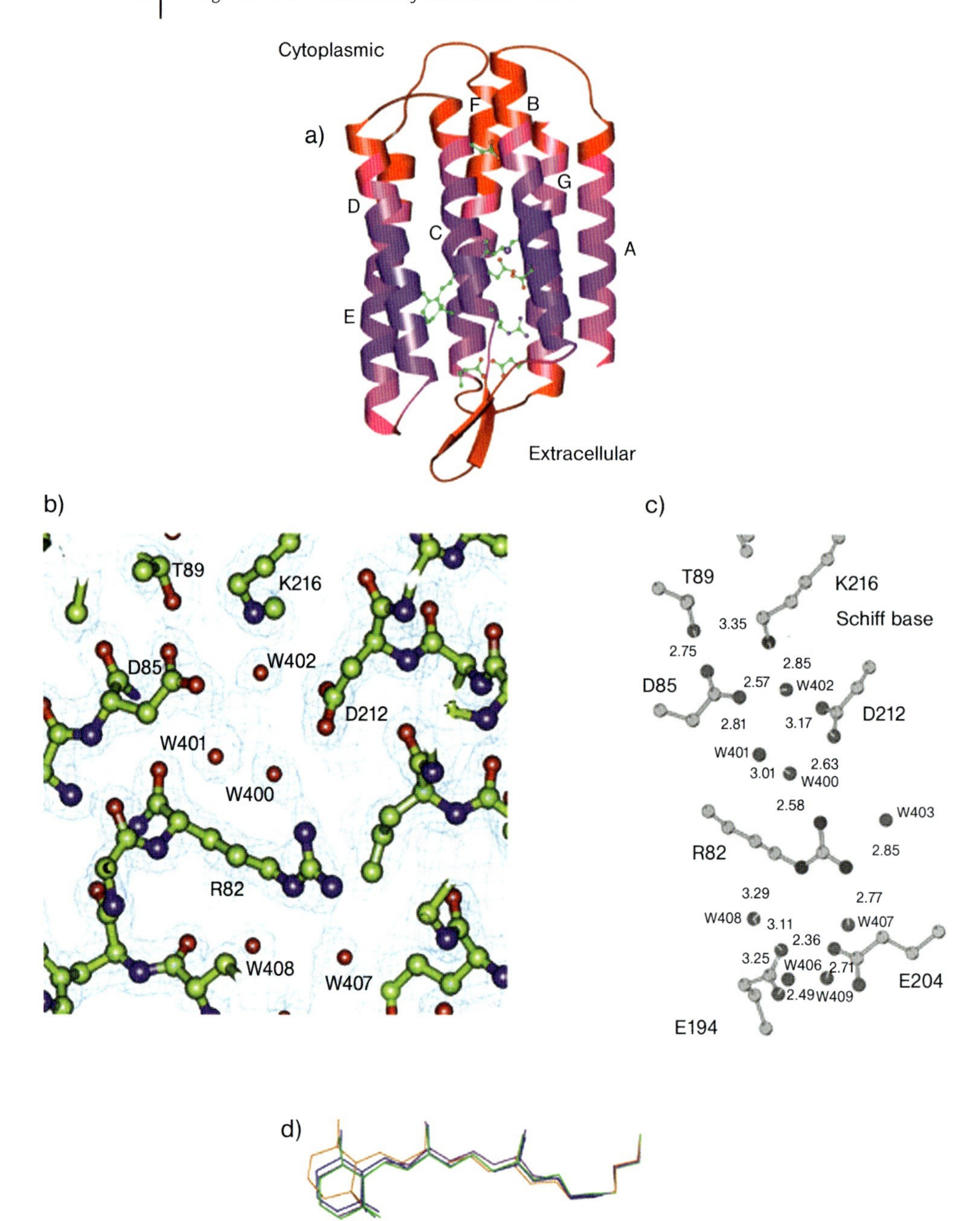

Fig. 1.5 Bacteriorhodopsin (BR). (A) Overall view of BR; the seven TMHs are labeled from A to G. The structure is colored according to the temperature factors (blue and red correspond to low and high factors, respectively). The retinal is positioned in the centre of the protein. (B) Electron density map at 1.9 Å resolution, contoured at 1 , in the vicinity of K216 to which the retinal is bound (not shown). Water molecules are shown as red spheres. (C) Hydrogen-bond network in the putative proton pathway from the Schiff base to the extracellular side of BR where the proton is expelled. (D) Comparison of the retinal structures in purple BR (1qhj), in orange sensory rhodopsin II (1h68) and green halorhodopsin (1e12), as determined from high-resolution X-ray crystallography. (Figs. 1.5A–C reproduced from Ref. [41].)

responsible for the import and export of several important metabolites [46]. AAC was the first protein of the mitochondrial carrier family (MCF) for which the structure was solved (Fig. 1.6). This protein is essential because it exports ATP, which is synthesized within the mitochondrial matrix, and also imports ADP from the cytosol. Because it is highly abundant in the heart mitochondria –representing up to 10% of the inner member proteins – it was extensively characterized and studied during the 1970s by using a variety of biochemical approaches [47, 48].

Although its characterization is an interesting issue in its own right, AAC also serves as a model for other members of the MCF family. For example, the amino acid sequences of these carriers were shown to be composed of three replicates each of about 100 amino acids, resulting most probably from a gene triplication (Fig. 1.6B). Each motif contains a specific amino-acid signature, the MCF motif, which today can be used to identify new MCF members from the newly sequenced genomes [49]. Almost 50 AAC members are present in humans and/or yeast, as well as in plants and other eukaryotic organisms. More recently, a putative MCF carrier was also identified from the mimivirus genome [50]. As yet, however, none of the searches for MCF motifs among all known genomes has ever identified a protein with a single motif, or even with a number of motifs other than three. These findings point to the importance of the three replicates for stability of the overall scaffold. The structure as elucidated by X-ray crystallography, shows that each motif consists of two TMHs linked by a loop containing a short amphipathic helix [51] (Fig. 1.6C). The six TMHs from the three motifs interact strongly, two by two, around a pseudo threefold axis (Fig. 1.6A). All of the TMHs are tilted with respect to the membrane, and the odd-numbered ones are kinked at the level of a proline residue belonging to the MCF motif. The structural analysis shows also that the basic and acidic residues from the MCF motifs are located on the C-terminal end of the odd-numbered helices, and that they interact two by two, forming salt bridges that bridge the three motifs (Fig. 1.6D). As a consequence of the tilts, the kinks and the arrangement of the six TMHs, the protein adopts a basket-like shape which is closed toward the mitochondrial matrix and open toward the inter membrane space, with a cone-shaped cavity ending at 10 Å of the mitochondrial matrix (Fig. 1.6E). It is also important to mention, that the protein extracted from the mitochondrial membrane is known to be unstable and was purified and crystallized in the presence of carboxyatractyloside (CATR), a well-known inhibitor that prevents nucleotide transport by blocking the entry site from the inter-membrane space. Therefore, CATR blocks the carrier in one well-defined conformation, which prevents the protein from being degraded during the purification and allows a homogeneous solution in terms of protein conformation, a prerequisite for the crystallization. The overall topology is a direct consequence of the presence of the three MCF motifs, and it seems most likely that all members of the MCF family share the same overall fold.

It is relatively easy to derive the overall structure of other carriers from AAC. However, due to the relatively low sequence identity (<20%), the models of other carriers cannot easily be used for direct structure–function interpretation. The high-resolution structure of AAC provides a detailed insight of the location of all

a)

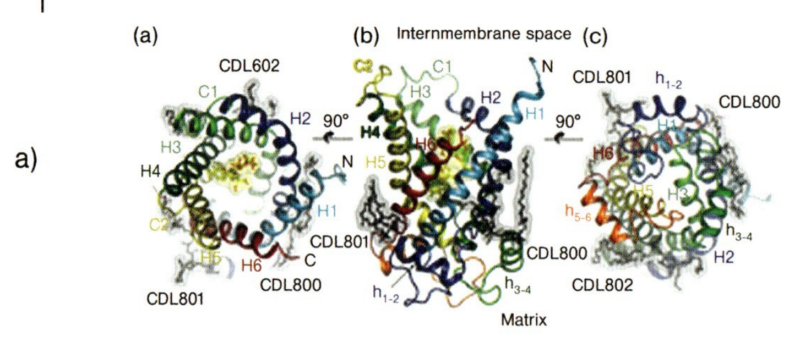
(a)
(b) Internmembrane space
(c)
CDL602
C1
H2
90°
H3
H4
N
H1
C2
H5
H6
C
CDL801
CDL800
Matrix
CDL802
h_{1-2}
h_{3-4}
h_{5-6}

b)

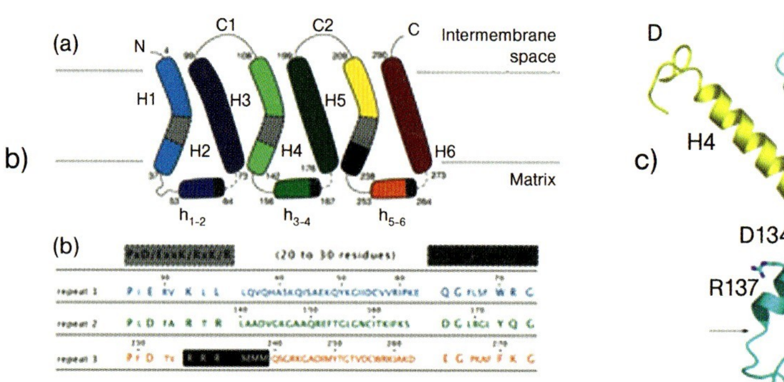
(a)
N
C1
C2
C
Intermembrane space
H1
H2
H3
H4
H5
H6
Matrix
h_{1-2}
h_{3-4}
h_{5-6}
(b)
(20 to 30 residues)
repeat 1
repeat 2
repeat 3
D
Motif 2
H4
H3
c)
D134
P132
R137
h3-4
M2

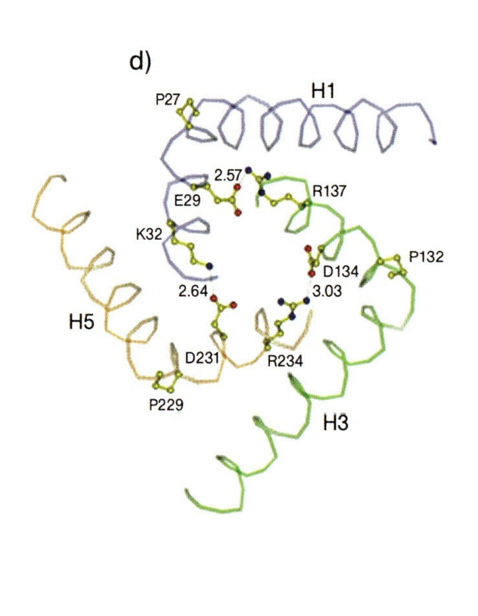
d)
H1
P27
2.57
E29
R137
K32
P132
D134
2.64
3.03
H5
D231
R234
P229
H3

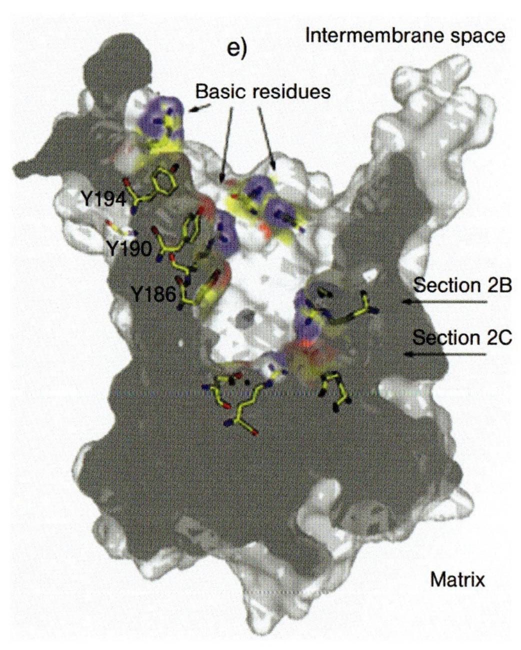
e)
Intermembrane space
Basic residues
Y194
Y190
Y186
Section 2B
Section 2C
Matrix

Fig. 1.6 The mitochondrial ADP/ATP carrier (AAC). (A) Overview of AAC, the ribbon diagram colored blue to red from the N-terminus to the C-terminus shows the six TMHs and three amphipathic helices located at the surface of the lipid membrane on the matrix side. Three cardiolipins are located around the protein. Because of the presence of three repeats in the amino-acid sequence of the protein, the backbone of AAC adopts a pseudo threefold symmetry. (B) Schematic representation of the topology of AAC. Regions containing the MCF motifs are colored in gray, the alignment of the three repeats is shown below with, on the top, the consensus MCF motif boxed in gray. (C) Structure of a single MCF motif showing the two TMHs, the first being sharply kinked at the level of a proline belonging to the MCF motif. Both helices are linked by a loop, which is partially structured as a small helix. (D) Salt bridges linking the three motifs two-by-two and based on acidic and basic residues from the MCF motifs. (E) Surface representation of the cavity. A longitudinal section through the protein highlights the cone-shaped cavity which enters deeply from the inter-membrane space toward the matrix. Conserved basic residues, as well as tyrosine residues, depicted on the Fig. are thought to attract ADP to the cavity. (Figs 1.6A and B reproduced from Ref. [68]; Fig. 1.6D and E reproduced from Ref. [53].)

its residues, which is particularly interesting for those residues located in the cavity. The presence of CATR at the bottom of the cavity not only allows an understanding of its inhibitory mechanism, but also highlights those residues which are important for the binding of ADP from the inter-membrane space – the first event to occur before ADP import. Initially, some of these residues were postulated as being important from functional studies of mutants [52], without knowing their location, whilst others highlighted by the structure were thought to act as selectivity filters [53]. Therefore, the structure represents a major tool in the design of mutants from functional hypotheses derived from the structure, in a rational manner by selecting the amino acids to mutate with respect to structural features, and by choosing those amino acids to be replaced but with minimal perturbation of the structure.

Other important functional issues are the conformational changes that the protein is able to undergo. Several hypotheses can be made from a single conformation structure, on the basis of structural properties of individual amino acids. In particular, proline residues are known to be responsible for kinks that can be modified, while glycine residues located at helical extremities are known to induce flexibility. When considering the conservation of specific proline and glycine residues among MCF carriers, certain helical movements that would allow the transport of metabolites to occur can be postulated. Although, it is exciting to realize these first clues for the transport mechanism, further experimental evidence is an absolute requirement to make progress, and in this respect X-ray crystallography may provide the structure for another conformation, if its crystallization were to be achieved. Other lower-resolution structural methods, such as electron microscopy on two-dimensional crystals [54] (see Chapter 2), might also bring different insights into the conformational landscape of AAC, and could be combined with high-resolution data on one conformation. Clearly, in the near future, NMR spectroscopy (Chapter 3) will become a method of choice for following the binding of various nucleotides and understanding the selectivity of AAC for adenonucleotides. Surface plasmon resonance (see Chapter 5) may also represent an exciting approach for measuring the binding affinities of nucleotides or other molecules. Dynamic data demonstrating the flexible regions of the helices or loops – either experimentally or computationally, as described in Chapters 7 and 8 – will help in our understanding of how conformational changes are possible. In addition, helical rearrangements can be followed by using IR or Raman spectroscopy (see Chapters 11 and 12, respectively). In summary, this example illustrates the importance of high-resolution structures and their implication in our understanding of structure–function relationships. Perhaps more importantly, they also show how such achievements can open new fields for exploration with several complementary, biophysical methods.

1.4.3
Oligomerization of Membrane Proteins in their Natural Environment

The topology and oligomerization of membrane proteins are interesting issues for which recent structures have opened new questions. For example, pseudo-symmetry is observed several times in membrane protein structures, as already illustrated by the pseudo threefold symmetry related to gene "triplication" of the ADP/ATP carrier described in Section 1.4.2. A twofold pseudo-symmetry was also described for the ammonium transporter AmtB [55], for which TMHs M1 to M5 are related to helices M6 to M10 within the monomer by a symmetry in the mid-plane of the membrane. However, in that case no gene duplication is obvious. Similar observations have been made for the lactose permease LacY [56], the SecY protein of the translocon [57], and aquaporins [58]. For the latter, the symmetry was already recognized from the gene analysis.

Several membrane proteins are known to function as oligomers and, indeed, they crystallize as such and their structures can be determined. For ion channels, the functional meaning of oligomerization is clear from the structure, as the heart of the channel is formed by several TMHs, each of which is derived from a different monomer (for a review, see Ref. [59]). In the case of bacteriorhodopsin, the trimer is known, from direct imaging by electron microscopy and other methods, to exist in the native membrane [9]. This trimer is present in the crystal and its structure has been determined, although the reason for such trimerization remains elusive. Bacterial rhodopsins also highlight the diversity in oligomerization among homologous proteins; for example, bacteriorhodopsin is trimeric, sensory rhodopsin II is probably dimeric [60], but the oligomerization of halorhodopsin remains unknown [44]. In other cases – such as the ADP/ATP carrier, the protein import machinery SecY and various GPCRs – dimerization was described as the result of an ensemble of biochemical or functional evidences. However, the high-resolution structures determined were monomeric, and the functional properties could be interpreted from monomers. Moreover, SecY 2D crystals, when analyzed by electron microscopy, showed the presence of a dimer [61], a finding which was supported by biochemical analyses. The X-ray structure determined from 3-D crystals highlighted the existence of a translocation pore in the center of a monomer [57].

Consequently, the authors postulated that SecY forms an oligomer *in vivo*, but that each monomer functions independently. The oligomerization of SecY in the native membrane may create a binding site to recruit other partners.

Several plausible hypotheses may be proposed to explain the discrepancies between high-resolution X-ray structures and biochemical data. For example, the oligomers may have been disrupted during purification, the previous biochemical observations may have been incorrect, or the protein may indeed function as a monomer but be present as an oligomer in the native membrane for better efficiency. A further possibility might be that for improved efficiency and/or stability, the protein must be packed into dense patches locally, without forming oligomers. In all cases, it will be necessary to conduct additional investigations, either on solubilized proteins (e.g., ultracentrifugation; see Chapter 4) or on native membranes or reconstituted systems by electron microscopy (Chapter 2) or atomic force microscopy (Chapter 6). Förster resonance energy transfer (FRET) experiments on proteoliposomes or intact cells in which the protein of interest is expressed with fluorescence labels, also represents an attractive method to ascertain oligomerization within a functional environment (Chapter 13).

1.5 Future Developments in X-Ray Crystallography of Membrane Proteins

Although many developments are currently under way, the crystallography of membrane proteins will clearly continue to improve in the near future. At present, the production of large quantities of pure, homogeneous membrane proteins remains a primary bottleneck which must be overcome. Whilst this problem is not specific to crystallography, and no "magic" method is expected in the short term, an array of positive and negative information should lead us towards a rational approach for exploring parameters in parallel. These include homologous proteins from different sources (possibly from bacteria), vectors and gene lengths, expression systems and conditions, detergents for solubilization, purification and crystallization, and crystallization methods. Today, new *in-vitro* expression systems have shown great promise despite not yet being extensively explored [62]. Notably, expression in mammalian cells is affordable due to the small quantities of protein needed for crystallization. These developments will all require a better understanding of protein–lipid or protein–amphiphile interactions in general, and will benefit from the progress made in protein stabilization. In order to increase the stabilization of proteins outside the natural environment, new amphiphilic molecules must be designed and, if used for crystallization, the protein–amphiphile complexes will need to be of well-defined size. Although the use of such molecules has already been suggested, few investigations have been completed. For example, "amphipols" are amphipathic polymers which adsorb onto transmembrane surfaces of the proteins by multiple attachment points [63]. Although initially designed for biochemical applications, the different molecules may be used for crystallization. In contrast to detergents, it is not necessary to use an excess of micelles in solution in order to maintain the stability of protein–amphipol complexes. Several hemifluorinated surfactants have also been partly investigated [64, 65]; these fluorocarbon surfactants resemble classical detergents, the hydrophobic tail of which is fluorinated rather than hydrogenated. These compounds are also lipophobic, act as poor solvents for lipids, and can therefore be expected to be less delipidating. Other groups have investigated hybrid molecules called "lipopeptides"; these consist of a polypeptide which forms an α-helix linked on both ends to an alkyl chain that will surround the hydrophobic surface of the protein [66]. All of these developments require adequate and efficient systems not only to characterize the proteins in solution but also to control the particles formed by proteins and amphiphiles. Such methods are currently undergoing improvement by adapting well-known techniques to membrane proteins, or by making their use compatible with small volumes, which in turn allows much larger screenings of conditions. The development of solution studies of proteins prior to crystallization will also improve the choice of favorable parameters (protein and detergent concentrations, additives such as lipids, ionic strength, etc.) for the protein solutions used in crystallization studies.

Clearly, the diversification of crystallization methods will create new possibilities, and both methods described in this chapter may prove to be more adequate for a protein of interest, though it cannot yet be predicted which method should be used – or even both in parallel! Today, the classical vapor diffusion technique has become so efficient by using nano-drop robots that almost 1000 conditions can be explored using less than 100 μL of protein solution. However, it also appears that *in fine* a limited number of precipitants are successful, and the number of initial conditions may therefore be restricted and their choice driven by a better knowledge of the physical chemistry of protein–detergent–precipitant solutions. This would favor the crystallization of those proteins expressed in smaller quantities in mammalian cells, and several laboratories are currently designing their own screenings based on such an approach. The lipidic cubic method will also undoubtedly improve and become more efficient by using robots to allow larger sampling of conditions and much smaller quantities of protein for each trial [33]. In parallel, the screening of precipitants must also refined in terms of detecting crystals in the lipidic phase and extracting them from the lipids for diffraction experiments. Several groups are working along these lines, and lipidic phases are showing great promise because the proteins are embedded in a medium close to their natural environment.

Due to the increasing number of known structures, molecular replacement will be used more often for phase determination. Different chemical compounds caging heavy atoms are also the subject of investigation, and may be adapted to membrane protein properties [67]. Studies are also under way to explore the possibility of obtaining diffraction data from micro-crystals [35]. This approach will create new possibilities when crystallization is limited to micro-crystals, or when only diffraction-quality micro-crystals are acceptable. The data must first be derived from a larger number of crystals (at least one image per crystal), and then scaled and merged to obtain a complete data set. Such improvements would not only lead to better experimental data and phase determinations but also provide access to more protein structures and automatic structure determinations.

1.6
Conclusions

Although X-ray crystallography provides a strong structural basis for our understanding of functional mechanisms, this knowledge is insufficient to fully understand functional processes at the level of the native membrane. Biophysical or computational approaches will provide complementary information which – when combined with the high-resolution structure – will lead to a coherent ensemble. Biophysical methods are not only useful for understanding functional mechanisms but are also essential for controlling the quality of a purified protein by analyzing the detergent and lipid content in the protein solution, and by characterizing the size of the particles and the protein oligomerization state. Each of these measurements can be performed while screening various parameters, leading to a rational manner of protein preparation. In addition, further spectroscopic analyses can be used to characterize the conformational state of a protein as a function of ligands, and these represent important methods by which to seek controlled crystallization conditions. Other biophysical methods will enable the study of membrane proteins within membranes, and in particular their spatial arrangements. Molecular dynamics can be monitored with both neutron scattering and NMR, both of which complement computational molecular dynamics. X-ray crystallography provides for high-resolution structures, which provides new insights, raises new questions, and provokes much discussion with regards to oligomerization and the conformational changes necessary for transport, channel, or signaling activities. Answers to all of these questions are important in order to understand the function and to propose alternative mechanisms. And – when used together – these methods will provide an integrated approach of structure and function of membrane proteins.

Acknowledgments

The author is grateful to Grenoble University, the CEA and the CNRS for constant support, and to Dr. Susanne Berthier-Foglar for helpful comments on the manuscript. The author's work is supported by the Agence Nationale pour la Recherche and by the European Community Specific Targeted Research Project grant "Innovative tools for membrane protein structural genomics."

References

1 G. Arselin, J. Vaillier, B. Salin, J. Schaeffer, M. F. Giraud, A. Dautant, D. Brethes, J. Velours, *J. Biol. Chem.* **2004**, *279*, 40392–40399.
2 T. Kirchhausen, *Annu. Rev. Biochem.* **2000**, *69*, 699–727.
3 R. S. Kiss, M. R. Elliott, Z. Ma, Y. L. Marcel, K. S. Ravichandran, *Curr. Biol.* **2006**, *16*, 2252–2258.
4 H. Lortat-Jacob, *Biochem. Soc. Trans.* **2006**, *34*, 461–464.
5 K. Jacobson, O. G. Mouritsen, R. G. Anderson, *Nat. Cell Biol.* **2007**, *9*, 7–14.

6 E. Pebay-Peyroula, J. P. Rosenbusch, *Curr. Opin. Struct. Biol.* **2001**, *11*, 427–432.
7 J. Deisenhofer, O. Epp, K. Miki, R. Huber, H. Michel, *Nature* **1985**, *318*, 618–624.
8 S. H. White, *Protein Sci.* **2004**, *13*, 1948–1949.
9 R. Henderson, P. N. Unwin, *Nature* **1975**, *257*, 28–32.
10 N. Grigorieff, T. A. Ceska, K. H. Downing, J. M. Baldwin, R. Henderson, *J. Mol. Biol.* **1996**, *259*, 393–421.
11 Y. Kimura, D. G. Vassylyev, A. Miyazawa, A. Kidera, M. Matsushima, K. Mitsuoka, K. Murata, T. Hirai, Y. Fujiyoshi, *Nature* **1997**, *389*, 206–211.
12 N. Unwin, *J. Mol. Biol.* **2005**, *346*, 967–989.
13 Y. Hiroaki, K. Tani, A. Kamegawa, N. Gyobu, K. Nishikawa, H. Suzuki, T. Walz, S. Sasaki, K. Mitsuoka, K. Kimura, A. Mizoguchi, Y. Fujiyoshi, *J. Mol. Biol.* **2006**, *355*, 628–639.
14 A. Arora, D. Rinehart, G. Szabo, L. K. Tamm, *J. Biol. Chem.* **2000**, *275*, 1594–1600.
15 C. Fernandez, C. Hilty, G. Wider, P. Guntert, K. Wuthrich, *J. Mol. Biol.* **2004**, *336*, 1211–1221.
16 P. M. Hwang, W. Y. Choy, E. I. Lo, L. Chen, J. D. Forman-Kay, C. R. Raetz, G. G. Prive, R. E. Bishop, L. E. Kay, *Proc. Natl. Acad. Sci. USA* **2002**, *99*, 13560–13565.
17 K. Oxenoid, J. J. Chou, *Proc. Natl. Acad. Sci. USA* **2005**, *102*, 10870–10875.
18 A. Watts, *Nat. Rev. Drug Discov.* **2005**, *4*, 555–568.
19 R. Grisshammer, *Curr. Opin. Biotechnol.* **2006**, *17*, 337–340.
20 C. G. Tate, *FEBS Lett.* **2001**, *504*, 94–98.
21 S. Miras, D. Salvi, M. Ferro, D. Grunwald, J. Garin, J. Joyard, N. Rolland, *J. Biol. Chem.* **2002**, *277*, 47770–47778.
22 S. B. Long, E. B. Campbell, R. Mackinnon, *Science* **2005**, *309*, 903–908.
23 S. Tornroth-Horsefield, Y. Wang, K. Hedfalk, U. Johanson, M. Karlsson, E. Tajkhorshid, R. Neutze, P. Kjellbom, *Nature* **2006**, *439*, 688–694.
24 M. Jidenko, G. lenoir, J. M. Fuentes, M. le Maire, C. Jaxel, *Protein Expr. Purif.* **2006**, *48*, 32–42.
25 N. Ben-Tal, A. Ben-Shaul, A. Nicholls, B. Honig, *Biophys. J.* **1996**, *70*, 1803–1812.
26 J. L. Popot, D. M. Engelman, *Annu. Rev. Biochem.* **2000**, *69*, 881–922.
27 A. Oberai, Y. Ihm, S. Kim, J. U. Bowie, *Protein Sci.* **2006**, *15*, 1723–1734.
28 Y. Liu, M. Gerstein, D. M. Engelman, *Proc. Natl. Acad. Sci. USA* **2004**, *101*, 3495–3497.
29 C. Dong, K. Beis, J. Nesper, A. L. Brunkan-Lamontagne, B. R. Clarke, C. Whitfield, J. H. Naismith, *Nature* **2006**, *444*, 226–229.
30 J. Drenth, *Principles of protein X-ray crystallography*, New York: Springer-Verlag, **1994**.
31 F. Reiss-Husson, D. Picot, in: *Crystallization of nucleic acids and proteins. A practical approach*, pp. 245–268, Oxford University Press, **1999**.
32 E. M. Landau, J. P. Rosenbusch, *Proc. Natl. Acad. Sci. USA* **1996**, *93*, 14532–14535.
33 V. Cherezov, A. Peddi, L. Muthusubramaniam, Y. F. Zheng, M. Caffrey, *Acta Crystallogr. D. Biol. Crystallogr.* **2004**, *60*, 1795–1807.
34 E. Pebay-Peyroula, G. Rummel, J. P. Rosenbusch, E. M. Landau, *Science* **1997**, *277*, 1676–1681.
35 C. Riekel, M. Burghammer, G. Schertler, *Curr. Opin. Struct. Biol.* **2005**, *15*, 556–562.
36 B. T. Wimberly, D. E. Brodersen, W. M. ClemonsJr, ., R. J. Morgan-Warren, A. P. Carter, C. Vonrhein, T. Hartsch, V. Ramakrishnan, *Nature* **2000**, *407*, 327–339.
37 P. Nollert, H. Qiu, M. Caffrey, J. P. Rosenbusch, E. M. Landau, *FEBS Lett.* **2001**, *504*, 179–186.
38 F. Meilleur, D. A. Myles, M. P. Blakeley, *Eur. Biophys. J.* **2006**, *35*, 611–620.
39 C. Hunte, *Biochem. Soc. Trans.* **2005**, *33*, 938–942.
40 E. Pebay-Peyroula, R. M. Garavito, J. P. Rosenbusch, M. Zulauf, P. A. Timmins, *Structure* **1995**, *3*, 1051–1059.
41 H. Belrhali, P. Nollert, A. Royant, C. Menzel, J. P. Rosenbusch, E. M. Landau, E. Pebay-Peyroula, *Structure* **1999**, *7*, 909–917.
42 H. Luecke, B. Schobert, H. T. Richter, J. P. Cartailler, J. K. Lanyi, *J. Mol. Biol.* **1999**, *291*, 899–911.
43 A. Royant, P. Nollert, K. Edman, R. Neutze, E. M. Landau, E. Pebay-Peyroula,

J. Navarro, *Proc. Natl. Acad. Sci. USA* **2001**, *98*, 10131–10136.

44 M. Kolbe, H. Besir, L. O. Essen, D. Oesterhelt, *Science* **2000**, *288*, 1390–1396.

45 S. Subramaniam, M. Lindahl, P. Bullough, A. R. Faruqi, J. Tittor, D. Oesterhelt, L. Brown, J. Lanyi, R. Henderson, *J. Mol. Biol.* **1999**, *287*, 145–161.

46 J. E. Walker, M. J. Runswick, *J. Bioenerg. Biomembr.* **1993**, *25*, 435–446.

47 P. Vignais, M. Block, F. Boulay, G. Brandolin, G. Lauquin, in: *Structure and properties of cell membranes*, pp. 139–179, CRC Press: Boca Raton, FL, **1985**.

48 M. Klingenberg, *Arch. Biochem. Biophys.* **1989**, *270*, 1–14.

49 P. Jezek, J. Jezek, *FEBS Lett.* **2003**, *534*, 15–25.

50 M. Monne, A. J. Robinson, C. Boes, M. E. Harbour, I. M. Fearnley, E. R. Kunji, *J. Virol.* **2007**, *81*, 3181–3186.

51 E. Pebay-Peyroula, C. Dahout-Gonzalez, R. Kahn, V. Trezeguet, G. J. Lauquin, G. Brandolin, *Nature* **2003**, *426*, 39–44.

52 D. R. Nelson, J. E. Lawson, M. Klingenberg, M. G. Douglas, *J. Mol. Biol.* **1993**, *230*, 1159–1170.

53 E. Pebay-Peyroula, G. Brandolin, *Curr. Opin. Struct. Biol.* **2004**, *14*, 420–425.

54 E. R. Kunji, M. Harding, *J. Biol. Chem.* **2003**, *278*, 36985–36988.

55 S. Khademi, J. O'ConnellIII, , J. Remis, Y. Robles-Colmenares, L. J. Miercke, R. M. Stroud, *Science* **2004**, *305*, 1587–1594.

56 J. Abramson, I. Smirnova, V. Kasho, G. Verner, H. R. Kaback, S. Iwata, *Science* **2003**, *301*, 610–615.

57 B. Van den Berg, W. M. Clemons, Jr., I. Collinson, Y. Modis, E. Hartmann, S. C. Harrison, T. A. Rapoport, *Nature* **2004**, *427*, 36–44.

58 D. Fu, A. Libson, L. J. Miercke, C. Weitzman, P. Nollert, J. Krucinski, R. M. Stroud, *Science* **2000**, *290*, 481–486.

59 I. R. Booth, M. D. Edwards, S. Miller, *Biochemistry* **2003**, *42*, 10045–10053.

60 V. I. Gordeliy, J. Labahn, R. Moukhametzianov, R. Efremov, J. Granzin, R. Schlesinger, G. Buldt, T. Savopol, A. J. Scheidig, J. P. Klare, M. Engelhard, *Nature* **2002**, *419*, 484–487.

61 C. Breyton, W. Haase, T. A. Rapoport, W. Kuhlbrandt, I. Collinson, *Nature* **2002**, *418*, 662–665.

62 C. Berrier, K. H. Park, S. Abes, A. Bibonne, J. M. Betton, A. Ghazi, *Biochemistry* **2004**, *43*, 12585–12591.

63 J. L. Popot, E. A. Berry, D. Charvolin, C. Creuzenet, C. Ebel, D. M. Engelman, M. Flotenmeyer, F. Giusti, Y. Gohon, Q. Hong, J. H. Lakey, K. Leonard, H. A. Shuman, P. Timmins, D. E. Warschawski, F. Zito, M. Zoonens, B. Pucci, C. Tribet, *Cell. Mol. Life Sci.* **2003**, *60*, 1559–1574.

64 C. Breyton, E. Chabaud, Y. Chaudier, B. Pucci, J. L. Popot, *FEBS Lett.* **2004**, *564*, 312–318.

65 S. S. Palchevskyy, Y. O. Posokhov, B. Olivier, J. L. Popot, B. Pucci, A. S. Ladokhin, *Biochemistry* **2006**, *45*, 2629–2635.

66 C. L. McGregor, L. Chen, N. C. Pomroy, P. Hwang, S. Go, A. Chakrabartty, G. G. Prive, *Nat. Biotechnol.* **2003**, *21*, 171–176.

67 E. Girard, E. Pebay-Peyroula, J. Vicat, R. Kahn, *Acta Crystallogr. D. Biol. Crystallogr.* **2004**, *60*, 1506–1508.

68 H. Nury, C. Dahout-Gonzalez, V. Trezeguet, G. J. Lauquin, G. Brandolin, E. Pebay-Peyroula, *Annu. Rev. Biochem.* **2006**, *75*, 713–741.

Part II
Structural Approaches

2
Membrane Protein Structure Determination by Electron Cryo-Microscopy

Christopher G. Tate and John L. Rubinstein

Electron cryo-microscopy (cryo-EM) is a versatile technique used to visualize and determine structures ranging in size from subcellular assemblies to small protein molecules. The cornerstone of its success is the ability to image biological structures preserved in their native conformation by freezing the sample so rapidly that the specimen is immobilized in a layer of amorphous, non-crystalline ice. Although originally used to define the structures of large viruses at an intermediate resolution, cryopreservation, coupled to improved imaging and image analysis, has now allowed the structure determination of a membrane protein embedded in a native membrane environment to a resolution of 1.9 Å [1].

This chapter will describe the technique of cryo-EM and its application to the structure determination of integral membrane proteins and membrane protein complexes at resolutions ranging from 1.9 to 30 Å. The divisions within the chapter reflect the different techniques that are used to determine structures from different samples, despite the similarity in the data collection using cryo-EM. The structures of proteins that have masses in excess of about 300 kDa can be determined from merging images of single particles which, in the most favorable cases, allows the assignment of secondary structure elements such as α-helices to the density (Section 2.2). Higher resolution is possible if the protein molecules are arranged in an ordered crystalline array, either as sheets (Section 2.3) or helical tubes (Section 2.4). The first section will introduce the electron microscope, because improvements in the ease of image acquisition and the quality of those images has played a large role in increasing the effectiveness of cryo-EM as a method for structure determination. Space does not permit a discussion about electron tomography techniques that are being developed to study protein complexes and subcellular structures such as mitochondria, and the interested reader is directed to an excellent recent review [2].

Biophysical Analysis of Membrane Proteins. Investigating Structure and Function. Edited by Eva Pebay-Peyroula

ISBN: 978-3-527-31677-9

2.1 Introduction

Electron microscopy (EM) is used extensively in the material sciences to study radiation stable structures at Ångstrom resolution. Unfortunately, the resolution attainable for biological molecules is much lower due to a number of factors, the most important being radiation-induced specimen damage. Electrons are scattered by a molecule in two ways: elastic scattering and inelastic scattering. Elastic scattering, where the electron leaves the specimen with the same energy that it had before interacting, provides most of the useful information about the biological structure in the image. Inelastic scattering leads to the deposition of energy in the specimen, resulting in its destruction during the imaging process. Therefore, the amount of energy deposited by inelastic interactions per useful elastic scattering interaction is the fundamental parameter in determining the utility of various particles (electrons, X-ray photons, neutrons) in imaging techniques. For imaging biological molecules, electrons deposit approximately one thousandth of the energy of comparable X-ray photons per useful scattering interaction [3]. X-ray crystallography of proteins is only possible because the diffraction pattern is an average of the pattern produced by the many (e.g., 10^{12}) molecules in the crystal. Consequently, the dose of radiation for any molecule in the crystal can be kept below the critical dose that would destroy the high-resolution information that is desired. The reduced radiation damage caused by electrons, combined with the ability to easily focus electrons using magnetic lenses, makes electrons an excellent choice of particle for imaging biological molecules. Nevertheless, in order to avoid destroying the specimen during imaging, the number of electrons allowed to interact with the specimen must be limited by using low-dose procedures, which leads to images that suffer from statistical noise. As a result, it is impossible to record an image of a single protein molecule with atomic detail. Instead, high-resolution information can be determined only by averaging many different images of identical molecules in the same orientation. If the images of the individual molecules can be properly aligned, the signal from each molecule will be correlated, while the noise will not be correlated; hence, when averaged, the signal at every resolution increases while the noise gradually cancels out. Due to the short wavelength of electrons, an electron microscope has an extremely large depth of focus. Consequently, the image produced by an electron microscope is, to a good approximation, the projection of the three-dimensional (3-D) structure of the specimen onto the two-dimensional (2-D) plane of the image. However, every infinitesimally small point in the image appears not as a perfect point but instead as the point-spread-function (PSF) of the microscope. Also, phenomena such as specimen charging, drift, and attenuation of the microscope signal at high resolution all contribute to EM images that are less than ideal. Image analysis is used to improve the signal-to-noise ratio in the images and correct for the PSF of the microscope. Well established mathematical procedures are used to build a 3-D model of the specimen from images obtained by EM [4].

2.1.1
The Electron Microscope

The quality of images obtained by cryo-EM is dependent on the quality and type of electron microscope used. There are two types of electron source employed in modern electron microscopes: thermionic; and field emission. Thermionic electron sources, such as tungsten or lanthanum hexaboride, are heated to eject electrons from a filament. These electrons are then accelerated towards the specimen by an electrostatic potential. Field emission guns (FEGs) use the quantum mechanical phenomenon of tunneling to extract electrons from a tip constructed from a specially engineered low work-function material. FEGs can emit orders of magnitude more electrons per unit surface area than thermionic sources with slightly less energy spread, resulting in a spatially coherent electron beam that can be used to produce higher-resolution images. Cryo-EM maintains the temperature of the sample in the microscope below –155 °C so that the ice surrounding the specimen remains unstructured [5]. Improvements in the design of cryo-holders has reduced the amount of drift or vibrational movement due to thermal gradients and external vibrations by placing the specimen in a cassette in the center of the microscope so that it is disconnected from the external environment. Some microscopes are designed to maintain the sample at about –270 °C by using liquid helium as a coolant [6]; this reduces the radiation damage of the sample by around a factor of two, but may introduce the problem of increased specimen charging due to decreased conductivity of the specimen. At present, the use of liquid helium specimen stages is controversial in cryo-EM. Developments improving microscope stability and user interfaces have resulted in commercially available instruments that are relatively easy to use and which are capable of producing high-quality images for structure determination.

Film is still used to record images of radiation stable specimens because it currently offers the best compromise between sensitivity and resolution. However, once an image is captured on film it must be digitized using a high-quality film scanner. Some charge-coupled devices (CCDs) and area detectors can record images with zero background noise [7]; once they are available in a sufficiently large format and with small enough pixels, they will undoubtedly replace film, even for high-resolution imaging.

2.2
Single-Particle Electron Microscopy

The simplest imaginable way that molecules can be subject to structural analysis with the electron microscope is to image randomly orientated individual molecules and to merge all the images to build a 3-D model. The study of protein molecules in this state is known as single-particle EM. Image analysis for this experiment is computationally intensive because all of the information about the location and orientation of the protein particles in the image must be computed from the image

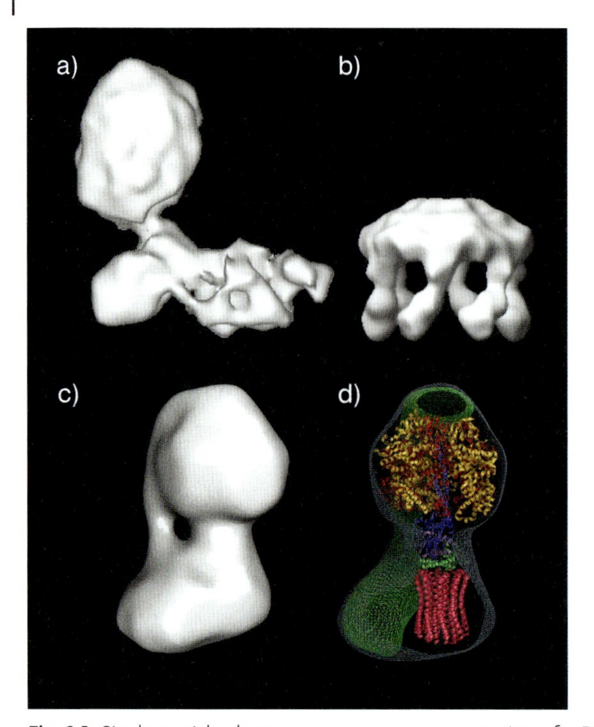

Fig. 2.1 Single-particle electron cryomicroscopy of membrane protein complexes that produce or use a proton motive force. (A) Bovine mitochondrial complex I is the first complex in the electron transport chain that generates the proton motive force (PMF) across the inner membrane of mitochondria. (Reprinted from *J. Mol. Biol.* **277**, 1033–1046, Grigorieff, N. Three-dimensional structure of bovine NADH: ubiquinone oxidoreductase (complex I) at 22 Å in ice; (c) 1998, with permission from Elsevier.) (B) The H^+-ATPase from *Neurospora crassa* couples the hydrolysis of ATP to the generation of a PMF across the cell's plasma membrane for secondary ion and nutrient transport. (Adapted by permission from Macmillan Publishers Ltd: *EMBO J.* **21**, 3582–3589, Rhee et al.; (c) 2002.) (C) The bovine mitochondrial ATP synthase uses the mitochondrial PMF for the generation of ATP. (D) Combining models from X-ray crystallography with the EM map allows for a more detailed understanding of the activity of the enzyme. [Parts (C) and (D) adapted by permission from Macmillan Publishers Ltd: *EMBO J.* **22**, 6182–6192, Rubinstein et al.; (c) 2003.)

itself. The method has been employed to study the structures of a number of large membrane-bound protein complexes, such as the complexes responsible for generating or using a proton motive force across a membrane (Fig. 2.1). Structures of complexes determined by cryo-EM have ranged in mass from several mega Daltons (e.g., [8,9]) to as small as the 350 kDa hetero-tetrameric insulin receptor [10]. The resolutions that have been obtained from these experiments range from ~10 Å [9] to 30 Å [11]. An exciting direction in single-particle EM is the combination of atomic models from X-ray crystallography or NMR spectroscopy within an envelope of an intact complex determined by cryo-EM. This experimental approach is only possible when atomic models of part of a complex and a model of the intact complex from either cryo-EM of negative stained specimens are available, and consequently it has not yet been applied extensively to membrane proteins. This approach has, however, been applied to the mitochondrial ATP synthase [11] and the voltage-gated potassium channel I_{to} [12]. Cryo-EM has also been used to elucidate the mechanism of cotranslational translocation of proteins across a membrane via a protein-conducting channel (PCC) [13]. Initial insight into this process was gained by cryo-EM of eukaryotic ribosome particles in complex with the detergent-solubilized PCC, Sec61 [8] and more recently, a structure was determined for the detergent-solubilized *Escherichia coli* PCC, SecYEG, in complex with a ribosome particle [14]. This experiment demonstrates the utility of cryo-EM to reveal the details of interaction of two protein complexes and was, in this instance, particularly informative because atomic models were available for the ribosome and for a homologue of SecYEG.

2.2.1
Sample Preparation and Requirements

Single-particle analysis of membrane proteins is usually performed on highly purified, detergent-solubilized samples, although there have been proposals to image individual membrane proteins that have been reconstituted into lipid vesicles [15]. The most significant difference between single-particle EM of soluble proteins and membrane proteins is the presence of a bound detergent–lipid micelle around the hydrophobic portion of the purified membrane protein complex and the requirement to maintain detergent in the buffer at a concentration above its critical micellar concentration to ensure that the membrane protein remains in solution in a monodisperse state.

Single-particle EM can be performed on samples that have either been preserved by negative staining or on unstained specimens in vitreous ice. The specimen supports used for either type of experiment are metal grids covered with continuous carbon films, or grids supporting perforated (also known as fenestrated or holey) carbon films. For negative staining, the protein particles are embedded in a film of heavy metal salt stain such as uranyl acetate or sodium phosphotungstate. Protein appears as "holes" in the layer of stain, and the shape of the holes provides information on the outer envelope of the protein structure. Negative staining offers the advantages of easy sample preparation, reasonable stability in the electron beam, and high-contrast images. However, the resolution of the image is limited by the granularity of the stain, and the molecular envelope produced is subject to a variety of distortions and artifacts, such as flattening and differential interaction of the stain with different regions of the protein. For cryo-EM of unstained specimens, a purified protein solution is rapidly frozen to form an amorphous ice containing protein particles [5] that approximates the aqueous environment of the macromolecular complexes in solution. Cryo-EM of unstained specimens offers the potential to produce high-resolution structures of the molecule, but the image of protein against an ice background has less contrast and a lower signal-to-noise ratio than images of protein embedded in stain.

2.2.1.1 Negative Staining of Specimens

The vast majority of negative stain experiments are performed using a continuous carbon film to support the protein particles. The carbon film is made hydrophilic by glow discharge in air, and a drop of protein solution is applied to the surface. Protein particles are allowed to adsorb for a few seconds or minutes, after which any excess solution is blotted away with filter paper. The grid can then be rinsed in different buffers or water to remove unbound protein and undesirable reagents such as glycerol, buffers and detergents that may interfere with staining. Finally, the grid is rinsed with a drop of stain solution, excess stain blotted away, and the remainder allowed to dry to form an amorphous film of heavy metal salt. There are also reports of negatively stained specimens embedded in an unsupported film of stain (e.g., 5% ammonium molybdate mixed with 0.1% trehalose) in a hole in a perforated carbon film-covered grid [16] which may reduce some of the distortions of negative-staining EM, but it also reduces the stability of the specimen in the electron beam.

2.2.1.2 Cryo-EM of Unstained Specimens

As with negative staining, cryo-EM can be performed on specimens supported by a continuous film of carbon. In this case, the grid preparation is identical to negative staining except for the final step where, instead of staining, the grid is plunged into a liquid cryogen (usually liquid ethane), with a guillotine device, to form the layer of vitreous buffer. The continuous carbon film offers the advantages that the detergent used to solubilize the protein, but which may interfere with the formation of amorphous ice, can be removed immediately prior to freezing. In addition, conditions for specimen preparation can be optimized using negative stain and then applied almost unchanged for the cryo-EM experiment. However, the continuous carbon film adds extra noise to the image and may induce preferred orientations of particles, which makes structure determination more difficult. For the preparation of specimens using a perforated carbon film, the film is usually made hydrophilic by glow discharge in air or an amylamine atmosphere. The use of amylamine makes the carbon film positively charged, which may prevent some protein particles from sticking to the carbon film and thus encourage them to populate the buffer that forms a thin film over the holes in the carbon. The excess protein buffer is blotted away and the grid plunged into the cryogen bath. It is still unclear whether continuous or perforated carbon support films can yield the highest-resolution models. A continuous carbon film has traditionally been used for high-resolution studies of the ribosome originating in the laboratory of Joachim Frank [14, 17], whereas unsupported layers have been used for the high-resolution studies of symmetrical particles such as the hepatitis-B virus capsid at a resolution of 7.4 Å [18] and the E2 core of pyruvate dehydrogenase at 8.7 Å resolution [19]. In the cryo-EM analysis of the type I ryanodine receptor (Fig. 2.2), resolution was improved by changing from continuous carbon film to perforated carbon

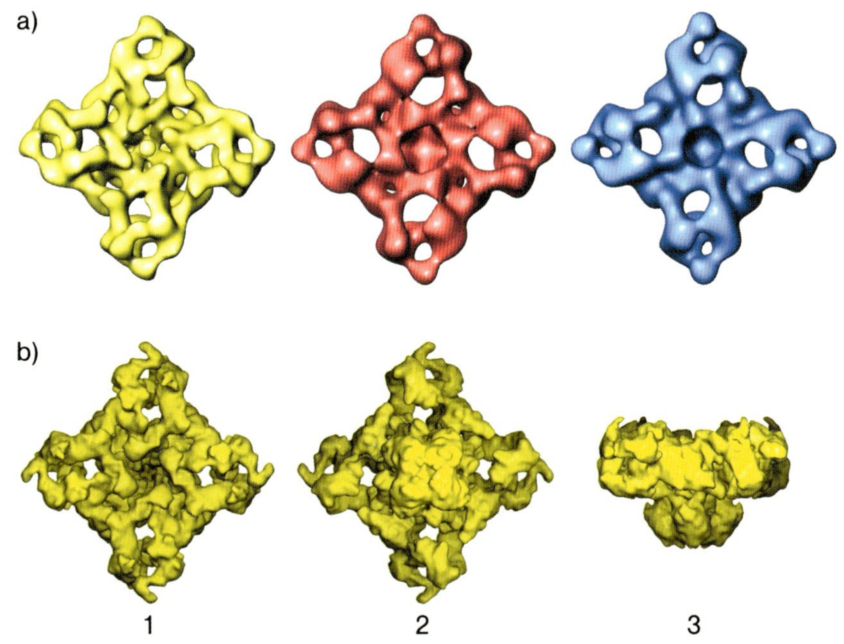

Fig. 2.2 Single-particle electron cryomicroscopy of the Ryanodine receptor. (A) Surface representations of the three isoforms of ryanodine receptor responsible for calcium release from endoplasmic reticulum/sarcoplasmic reticulum. Although expressed more widely than indicated by their nomenclature, the isoforms are RyR1 (yellow), found in skeletal muscle, RyR2 (red) from heart muscle, and RyR3 (blue), often referred to as the brain isoform. The three receptors are shown as viewed from the cytoplasmic face. (Adapted by permission from Unipress Padua: *Basic Appl. Myol.* **14**, 299–306, Sharma and Wagenknecht; (c) 2004.) (B) Cytoplasmic (1), sarcoplasmic reticulum (2) and side (3) views of the RyR1 receptor at ~10 Å resolution. Improved resolution was achieved by improved specimen preparation, use of a dataset containing more particle images and use of more sophisticated image analysis. (Adapted by permission from Macmillan Publishers Ltd: *Nat. Struct. Mol. Biol.* **12**, 539–544, Samsó et al.; (c) 2005.)

film-covered grids [9]. One way in which this transition benefited the model was that it removed much of the tendency of the particles to take on a preferred orientation on the grid, and consequently made the resolution of the model more isotropic. A similar problem with preferred orientation when performing cryo-EM with a continuous carbon film-covered grid was observed for the ATP synthase [11].

Evaporation during grid preparation may cause problems for preparing specimens of frozen hydrated proteins, particularly if detergent is present. During a typical freezing experiment in a non-humidity controlled environment, concentrations can increase by 50% or more [20]. The concentration of detergents can lead to a buffer composition that is incompatible with forming smooth, featureless amorphous ice. This effect can be minimized by performing the freezing in a humidity-controlled freezing chamber [21]. Evaporation can also be reduced by freezing grids in a cold-room or refrigerated cabinet since the vapor pressure of water decreases at lower temperature. Automated freezing devices with control over environmental temperature and relative humidity are now commercially available.

2.2.1.3 Choice of detergent

Several different detergents have been successfully employed in single-particle EM. The list includes the zwitterionic 3-[(3-cholamidopropyl)dimethylammonio]-1-propanesulfonate (CHAPS) [9, 22] and non-ionic polyoxyethylene ether (Brij-35) [11], *n*-dodecyl-β-D-maltoside (DDM) [23, 24] and digitonin [25]. Usually, the detergent used for preparing EM grids is the same one that was chosen for the purification of the molecule. However, it may be necessary or easiest to switch to a different detergent. When working with ATP synthase, it was observed that while there was no problem using DDM for negative staining of the complex [26], the DDM, possibly in combination with other reagents in the protein buffer, led to features in the ice comparable in size to protein particles. The problem was solved by substituting DDM with Brij-35 immediately before preparing cryo-EM grids [11]. Control experiments with a protein of known structure can be used to easily assess the suitability of different detergents for preparing samples of unstained protein particles.

An alternative to the use of detergents for keeping membrane proteins in solution is the use of amphipols [27]. Amphipols are long amphipathic polymers that bind irreversibly to the transmembrane portion of membrane proteins and can stabilize some membrane proteins. The low concentration of free amphipol that is required to maintain the membrane protein in a monodisperse state may be a significant benefit for single-particle EM. Amphipols have been used in the single-particle cryo-EM study of the ATP synthase from *E. coli* [28].

2.2.2 Image Analysis

2.2.2.1 Classification of Images

In the initial stages of image analysis, it is often useful to increase the signal-to-noise ratio of individual images by identifying and averaging noisy images that are the same view of different molecules. Multivariate statistical analysis (MSA) has proven to be an invaluable tool for this task. However, before MSA can be applied, images must be aligned with respect to each other. This alignment is described by one rotation angle and two shifts. There are many ways to carry out the alignment process. A typical approach is to average all the images in the data set and then to align each image to the averaged image. The newly aligned images are averaged again and the process iterated until the position of each image ceases to change. At this point, MSA can be performed in order to group together and average similar images. The resulting class-averages contain a mixture of views of the molecule and incorrectly classified and incoherently averaged images. Often, the *bona fide* views of the molecule are identified interactively (a risky but necessary process if there is no prior information on the structure of the molecule). In some situations, it is possible to use simple logic and symmetry rules to identify those class-averages that are most likely to represent true views of the molecule [11].

2.2.2.2 Model Building and Refinement

In order to build a 3-D model from images of randomly oriented particles, there are five parameters that must be determined for each molecule: the exact position of the center of the molecule on the piece of film or detector (described by two shifts or translations), and three angles that describe the orientation of the molecule with respect to some standard orientation. If these parameters were known exactly, calculations from first principles have shown that atomic resolution structures could be determined for proteins from as few as 500 electron microscope images of randomly oriented frozen-hydrated molecules [19]. At present, it has not been possible to determine atomic resolution maps even from hundreds of thousands of images of individual molecules. It is this requirement of being able to glean enough information from the raw images to determine the orientation of each molecule that places a lower limit on the mass of complexes that can be studied. Contrast in the images is approximately proportional to the mass of the complex. In order to have enough contrast to orient the image of a frozen-hydrated molecule it must be at least 200 000 or 300 000 Daltons. Images may be aligned for negatively stained complexes of lower mass due to the higher contrast in images of protein in negative stain. The process of analyzing images of single particles when building a 3-D map is usually divided into: (i) building a low-resolution initial model; and (ii) refining the model to try to extract as much high-resolution information as possible.

There are two main approaches for determining 3-D structures from single particles. Usually, these approaches require the signal-to-noise ratios that can be achieved by using class-average images rather than raw images.

The random conical tilting (RCT) method [29] becomes useful in cases where a homogeneous population of structures adopt a single orientation on the EM grid, or a subpopulation can be identified that lie in the same orientation. The most frequently used implementation of the RCT method is in the program SPIDER [30]. In practice, particles are picked from a pair of micrographs recorded from the same area of the EM grid. For the first micrograph, the specimen is tilted to a known angle, while for the second micrograph the specimen is kept untilted with the plane of the carbon grid perpendicular to the electron beam. The random azimuthal angle of particles presenting the same view may be determined from the untilted micrograph using the 2-D techniques of alignment described above. The images in the tilted micrograph then comprise a conical projection series with the azimuth known from the 2-D alignment and the cone angle determined by the tilt applied to the specimen. The first micrograph can be taken under strict low-dose conditions. The second micrograph need not be used in the reconstruction. A variation of the RCT method that works well for randomly oriented particles is the orthogonal tilt reconstruction method [31].

The common lines method [32] was originally applied to spherical virus particles with icosahedral symmetry, and remains the method of choice for this type of structural problem [18]. Unfortunately, this robust method finds little application in the study of membrane proteins as there are no membrane–protein complexes with icosahedral symmetry. The common lines strategy was restated in real-space for a more general symmetry case using Radon transforms with the IMAGIC software package [33]. IMAGIC is a user-friendly package that is capable of determining the structures of protein complexes with some degree of symmetry. Indeed, symmetry, by providing redundant information about a protein structure, generally makes image analysis significantly easier. It is occasionally possible to devise an unrelated initial model building method for special cases, such as class-averages that are all approximately views about a single axis of molecular rotation [11].

The most powerful and widely used approach for refining a 3-D model is the projection matching method [34]. Although the algorithm used to carry out this process can vary greatly between implementations, the process is always equivalent to generating a large number of projections of the 3-D model and comparing these projections to the experimental images. By identifying the projection that most closely resembles the image, and the shifts necessary to bring the image into register with the projection, it is possible to identify the five parameters necessary to relate the orientations of different images. A new model can then be built from the images with the newly assigned orientation parameters. The process of projection matching and model building is iterated until the orientations of individual particles stops changing or statistics that describe the quality of the model cease to improve.

2.2.2.3 **Assessing Resolution**

Unlike 2-D, 3-D, or helical crystallography, there are no diffraction spots or layer lines to inform the experimentalist on the resolution of a single-particle EM model. Consequently, the measurement of resolution in single-particle EM consistently generates controversy. The most common approach for determining resolution is the use of a Fourier ring correlation (FRC) chart for 2-D averages or a Fourier shell correlation (FSC) chart for 3-D maps [35, 36]. To generate a Fourier correlation chart, the data set that was used to form the average is divided into two separate data sets. Averages are then calculated from each data set and the correlation between the two averages is calculated as a function of resolution. Typically, the averages will correlate well at low resolution, but the correlation will drop to insignificance at higher resolution. The resolution at which the correlation ceases to be significant is taken to be the resolution at which the 2-D projection or 3-D model should be truncated. A cut-off of 0.5 provides a conservative estimate of the resolution at which correlation remains significant. More recently, a value of 0.143 was shown to be a more appropriate threshold, because it accounts for the fact that the final model contains twice as many particle images as the models used for calculating the correlation [19]. However, it is widely recognized that a FSC or FRC plot has the inherent problem that the two half models are not truly independent, being generated by particles that were aligned to the same model. Therefore, these plots by themselves do not guarantee the resolution or accuracy of a single-particle model.

2.2.3
Future Perspectives

Although the number of membrane–protein complexes analyzed by single-particle EM is increasing, it does not appear to be increasing at the same rate as the number of structures of soluble protein complexes analyzed. Rather than a lack of interest, this difference probably has more to do with the difficulty associated with obtaining purified and detergent-solubilized membrane protein complexes in a stable, monodisperse state. As EM methods develop, they will continue to provide information on all classes of biological macromolecule. However, it is in the area of membrane–protein complexes that some of the greatest advances will likely be achieved, as it is this area that has traditionally resisted structural analysis by other methods. In particular, cryo-EM of unstained specimens has great potential. Currently, maps of protein complexes at intermediate resolution (10–30 Å) can be combined with atomic models of the structures of subunits determined by X-ray crystallography and NMR spectroscopy. By docking atomic models into an EM map, it is possible to build up a high-resolution model of an intact assembly (Fig. 2.1). With membrane–protein complexes that are particularly difficult to crystallize, this approach is often the only means by which information on subunit–subunit interactions can be obtained. Calculations from first principles [3] suggest that it should be possible to build 3-D maps of protein complexes at resolutions sufficient to construct atomic models of protein complexes without using high-resolution models determined by other techniques. The ability to determine the atomic-resolution structure of proteins by single-particle cryo-EM would revolutionize the study of membrane proteins.

2.3
Structure Determination from 2-Dimensional Crystals

The first structure determined for any membrane protein was that of bacteriorhodopsin at 7 Å resolution from 2-D crystals using EM, in the pioneering investigations of Henderson and Unwin in 1975 [37]. It took Henderson and colleagues another 15 years to solve the technical problems in data analysis before an atomic model could be built from data at 3.5 Å resolution [38]. Bacteriorhodopsin is unusual for a membrane protein in many respects, not least that it is found in the archaebacterium *Halobacteria salinarum* as an ordered 2-D array within the cytoplasmic membrane. This naturally occurring 2-D crystal probably arose in evolution because a crystal is the most efficient way of packing together large numbers of molecules in the smallest area for, in this instance, the capture of light by its prosthetic group, retinal. It has now been found that many membrane proteins can be crystallized into 2-D arrays, and the structures have been solved of many channels, transporters and receptors, usually at an intermediate resolution of about 5 to 8 Å [39]. At this resolution it is possible to discern the major secondary structure elements which, for the majority of membrane proteins, is predominantly α-helix. A few membrane proteins form 2-D crystals that diffract to high resolution, allowing the determination of atomic models. It is interesting to note that the two highest resolution structures were determined for bacteriorhodopsin [40] and the lens major intrinsic protein (aquaporin 0, AQP0; Fig. 2.3) [1], and both of these proteins are found in native cells as crystalline arrays.

One major advantage of determining structures of membrane proteins from 2-D crystals is that the membrane protein is present in an environment that is extremely similar to its native environment within the cell. Two-dimensional crystals are grown by reconstituting purified membrane protein into a lipid bilayer, usually composed of a single synthetic lipid such as dimyristoylphosphatidylcholine. In

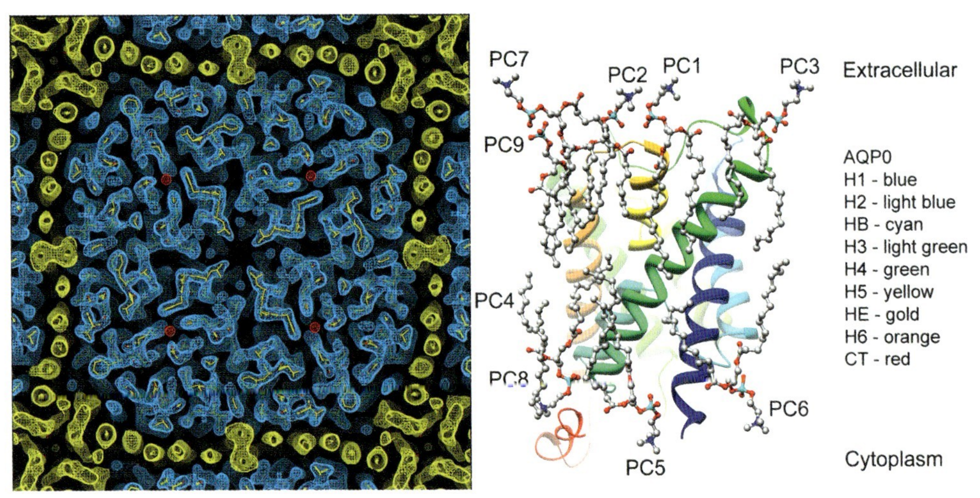

Fig. 2.3 Structure of the major intrinsic lens protein, aquaporin 0 (AQP0) at 1.9 Å resolution. Left: Section parallel to the membrane plane through the AQP0 tetramer (blue); the center of each pore in a monomer is indicated by the position of a water molecule colored red and the lipids surrounding each tetramer are in yellow. Right: Ribbon diagram of an AQP0 monomer viewed parallel to the membrane plane, with bound lipids depicted in a ball-and-stick representation. (Fig.s reprinted by permission from Macmillan Publishers Ltd: *Nature* **438**, 633–638, Gonen et al.; (c) 2005.)

contrast, structure determination by X-ray crystallography determines the structure from a 3-D crystal of a membrane protein in detergent solution (see Chapter 1). In most cases, the structures of membrane proteins determined by X-ray crystallography are considered to be an accurate model of a conformation of the protein as found in nature. When 3-D structures have been determined by both techniques, there is an obvious similarity between the two structures, and it is possible to dock the high-resolution X-ray structure into the α-helical densities from the cryo-EM structure (e.g., see Refs. [41–43]). However, there are three cases where the structures determined by X-ray diffraction (XRD) are thought not to represent an accurate representation of a biologically relevant conformation [44–46]. In the case of EmrE, a 3-D cryo-EM structure determined from 2-D crystals is considerably different from two X-ray structures [46]. In another instance, the distortion of the voltage sensor domain in an X-ray crystallographic structure of the potassium channel KvAP was directly attributed to the instability of this domain in detergent solution compared to the lipid bilayer [47]. Two-dimensional crystallography provides a solution to these difficulties by directly determining the structure of a membrane protein in a lipid bilayer.

Conformational changes in membrane proteins have been defined by comparing the differences in structure after the addition of substrates to 2-D crystals. In the case of bacteriorhodopsin, projection maps of photointermediates throughout the photocycle were compared and the structural changes defined as the movement of two α-helices [48, 49]. These conformational changes did not affect the packing of bacteriorhodopsin trimers in the 2-D crystal and, similarly, packing within 2-D crystals of NhaA was unaffected by the small conformation changes upon increasing the concentration of the H^+ substrate [50]. In contrast, the addition of substrates such as tetraphenylphosphonium and ethidium to 2-D crystals of the multidrug transporter EmrE induced a conformational change that destroyed the order of the 2-D array and induced repacking of the functional dimers into a crystal with a different space group [51]. In the case of bovine rhodopsin, the conformational change between metarhodopsin I and metarhodopsin II does not occur fully in the 2-D crystal, possibly either due to the strength of crystal packing interactions or the lack of native unsaturated lipids [52]. Thus, although a 2-D crystal provides an environment for the membrane protein that is more similar to its native environment than found within 3-D crystals, it still must be proven whether the membrane protein can attain the full range of native conformations required for its function.

In addition to determining the structure of membrane proteins, 2-D crystals have provided two examples where the structures of the lipids surrounding the membrane proteins has also been determined, yielding a unique snapshot of the structure of an entire membrane. In the case of bacteriorhodopsin, the membrane is composed entirely of native lipids that show many specific interactions with bacteriorhodopsin [40, 53]; in contrast, the packing interactions between AQP0 tetramers within the 2-D crystal are mediated solely by the synthetic lipid used to make the crystals (Fig. 2.3), and thus provides a wonderful example of non-specific protein–lipid interactions [1].

2.3.1
Two-Dimensional Crystallization of Membrane Proteins

A prerequisite for any crystallization process, whether to make 2-D or 3-D crystals, is a few milligrams of purified membrane protein, which is fully functional and in a monodisperse homogeneous preparation. This is often difficult to achieve. Membrane proteins must first be removed from the membrane by the use of detergents, which disrupt the membrane and maintain the membrane protein in a soluble form by creating a "lifebelt" of detergent molecules around the hydrophobic region [54]. Unfortunately, it is not possible to predict which detergents will be best for the solubilization and purification process, so this has to be determined by trial and error. In addition, lipids can also be important to maintain the activity of membrane proteins [55]; these are often lost during purification and can lead to irreversible inactivation of the membrane protein [56, 57]. Dodecylmaltoside (DDM) has proven to be one of the most successful detergents for maintaining the integrity of membrane proteins, and has also been used for the crystallization of some of them (http://www.mpibp-frankfurt.mpg.de/michel/public/memprotstruct.html). Two simple criteria can be used to assess the quality of a purified membrane protein: a silver-stained, sodium dodecylsulfate–polyacrylamide gel electrophoresis (SDS–PAGE) will determine the purity of the protein, while size-exclusion chromatography (SEC) will assess the homogeneity of the particle size [58]. Both techniques are, however, affected by the hydrophobic properties of the membrane protein; membrane proteins often migrate faster on SDS–PAGE than soluble proteins, probably because they bind more SDS per unit mass than soluble proteins, whereas membrane proteins will migrate more rapidly on SEC due to the additional mass of detergent and lipid bound to the protein. An extreme example of the latter is EmrE; the tagged version of EmrE (M_r 15 kDa) migrates on an SDS–PAGE gel at an apparent M_r of 13 kDa, but in the DDM-solubilized form, which is a dimer, it migrates on SEC at an apparent M_r of 137 kDa due to the presence of 10 lipid and 210 DDM molecules that are bound specifically to each dimer [59].

Once a suitable membrane protein preparation has been obtained, 2-D crystallization trials can begin. There are many parameters that can affect whether 2-D crystals are formed, or not [60, 61], the most important being the quality of the protein, whether the substrate is bound or not, and the lipid-to-protein ratio (LPR). If too much lipid is present, then upon reconstitution large vesicles are formed that usually do not contain any crystalline areas; in contrast, if too little lipid is added, then upon detergent removal, the membrane protein precipitates. At an intermediate LPR, crystals may form (Fig. 2.4). The LPR for crystal formation is dependent upon each individual protein and on how it is purified. Changes in the number of purification steps, the detergent concentration, or even the ratio of protein to column bed size, can all affect the amount of lipid which copurifies with the membrane protein and hence the LPR for crystal formation. The type of lipid is also extremely important for crystal formation, as are the buffer pH and type,

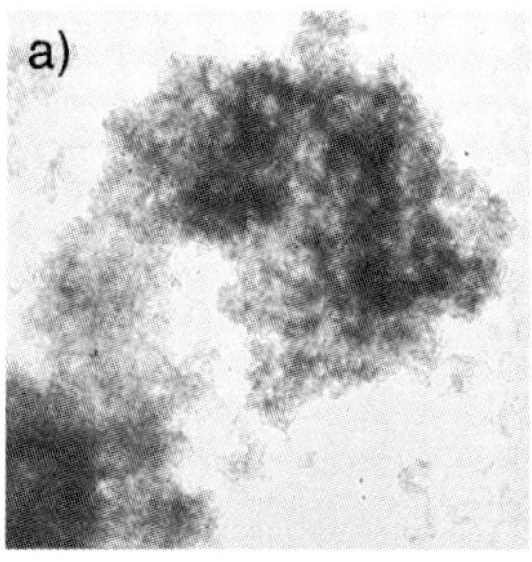

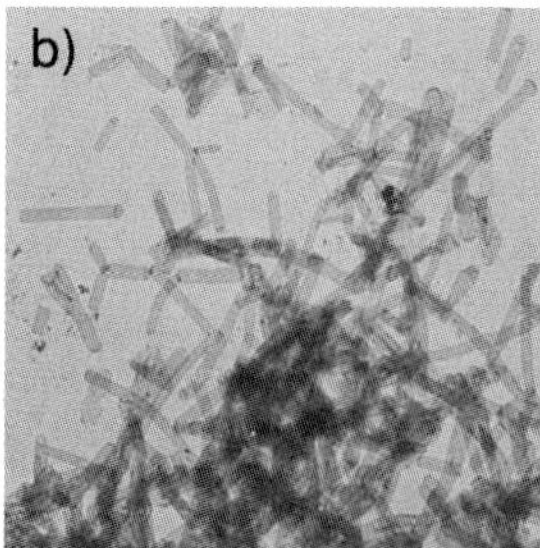

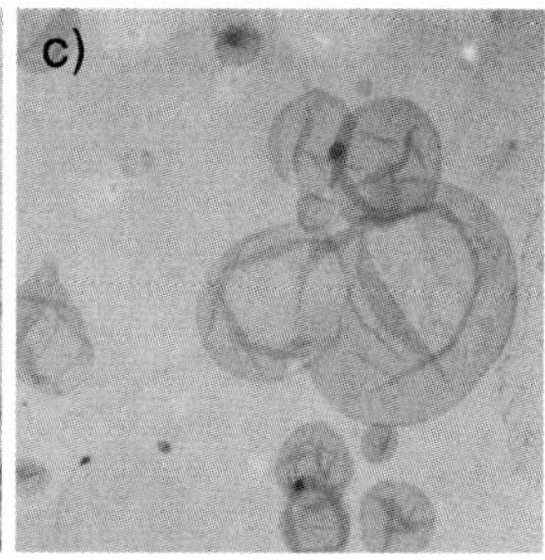

Fig. 2.4 The effect of the lipid-to-protein ratio (LPR) on the formation of 2-D crystals. Crystallization trials were performed with EmrE at a final concentration of $0.5\,mg\,mL^{-1}$ purified in dodecylmaltoside (DDM). After addition of different amounts of the lipid dioleioylphosphatidylglycerol, the samples were dialyzed for two weeks until all detergent was removed. The LPR in each sample, expressed in molar terms (lipid molecules per dimer) or in mass terms (mg lipid mg^{-1} EmrE, in parentheses). At low LPR [panel A, 5 (0.2)) membranous aggregates form; at an intermediate LPR (panel B, 7 (0.4)) crystalline tubes appear; at higher LPRs large non-crystalline vesicles form (panel C, 10 (0.6)). For an idea of scale, the tubes in panel B are approximately 0.2 μm wide. The samples were negatively stained with phosphotungstate, pH 7.

salt concentration, protein concentration, the detergent used to prepare the protein, the presence of additives (other detergents, amphiphiles, divalent cations), the temperature (fixed or ramped), and the rate of detergent removal. The most common way to induce reconstitution is to remove the detergent slowly by dialysis [62], although absorption of the detergent using hydrophobic beads and dilution have also both been used [63, 64]. Whichever method is used, however, there are major problems compared to 3-D crystallization trials in making the method high-throughput in nature. There is no comparable system to the robotics used in 3-D crystallography to set-up 1500 crystallization conditions in a matter of hours, although the dilution method is more amenable to scale-up compared to the dialysis method. In addition, the outcome of the 2-D crystallization trial must be assessed by the EM of negatively stained samples. Unfortunately, this is a severe bottleneck as even with automated EM techniques, only 96 samples were screened in 28.5 h [65]. Despite these disadvantages, it appears that many membrane proteins form 2-D crystals quite readily. Two-dimensional crystals are often observed in negatively stained samples as membranes with straight edges, usually as collapsed tubes with two parallel straight edges. However, crystalline sheets that contain only a single membrane layer may not have straight edges, which is the case for crystals of bacteriorhodopsin. The only way to test whether a potential crystal contains ordered protein molecules is to take a high-resolution image and to use a laser diffractometer to observe whether there is a diffraction pattern. Multiple rounds of 2-D crystallization trials and EM are usually required to obtain highly ordered, reproducible 2-D crystals.

2.3.2 Image Acquisition and Structure Determination

Once 2-D crystals have been identified by negative staining, high-resolution images are collected under cryo conditions to enable structure determination. Two methods are used to preserve the protein structure for cryo-EM: (i) by freeze-plunging in liquid ethane to preserve the sample in vitreous ice; or (ii) by using a thin film of preserving agent such as glucose, trehalose, or tannin [66]. The latter agents tend to give better contrast and so allow easier visualization of the crystals under very low-dose conditions. In addition, the blotting technique and drying may form a flatter sample more reproducibly than when preserving crystals in ice [66], although sandwiching the crystals between two layers of carbon has been reported to improve sample flatness further [67]. High-resolution images are taken under low-dose conditions so that the total electron dose is about 10 to 15 e^{-} Å^{2} using a 200 to 300-keV FEG microscope. The first step is take images of crystals that are untilted in the microscope to determine the unit cell dimensions and planar space group to allow the calculation of a projection map. The images are first assessed for quality using a laser diffractometer, which shows the quality of the crystal by displaying the diffraction pattern from the EM negative. The quality of the Thon rings can also be used as a guide as to whether the data will be affected by specimen movement (drift). Good images are scanned in order to digitize them for image processing.

Image processing for structure determination is based upon the MRC package of programs [68]. At this point, limitation of space prevent any description in detail of how these programs work, but a brief outline will be presented to show the advantage of working with a crystalline specimen compared to single-particle imaging. The key to the processing is being able to work in reciprocal space using a fast Fourier transform (FFT) derived from the digitized image of the crystal. As with X-ray crystallography, the first step is to index the diffraction spots; because the diffraction pattern is a regular predictable array, this information can be used in two processes. First, the noise in the FFT that is present between the diffraction spots is removed, along with regions outside the crystal or parts of the crystal that are highly disordered. Second, there is an unbending procedure, which corrects lattice distortions that probably arise from inherent disorder in the crystal and from the interaction of the 2-D crystal with the carbon support layer on the grid. A small area of the image is cross-correlated with the whole of the image, and a set of distortion vectors are then derived that account for the displacement of the cross-correlation peaks from the ideal lattice positions; the original image is then corrected by reinterpolation using these distortion vectors [69]. Two or more rounds of unbending result in a corrected FFT of the whole crystal that significantly improves the resolution of the structural data, and additional unbending using a model improves the data further [70]. Both these steps significantly improve the signal-to-noise ratio. Finally, a reverse transform of the unbent and de-noised FFT and the imposition of symmetry averaging as defined by the planar space group [71], allows display of the projection density map. If the images are taken of a sample that is not tilted in the electron microscope, then the projection map represents a view of the membrane protein perpendicular to the membrane plane. The contours of the projection map represent the sum of density throughout the thickness of the membrane, with positive contours representing density greater than the mean density of the whole sample.

The interpretation of a projection structure can be difficult, because the projection density represents the sum of density throughout the membrane. Helices that are nearly perpendicular to the membrane are distinguished by a series of circular features at 7 Å resolution but, if the helices are tilted in the membrane, then the projection densities can overlap one another to produce an arc-shaped feature (Fig. 2.5). Despite these difficulties, important biological inferences can be made from these data. For example, in the case of the multidrug transporter EmrE, the projection map showed that the molecule formed an asymmetric dimer rather than a trimer as previously thought [72].

Once a projection map has been determined, the next step is to obtain a 3-D model. This task is accomplished by taking images of 2-D crystals that have been tilted in the microscope, processing the images essentially as described above, and merging the data to give a 3-D map [73]. For mainly technical reasons, this procedure is far more demanding than the short previous sentence suggests, and it may take over six months to acquire the necessary images and to merge the data. Most of the difficulties arise in image acquisition for two reasons. First, the stability of the cryo-holder in a tilted position is considerably worse than when in the horizontal position, which means that a significant amount of time is required to take

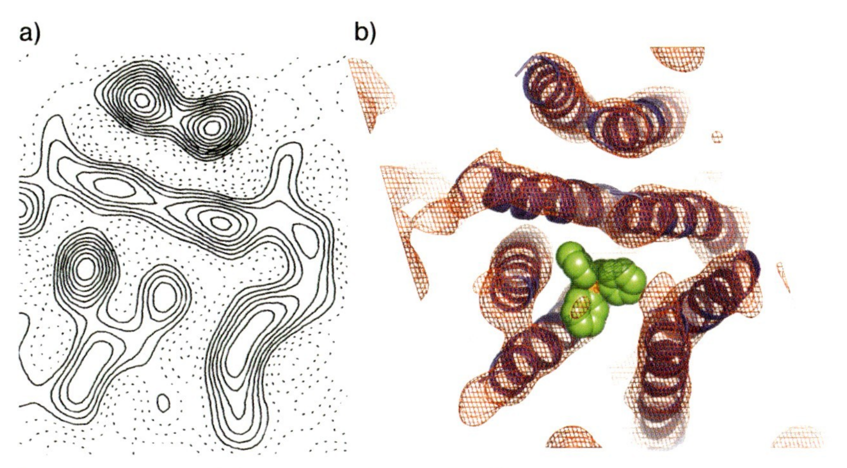

Fig. 2.5 Comparison of the projection structure of EmrE with the corresponding 3-D structure at 7.5 Å resolution. (A) Projection structure of EmrE at 7 Å resolution. (B) 3-D structure of EmrE with a molecule of the substrate tetraphenylphosphonium in the binding site. (Figures are reprinted with permission from Elsevier from Refs. [51] and [46], respectively.)

each image; this has been significantly improved with newer microscopes where the grid is loaded in a cassette that is separate from the holder. In addition, it is essential that the sample is flat on the carbon support, because if it is not, then the crystal is no longer ordered in the direction perpendicular to the membrane plane; this results in a Fourier transform where the data are anisotropic with considerable loss of data perpendicular to the tilt axis [67]. These difficulties increase as the samples are tilted to higher angles in the microscope. The highest angle that samples can be tilted to in the microscope is ~70°, thus there will inevitably be a missing cone of data due to the impossibility of data collection up to a 90° angle. The anisotropy of tilted data and the missing cone of data result in 3-D structures of membrane proteins that have a lower resolution perpendicular to the membrane plane than within the membrane plane.

A number of membrane proteins have had their 3-D structures determined from 2-D crystals to an in-plane resolution of about 8 Å, which is sufficient to determine the pathway of the α-helices and define the fold of the protein. At this resolution it is not possible to assign directly the polypeptide chain to individual helical densities. However, a combination of evolutionary sequence conservation and biochemical data is often sufficient to allow the deduction of which density represents which α-helix [74]. This approach can lead to the construction of models that are sufficiently detailed and accurate to guide further experimentation. This method was pioneered by Baldwin [75], who used the cryo-EM structure of rhodopsin to correctly predict the structure of the α-helical domains with an accuracy of 3 Å rmsd some three years before the structure was solved using X-ray crystallography. This technique has recently been applied to the gap junction structure, which has allowed the rationalization of many mutations that lead to the malfolding of the protein [76].

Many integral membrane proteins crystallize into tubes that are 0.2 to 1.0 μm wide, and they often give structural data to about 7 Å resolution. However, persuading membrane proteins to form crystals that diffract to better than 4 Å resolution is extremely difficult. All of the high-resolution structures determined to date by cryo-EM have been determined from crystals that form sheets. Unfortunately, most 2-D crystals seem to form wide tubes that are then flattened onto the carbon support of the grid. This results in a sample that is often not flat enough for the collection of good data during tilting, possibly due to the trapping of liquid within the tube during blotting; they are often also too small for electron diffraction. In contrast, sheets form flat layers on the carbon support that can allow the collection of excellent data even to 70° tilts. In addition, the sheets are often over 1 μm in diameter, which allows the collection of electron diffraction patterns directly in the microscope. As with XRD data, the electron diffraction data contains only information on amplitudes of the structure factors, whereas the image data contain both amplitude and phase information. However, because the collection of diffraction data is largely unaffected by specimen movement, it can be used to determine accurately the amplitudes at high tilts, which helps to improve the quality of the reconstruction. Both, data from images and electron diffraction patterns were merged to determine the structures of bacteriorhodopsin [38, 40, 77], light harvesting complex II [78], and aquaporin AQP1 [79] to produce atomic models. The recently solved structure of AQP0 [1] was solved from well-ordered crystals solely from electron diffraction data and by using molecular replacement with structure factor amplitudes and phases from the related crystal structure of AQP1.

2.3.3 Future Perspectives

The structure determination of membrane proteins from 2-D crystals is a complementary technique to X-ray crystallography in many ways. In the past, X-ray crystallography has undergone a far greater development than electron crystallography, such that when the crystals have been obtained it is far easier and quicker to solve the structure of a membrane protein, provided that the phases can be determined. However, the growth of good 3-D crystals that diffract isotropically to high resolution is by no means easy. In addition, it is sobering to realize that the X-ray structures of the membrane proteins that have been solved so far are in fact those that were most straightforward to crystallize, just as the first soluble protein structures determined during the 1970s were of those proteins that most readily formed crystals. The future challenge will be to solve the structures of mammalian integral membrane proteins that are often far less stable in detergent than their bacterial counterparts, and it is this area where 2-D crystallography may have its greatest impact. The ability to grow 2-D crystals from virtually any detergent, resulting in the membrane protein being inserted back into its most stable environment – the membrane itself – may allow the structural analysis of many unstable human receptors, ion channels and transporters that are important drug targets and may be intimately involved in human disease. Even low-resolution structures will allow detailed models to be built based upon high-resolution structural data from bacterial homologues, biochemical experiments and evolutionary sequence conservation data. However, as many human membrane proteins do not have a bacterial homologue, it will be essential to solve high-resolution structures of these by electron crystallography. The challenge is to persuade these difficult membrane proteins to produce crystalline sheets that diffract to high resolution.

2.4 Helical Analysis of Tubes

During attempts to produce 2-D crystals, it is sometimes the case that the membrane protein arranges to form tubular crystals with helical symmetry that have a diameter of less than 0.1 μm. These crystals are not ideal for structure determination using the methodology described above for 2-D crystals because they are too narrow and are unlikely to form flat, well-ordered 2-D arrays when placed upon a carbon support film. The method of choice for narrow tubes is to use the helical periodicity of the molecules in the tube for structure determination. Tubes are suspended over holes in a perforate carbon film-covered grid and frozen in

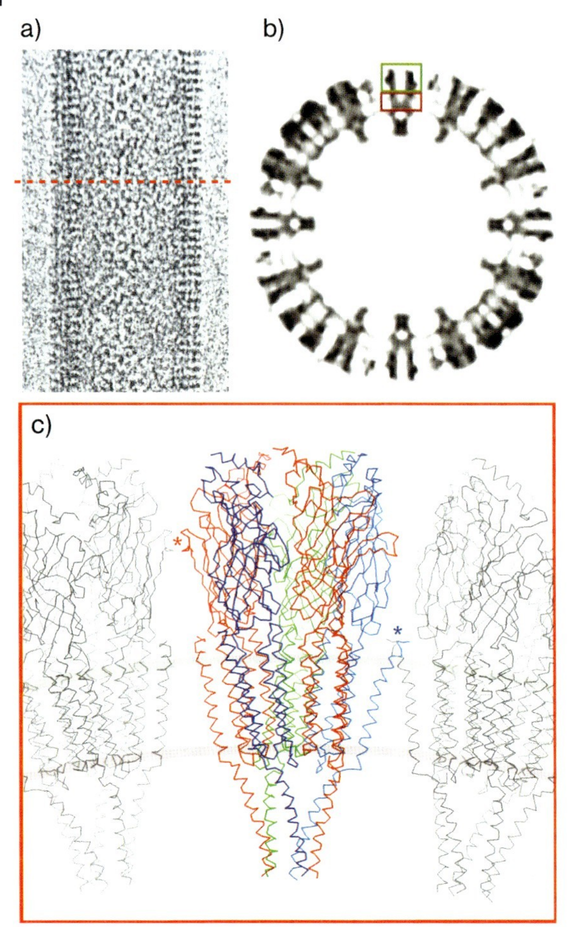

Fig. 2.6 Structure of the nicotinic acetylcholine receptor (nAChR) at 4 Å resolution. (A) Image of a tube in ice. (B) Cross-section of the tube [dashed line in (A)] showing the nAChRs embedded in the lipid bilayer. The green box indicates the extracellular domain, and the red box highlights the transmembrane domain. (C) Ribbon diagram showing the arrangement of α-helices in the membrane and the β-sheet domain where acetylcholine binds. (Panel A reprinted by permission from Macmillan Publishers Ltd: *Nature* **373**, pp. 37–43; (c) 1995. Panel B reprinted by permission from Macmillan Publishers Ltd: *Nature* **423**, 949–955; (c) 2003. Panel C reprinted by permission from Elsevier: *J. Mol. Biol.* **346**, 967–989; (c) 2005.)

vitreous ice. Images are taken under low-dose conditions (Fig. 2.6), digitized, and then processed in a manner analogous to 2-D crystals, with Fourier transforms of the crystalline areas used in a series of steps to correct for lattice distortions caused by the tubes being bent, tilted, twisted or stretched to different degrees along the length of each tube [80]. One significant advantage of tubes over 2-D crystals is that each image of a tube contains views of the same proteins from many different orientations, which means that reconstructions from the averaged de-noised amplitudes and phases results in a structure that has the same resolution perpendicular to the membrane plane as in the membrane plane. The major difficulty with solving structures at high resolution from helical arrays is that, because the tubes are very narrow, the signal-to-noise is low and exceptional cryo-EM images are required.

Two membrane protein structures have been determined from tubular arrays. The resolution of the sarcoplasmic reticulum Ca^{2+}-ATPase was taken to 8 Å [81] before the structure was solved by X-ray crystallography [82]. In contrast, it has not been possible to grow well-diffracting 3-D crystals of the nicotinic acetylcholine receptor, and so this structure has been determined to 4 Å resolution solely from cryo-EM images of tubes [83]: in total, 359 images of tubes, which are 760 to 832 Å in diameter, were recorded at liquid helium temperatures, processed and merged, which represents approximately one million receptors. Modeling of the extracellular domain of the nicotinic acetylcholine receptor, which is predominantly β-sheet, was aided by the structure of the soluble acetylcholine binding protein from a mollusk synapse [84].

2.5 Conclusions

The structure determination of integral membrane proteins by cryo-EM has provided valuable insights into membrane proteins that have proved difficult to crystallize in three dimensions. Even when membrane proteins structures have been solved by XRD, cryo-EM structures continue to provide valuable information on membrane protein conformational changes that are often constrained when the protein is crystallized in a 3-D lattice. In addition, it is now apparent that a few membrane proteins are distorted when they crystallize in three dimensions, resulting in a structure that is not biologically relevant. The ability to solve the structure of membrane proteins in a near-native environment from 2-D crystals and helical arrays provides a solution to these problems. It is likely that many membrane proteins, especially those of eukaryotic origin, are extremely unstable in detergent solution, particularly the harsh detergents that are most favored for obtaining well-diffracting 3-D crystals. The ability to grow well-ordered 2-D crystals from virtually any detergent might provide an ideal avenue for structure determination. However, a problem remains in obtaining structural data to beyond 4 Å resolution from both 2-D crystals and helical arrays, which could be associated with the relatively weak hydrophobic packing interactions between the molecules in the array. A systematic screening approach to 2-D crystallization similar to that for 3-D crystallization may well allow the rapid determination of conditions that provide well-ordered sheets, which are the ideal sample for high-resolution structure determination.

One major strength of cryo-EM is being able to visualize individual molecules, and therefore it provides a route to determine structures from extremely large and dynamic protein complexes that are refractory to either 2-D or 3-D crystallization.

Currently, these techniques provide molecular envelopes at low resolution, which allows the docking of components that have been solved by X-ray crystallography, NMR, or cryo-EM. It has been calculated, however, that there is theoretically sufficient structural information in cryo-EM images to determine structures beyond 4 Å resolution, and it is likely that this will be an area where further progress will be made in the future [3, 39].

References

1 T. Gonen, Y. Cheng, P. Sliz, Y. Hiroaki, Y. Fujiyoshi, S. C. Harrison, T. Walz, *Nature* **2005**, *438*, 633–638.

2 V. Lucic, F. Forster, W. Baumeister, *Annu. Rev. Biochem.* **2005**, *74*, 833–865.

3 R. Henderson, *Q. Rev. Biophys.* **1995**, *28*, 171–193.

4 D. J. De Rosier, A. Klug, *Nature* **1968**, *217*, 130–134.

5 J. Dubochet, M. Adrian, J. J. Chang, J. C. Homo, J. Lepault, A. W. McDowall, P. Schultz, *Q. Rev. Biophys.* **1988**, *21*, 129–228.

6 Y. Fujiyoshi, T. Mizusaki, K. Morikawa, H. Yamaguchi, Y. Aoki, H. Kihara, Y. Harada, *Ultramicroscopy* **1991**, *38*, 241–251.

7 G. McMullan, D. M. Cattermole, S. Chen, R. Henderson, X. Llopart, C. Summerfield, L. Tlustos, A. R. Faruqi, *Ultramicroscopy* **2007**, *107*, 401–413.

8 R. Beckmann, C. M. Spahn, N. Eswar, J. Helmers, P. A. Penczek, A. Sali, J. Frank, G. Blobel, *Cell* **2001**, *107*, 361–372.

9 M. Samso, T. Wagenknecht, P. D. Allen, *Nat. Struct. Mol. Biol.* **2005**, *12*, 539–544.

10 C. N. Woldin, F. S. Hing, J. Lee, P. F. Pilch, G. G. Shipley, *J. Biol. Chem.* **1999**, *274*, 34981–34992.

11 J. L. Rubinstein, J. E. Walker, R. Henderson, *EMBO J.* **2003**, *22*, 6182–6192.

12 L. A. Kim, J. Furst, D. Gutierrez, M. H. Butler, S. Xu, S. A. Goldstein, N. Grigorieff, *Neuron* **2004**, *41*, 513–519.

13 K. Mitra, J. Frank, *FEBS Lett.* **2006**, *580*, 3353–3360.

14 K. Mitra, C. Schaffitzel, T. Shaikh, F. Tama, S. Jenni, C. L. Brooks, III, N. Ban, J. Frank, *Nature* **2005**, *438*, 318–324.

15 Q. X. Jiang, D. W. Chester, F. J. Sigworth, *J. Struct. Biol.* **2001**, *133*, 119–131.

16 J. R. Harris, *Microscopy and Analysis* **2006**, *20*, 5–9.

17 M. Halic, T. Becker, M. R. Pool, C. M. Spahn, R. A. Grassucci, J. Frank, R. Beckmann, *Nature* **2004**, *427*, 808–814.

18 B. Bottcher, S. A. Wynne, R. A. Crowther, *Nature* **1997**, *386*, 88–91.

19 P. B. Rosenthal, R. Henderson, *J. Mol. Biol.* **2003**, *333*, 721–745.

20 J. Trinick, J. Cooper, *Microscopy* **1990**, *159*, 215–222.

21 J. R. Bellare, H. T. Davis, L. E. Scriven, Y. Talmon, *J. Electron Microsc. Tech.* **1988**, *10*, 87–111.

22 M. R. Sharma, L. H. Jeyakumar, S. Fleischer, T. Wagenknecht, *J. Biol. Chem.* **2000**, *275*, 9485–9491.

23 N. Grigorieff, *J. Mol. Biol.* **1998**, *277*, 1033–1046.

24 K. H. Rhee, G. A. Scarborough, R. Henderson, *EMBO J.* **2002**, *21*, 3582–3589.

25 M. Wolf, A. Eberhart, H. Glossmann, J. Striessnig, N. Grigorieff, *J. Mol. Biol.* **2003**, *332*, 171–182.

26 J. Rubinstein, J. Walker, *J. Mol. Biol.* **2002**, *321*, 613–619.

27 C. Tribet, R. Audebert, J. L. Popot, *Acad. Sci. USA* **1996**, *93*, 15047–15050.

28 S. Wilkins, *J. Bioenerg. Biomembr.* **2000**, *32*, 333–339.

29 M. Radermacher, *J. Electron Microsc. Tech.* **1988**, *9*, 359–394.

30 J. Frank, M. Radermacher, P. Penczek, J. Zhu, Y. Li, M. Ladjadj, A. Leith, *J. Struct. Biol.* **1996**, *116*, 190–199.

31 A. E. Leschziner, E. Nogales, *J. Struct. Biol.* **2006**, *153*, 284–299.

32 R. A. Crowther, L. A. Amos, J. T. Finch, D. J. DeRosier, A. Klug, *Nature* **1970**, *226*, 421–425.

33 M. van Heel, G. Harauz, E. V. Orlova, R. Schmidt, M. Schatz, *J. Struct. Biol.* **1996**, *116*, 17–24.

34 G. Harauz, F. P. Ottensmeyer, *Science* **1984**, *226*, 936–940.

35 W. O. Saxton, W. Baumeister, *J. Microsc.* **1982**, *127*, 127–138.

36 G. Harauz, M. van Heel, *Optik* **1986**, *73*, 146–156.

37 R. Henderson, P. N. Unwin, *Nature* **1975**, *257*, 28–32.

38 R. Henderson, J. M. Baldwin, T. A. Ceska, F. Zemlin, E. Beckmann, K. H. Downing, *J. Mol. Biol.* **1990**, *213*, 899–929.

39 R. Henderson, *Q. Rev. Biophys.* **2004**, *37*, 3–13.

40 N. Grigorieff, T. A. Ceska, K. H. Downing, J. M. Baldwin, R. Henderson, *J. Mol. Biol.* **1996**, *259*, 393–421.

41 J. J. Ruprecht, T. Mielke, R. Vogel, C. Villa, G. F. Schertler, *EMBO J.* **2004**, *23*, 3609–3620.

42 B. Van den Berg, W. M. Clemons, Jr., I. Collinson, Y. Modis, E. Hartmann, S. C. Harrison, T. A. Rapoport, *Nature* **2004**, *427*, 36–44.

43 E. Screpanti, E. Padan, A. Rimon, H. Michel, C. Hunte, *J. Mol. Biol.* **2006**, *362*, 192–202.

44 F. Bezanilla, *Trends Biochem. Sci.* **2005**, *30*, 166–168.

45 A. L. Davidson, J. Chen. *Science* **2005**, *308*, 963–965.

46 C. G. Tate, *Curr. Opin. Struct. Biol.* **2006**, *16*, 457–464.

47 S. Y. Lee, A. Lee, J. Chen, R. MacKinnon, *Acad. Sci. USA* **2005**, *102*, 15441–15446.

48 S. Subramaniam, M. Gerstein, D. Oesterhelt, R. Henderson, *EMBO J.* **1993**, *12*, 1–8.

49 S. Subramaniam, M. Lindahl, P. Bullough, A. R. Faruqi, J. Tittor, D. Oesterhelt, L. Brown, J. Lanyi, R. Henderson, *J. Mol. Biol.* **1999**, *287*, 145–161.

50 K. R. Vinothkumar, S. H. Smits, W. Kuhlbrandt, *EMBO J.* **2005**, *24*, 2720–2729.

51 C. G. Tate, I. Ubarretxena-Belandia, J. M. Baldwin, *J. Mol. Biol.* **2003**, *332*, 229–242.

52 I. Szundi, J. J. Ruprecht, J. Epps, C. Villa, T. E. Swartz, J. W. Lewis, G. F. Schertler, D. S. Kliger, *Biochemistry* **2006**, *45*, 4974–4982.

53 H. Luecke, B. Schobert, H. T. Richter, J. P. Cartailler, J. K. Lanyi, *J. Mol. Biol.* **1999**, *291*, 899–911.

54 M. Roth, A. Lewitt-Bentley, H. Michel, J. Deisenhofer, R. Huber, D. Oesterhelt, *Nature* **1989**, *340*, 659–662.

55 P. K. Fyfe, K. E. McAuley, A. W. Roszak, N. W. Isaacs, R. J. Cogdell, M. R. Jones, *Trends Biochem. Sci.* **2001**, *26*, 106–112.

56 M. J. Lemieux, R. A. Reithmeier, D. N. Wang, *J. Struct. Biol.* **2002**, *137*, 322–332.

57 Z. Yao, B. Kobilka, *Anal. Biochem.* **2005**, *343*, 344–346.

58 M. Auer, M. J. Kim, M. J. Lemieux, A. Villa, J. Song, X. D. Li, D. N. Wang, *Biochemistry* **2001**, *40*, 6628–6635.

59 P. J. Butler, I. Ubarretxena-Belandia, T. Warne, C. G. Tate, *J. Mol. Biol.* **2004**, *340*, 797–808.

60 W. Kuhlbrandt, *Q. Rev. Biophys.* **1992**, *25*, 1–49.

61 B. K. Jap, M. Zulauf, T. Scheybani, A. Hefti, W. Baumeister, U. Aebi, A. Engel, *Ultramicroscopy* **1992**, *46*, 45–84.

62 J. Rigaud, M. Chami, O. Lambert, D. Levy, J. Ranck, *Biochim. Biophys. Acta.* **2000**, *1508*, 112–128.

63 J. L. Rigaud, G. Mosser, J. J. Lacapere, A. Olofsson, D. Levy, J. L. Ranck, *J. Struct. Biol.* **1997**, *118*, 226–235.

64 H. W. Remigy, D. Caujolle-Bert, K. Suda, A. Schenk, M. Chami, A. Engel, *FEBS Lett.* **2003**, *555*, 160–169.

65 C. S. Potter, J. Pulokas, P. Smith, C. Suloway, B. Carragher, *J. Struct. Biol.* **2004**, *146*, 431–440.

66 J. Vonck, *Ultramicroscopy* **2000**, *85*, 123–129.

67 N. Gyobu, K. Tani, Y. Hiroaki, A. Kamegawa, K. Mitsuoka, Y. Fujiyoshi, *J. Struct. Biol.* **2004**, *146*, 325–333.

68 R. A. Crowther, R. Henderson, J. M. Smith, *J. Struct. Biol.* **1996**, *116*, 9–16.

69 R. Henderson, J. M. Baldwin, K. H. Downing, J. Lepault, F. Zemlin, *Ultramicroscopy* **1986**, *19*, 147–178.

70 E. R. Kunji, S. von Gronau, D. Oesterhelt, R. Henderson, *Acad. Sci. USA* **2000**, *97*, 4637–4642.

71 J. M. Valpuesta, J. L. Carrascosa, R. Henderson, *J. Mol. Biol.* **1994**, *240*, 281–287.

72 C. G. Tate, E. R. Kunji, M. Lebendiker, S. Schuldiner, *EMBO J.* **2001**, *20*, 77–81.

73 L. A. Amos, R. Henderson, P. N. Unwin, *Prog. Biophys. Mol. Biol.* **1982**, *39*, 183–231.

74 S. J. Fleishman, V. M. Unger, N. Ben-Tal, *Trends Biochem. Sci.* **2006**, *31*, 106–113.

75 J. M. Baldwin, G. F. Schertler, V. M. Unger, *J. Mol. Biol.* **1997**, *272*, 144–164.

76 S. J. Fleishman, V. M. Unger, M. Yeager, N. Ben-Tal, *Mol. Cell* **2004**, *15*, 879–888.

77 K. Mitsuoka, T. Hirai, K. Murata, A. Miyazawa, A. Kidera, Y. Kimura, Y. Fujiyoshi, *J. Mol. Biol.* **1999**, *286*, 861–882.

78 W. Kuhlbrandt, D. N. Wang, Y. Fujiyoshi, *Nature* **1994**, *367*, 614–621.

79 K. Mitsuoka, K. Murata, T. Walz, T. Hirai, P. Agre, J. B. Heymann, A. Engel, Y. Fujiyoshi, *J. Struct. Biol.* **1999**, *128*, 34–43.

80 R. Beroukhim, N. Unwin, *Ultramicroscopy* **1997**, *70*, 57–81.

81 P. Zhang, C. Toyoshima, K. Yonekura, N. M. Green, D. L. Stokes, *Nature* **1998**, *392*, 835–839.

82 C. Toyoshima, M. Nakasako, H. Nomura, H. Ogawa, *Nature* **2000**, *405*, 647–655.

83 N. Unwin, *J. Mol. Biol.* **2005**, *346*, 967–989.

84 K. Brejc, W. J. van Dijk, R. V. Klaassen, M. Schuurmans, J. Van Der Oost, A. B. Smit, T. K. Sixma, *Nature* **2001**, *411*, 269–276.

3
Introduction to Solid-State NMR and its Application to Membrane Protein–Ligand Binding Studies

Krisztina Varga and Anthony Watts

3.1
Introduction

3.1.1
Membrane Proteins: A Challenge

Membrane proteins still represent a challenge to structural biologists, as evidenced throughout this volume. The statistics of the Protein Data Bank (PDB) [1] reflect the remarkable progress in structural biology; the number of structures solved every year has grown exponentially since the first two structures were deposited in 1972, and as of March 2007, there are 42 082 structures deposited. However, membrane proteins represent only a small fraction of the known structures (less than 2%), even though they constitute 20 to 40% of all genomes. Only the bold and adventurous (or happily oblivious) venture into the area of studying these proteins, in spite of their significance in biology. Their biological function is diverse; they function as receptors, channels, and pumps.

Membrane proteins are also major drug targets. It is estimated that receptors and ion channels constitute at least half of all drug targets [2], and thus elucidation of their three-dimensional (3-D) structure and ligand binding will advance drug design [3]. During the past two decades, drug design has been shifting from the "trial and error" approach to drug design based on structural information and modeling studies of the target protein and small molecule ligands [4]. Since structural data for membrane proteins is still scarce, they are challenging targets for conventional methods – solution NMR and X-ray crystallography. Solid-state NMR is emerging as a promising alternative technique, with the advantage that it allows for the study of membrane proteins in various membrane mimetic environments.

In this chapter, first, a description of the fundamentals of solid-state NMR techniques is given, after which applications of the technique in the study of ligand–receptor binding interactions and dynamics will be discussed.

Biophysical Analysis of Membrane Proteins. Investigating Structure and Function. Edited by Eva Pebay-Peyroula

ISBN: 978-3-527-31677-9

3.1.2
Why Solid-State NMR?

Over the past few years, the remarkable advances in solid-state NMR approaches have contributed significantly to the characterization of protein structure and function [5–11]. Even so, it is a less widely known technique than X-ray diffraction and solution NMR, both of which have been the most prominent techniques for obtaining high-resolution protein structures. X-ray diffraction and solution-state NMR have therefore proved to be imperative in structural biology, and their remarkable success has contributed to the development and understanding of many other fields of biology. X-ray crystallography is a long-range method more suited to determining the structures of large complexes, while NMR is ideal for smaller proteins (<40 kDa) as it is more adept at measuring short-range distances. In addition, there is often overlapping and overcrowding of information in NMR spectra, and hence simplification is required.

Membrane proteins pose their own challenges when they are studied using these techniques. Solution NMR has limitations in the study of insoluble macromolecules and large complexes (due to their slow tumbling). Although there are notable accomplishments in assignments (e.g., KcsA [12]) and structure determination of (usually small) membrane proteins by solution NMR (e.g., β-barrel proteins [13–19], ATP synthase subunit *c* [20], crambin [21]), it is difficult to keep the membrane protein solubilized and at a sufficiently high concentration to obtain spectra with high signal-to-noise and resolution. Integral membrane proteins are stable in the membrane, and there has been increasing evidence that lipid species play an important structural and functional role. Some of these lipids may have relevance to the function of the protein [22–24], whereas others promote the formation of highly ordered crystals [25–28]. The purification of integral membrane proteins requires that they must be extracted from the membrane and rendered water-soluble. Following detergent solubilization, membrane proteins often become unstable and quickly lose activity. Thus, a wide variety of investigations have been undertaken to study membrane proteins using NMR in non-detergent environments which mimic the membrane to different extents; these include bicelles, lipid vesicles, nanodiscs, non-detergent surfactants, and organic solvents. For recent reviews on solution NMR membrane protein studies, the reader is referred to Refs. [19, 29, 30].

Solid-state NMR is a rapidly developing technique which can be applied to explore membrane proteins and other non-soluble or large biological structures that are difficult to study with solution NMR and X-ray crystallography. Some advantages of solid-state NMR are that:

- Insoluble macromolecules which cannot be crystallized are amenable to study. For instance, solid-state NMR techniques were used to determine structural models of amyloid fibrils [31–33] which are formed by protein aggregates *in vivo* and are the basis of many conditions, such as Alzheimer's and Parkinson diseases.

- Protein size is not a limiting factor – in theory. However, in practice large proteins are more difficult to study because of the current limitations of spectral resolution and overload of spectral lines. Therefore, for large proteins selective isotope labeling may be employed to elucidate questions about specific protein sites.
- Protein can be studied in a variety of environments, such as oriented bilayers, crystals, or bicelles. The temperature and pH conditions also may be varied within limits, usually determined by the protein.
- Protein dynamics can be studied directly from nuclear relaxation behavior.

Each of these aspects will be elaborated here and, because of the advantage that solid-state NMR can offer in membrane protein studies, it holds the promise to become a more commonly used method to obtain high-resolution structures of membrane proteins and macromolecular systems. In the following, a brief description of the fundamentals of solid-state NMR techniques is first provided, followed by some details and discussion of the applications of solid-state NMR techniques to study ligand–receptor binding interactions and dynamics.

3.2
Solid-State NMR

Nuclear magnetic resonance (NMR) is a spectroscopic technique which exploits the interaction of the magnetically active nuclei with an applied magnetic field, with solution-state NMR finding widespread applications in structural biology. The major difference between solution- and solid-state NMR samples is the motion of the molecules under study. Molecules tumble rapidly in solution – small molecules on the picosecond, macromolecules on the nanosecond timescale. In general, NMR line widths are proportional to the tumbling rate (τ_c^{-1}) and thus molecular size: the larger the molecule, the slower the tumbling rate ($\tau_c^{-1} \sim 0.75\, kT/\pi r^3 \eta$, where r is the molecular radius and η the viscosity), the line widths become larger.

Spectral quality can be assessed by its signal-to-noise and resolution, both of which are influenced by the strength of the applied field of the magnet. In short, higher-field magnets yield spectra with substantially higher signal-to-noise and resolution. Currently, the highest field superconducting commercially available solid-state NMR magnets have a field of 21.1 Tesla (900 MHz), although there are a few "home-built" resistive or hybrid magnets with higher field (e.g., see the solid-state NMR home page of National High Magnetic Field Laboratory in Tallahassee, Florida, USA [34]).

3.2.1
Sample Preparation: What is an Ideal Sample?

The primary reason for choosing a protein should be the biological interest and appeal of the system, with an aim of addressing an important biological question. Certain characteristics are required for a sample for solid-state NMR studies, as discussed here.

3.2.1.1 Availability

For most membrane proteins the overexpression and purification remains a challenge, and much method development has been carried out with readily available wild-type proteins. A few hundred nanomols (milligram quantities) of (usually isotopically enriched) protein is required. Most expression vectors have been exploited, but labeling can reduce the options on cost and efficiency grounds, with *Escherichia coli* and *Archae* being most commonly used, although inclusion body formation and refolding technology may produce functionally inactive protein. Solid-phase peptide synthesis is also commonly used for smaller peptides, and labeling at specific sites can be highly informative.

3.2.1.2 Stability

Those proteins which have good stability are best for solid-state NMR studies, in common with other methods. With the currently available techniques, a typical two-dimensional (2-D) experiment may take from a few hours up to a few days, while the data for a 3-D experiment may be collected for 7 to 8 days. Usually, several spectra are collected on the same sample, which may be used for weeks or months at a time if sample survival is assured.

Sample heating is another general consideration in solid-state NMR experiments, as samples may heat up significantly during experiments. This occurs mainly because of strong proton decoupling applied during acquisition, and thus the length of the decoupling pulses and power must be kept to lower safe levels. The heating can be calibrated by the compound $Sm_2Sn_2O_7$, where the ^{119}Sn NMR chemical shift can be used as a temperature indicator [35]. The amount of heating is also related to the hydration level and salt content; high salt-containing protein samples are more prone to heating. In experiments in which sample spinning is required (see below), the samples heat up additionally (an average of 10 °C) due to the frictional forces caused by spinning at high speed (~10–13 kHz). Higher-resolution spectra can be obtained with higher field magnets, although sample heating becomes a greater problem at higher fields with necessary higher decoupling powers. Sample temperatures are controlled during experiments with a continuous air flow, and sometimes are cooled to subzero temperatures (typically −5 to −50 °C), although the actual temperature of the sample during pulsing may be tens of degrees higher. In spite of all the precautions taken, it is not uncommon that less-stable protein samples "cook" during the experiments.

Another concern is dehydration of the sample. At high spinning frequencies and over time, the water may escape from the sample through tiny holes around the spacers. Various methods have been suggested to prevent dehydration, including the application of O-rings and a thin film of a fluorolube wax [36]. Protein stability can be monitored by observing chemical shift perturbations of specific peaks; a shift usually indicates a change in the secondary structure. Broadening of line widths is also a general indication of denaturation, making the information gained irrelevant.

3.2.1.3 **Secondary Structure**

There is a strong correlation between protein secondary structure and NMR chemical shifts [37–39]. All α-helical or β-sheet secondary structure elements produce severe spectral overlap in NMR. Mixed secondary structure elements facilitate assignments and structural studies, since chemical shift dispersion will help to resolve the secondary structural features one from another. Thus, proteins of mixed secondary elements are more accessible for solid-state NMR studies. Unfortunately, membrane proteins tend to be dominated by one or the other structural element (e.g., transmembrane helices or β-barrels).

3.2.1.4 **Sample Form: Local Order**

What does a solid-state NMR sample look like? As the term solid state implies, the sample must be some form of a *solid*, and a wide variety of solid-state NMR studies have used proteins as lyophilized powders, as microcrystals precipitated by organic molecules, membrane proteins in membranes, and as frozen protein solutions.

Sample preparation is perhaps the most important factor for obtaining high-quality protein spectra. The structural homogeneity of the sample determines line widths, which in turn affects the spectral resolution. Peak doubling, or unexpected multiplets and broadenings, indicate structural heterogeneity in a particular formulation. In recent years increasing consideration has been given to different sample preparation conditions in comparison with resolution [40–46].

In the earlier protein studies with solid-state NMR (up to the late 1990s), the apparently simplest method of preparing a solid protein sample – lyophilization (freeze-drying) – was used in many cases. Although this method has long been used to increase the long-term stability of many proteins, both the freezing and drying steps may cause damage to the proteins and alter their secondary structure. Lyophilized proteins may exhibit poor resolution (typically 1–2 ppm), which is not sufficient for most structural studies [42]. Even if the protein backbone is correctly folded, the side-chain conformations may be quite variable and this results in sample inhomogeneity. This is consistent with the observation that the hydration of a sample improves ^{13}C resolution in various proteins (e.g., lysozyme [43], bovine serum albumin [44], and SH3 [40]).

A major advance in recent years has been the recognition that microcrystalline protein samples yield narrow resonance lines for soluble proteins [40, 47, 48]. There is accumulating evidence that protein precipitates that have been created in a controlled manner yield equivalent spectral resolution as do medium-quality or X-ray quality crystals [40, 42, 49]. The precipitating conditions can strongly influence the apparent line width, and several precipitation conditions may have to be screened for a high-resolution NMR sample. For example, although ubiquitin can be crystallized both from polyethylene glycol (PEG) 8000 and 2-methyl-2,4-pentanediol (MPD), the MPD-induced crystals have narrower lines for natural-abundance ubiquitin [48]. The prospect of studying proteins as precipitates is particularly a significant benefit for membrane proteins, which are very difficult to crystallize to X-ray quality due to the presence of detergents. As high-quality crystals are not required for structural studies, membrane proteins can be studied in high resolution and, under variable conditions, can be precipitated in detergent micelles or bicelles for solid-state NMR.

There are, however, very few studies available which refer to the optimal solid-state NMR sample preparation method for membrane proteins and, in common with many membrane protein studies, each protein seems to be different, with few generalizations. Detergent-purified membrane proteins can either be precipitated as microcrystals or reconstituted into lipid bilayers to form proteoliposomes or 2-D crystals. The results of the available (albeit limited) studies suggest that proteoliposomes yield the worst resolution spectra among these methods. Oschkinat and colleagues [46] reconstituted the purified *E. coli* outer membrane protein G (OmpG) into a *E. coli* total lipid extract in 1:2 and 3:2 (w/w) lipid-to-protein ratios to yield 2-D crystals and proteoliposomes, respectively. Spectra obtained from the 2-D crystals showed better resolved lines than spectra from proteoliposomes. In another study, Glaubitz and coworkers [45] reported that microcrystalline diaglycerol kinase (DGK) yielded better resolution spectra than proteoliposomes. In the study of McDermott and colleagues [50], U-^{13}C, ^{15}N KcsA potassium ion channel was precipitated using polyethylene glycol (PEG) in detergent micelles, where the linewidth of a single ^{13}C peak was about 80 to 100 Hz, similar to that of 2-D crystals of OmpG [46] and of bacteriorhodopsin (unpublished results).

3.2.2
NMR Active Isotopes and Labeling

Almost every element in the Periodic Table has an NMR active isotope. Nuclei with an intrinsic magnetic moment have a detectable NMR signal. In biological solution and solid-state NMR, the most frequently used nuclei in biology are ^{1}H, ^{2}H, ^{13}C, and ^{15}N. The natural abundance of ^{1}H is almost 100 %; on the other hand, the natural abundance of ^{2}H, ^{13}C, and ^{15}N is much lower: 0.015%, 1.1%, and 0.37%, respectively. In order to enhance the signal-to-noise ratio of the NMR signal, it is general practice to enrich isotopically the biomolecule of interest. In the recent years, the availability of various labeled amino acids has increased considerably; moreover, with increasing availability isotopically labeled compounds (amino acids, glucose, and ammonium salts) have also become more affordable. Small peptides can be synthesized from isotopically enriched amino acids. In the case of proteins, the cells that produce them are grown in isotopically enriched media.

Several methods have been developed to optimize protein expression in isotopically enriched medium with the minimum cost [51–53]. As an alternative to protein expression *in vivo* in mostly bacterial hosts, cell-free expression (or *in-vitro* expression) systems are becoming increasingly popular for producing labeled proteins – including membrane proteins – for NMR studies (for recent reviews, see Refs. [54, 55]). Another increasingly popular technique in solution NMR is to label some parts of the protein by intein-mediated protein ligation (reviewed recently in [56]), though this has yet to be applied to proteins studies using solid-state NMR.

Peptides and proteins can be either uniformly or partially labeled (for a review, see Ref. [57]), and the labeling scheme must be carefully designed in advance and tailored to the NMR experiment. In the case of uniform labeling, all residues must be labeled with NMR active isotopes. In solution NMR, it has long been common practice to measure proteins with uniform labeling schemes, including one or a combination of the most popular isotopes of ^{13}C, ^{15}N, and ^{2}H. The advantage of uniform labeling is that it provides the best potential signal-to-noise of the NMR signal; however, because all the sites are labeled, a possible predicament is the overlap of the signals, especially in large proteins.

Alpha-helical intrinsic membrane proteins have heavily congested spectra, and thus are very challenging targets for assignment, particularly in solid-state NMR. Partial labeling can be advantageous when resolving overlapping signals or simplifying crowded spectra. Partial labeling can range from a few selected sites to plentiful labeling, but in either case as not all the residues contain NMR active isotopes, partial labeling results in fewer resonances relative to uniform labeling, which in turn may facilitate site-specific assignments. One approach is to label a only a few, selected sites of interest of a protein or peptide. If the labeling is sparse enough, the resonances may be resolved and unambiguously assigned, even in one-dimensional (1-D) spectra. For short peptides synthesized in the laboratory, it is relatively simple to label selected sites if the labeled amino acid is available. For proteins expressed in bacteria, the growth medium is supplemented with the labeled amino acids or amino acid biosynthetic precursors which are incorporated by the bacteria into the protein. Although amino acid scrambling may be a major obstacle of this technique, selective labeling techniques have been extensively used in ligand–receptor binding studies to measure distances, determine torsion angles, and to observe the dynamics of the bound and unbound ligand.

Another partial labeling approach is to label an abundance of sites in a specific pattern, often referred to as "extensive labeling". An alternative to uniform labeling has been the increasingly popular biosynthetic site labeling approach, in which the growth medium is supplemented with [2-^{13}C]glycerol or [1,3-$^{13}C_2$]glycerol [58, 59]. These labeling schemes result in extensive, systematic ^{13}C labeling of certain sites at high level, while others remain unlabeled in a specific pattern whereby most labeled carbons have unlabeled neighbors, similar to a "chessboard pattern". As most chemically bonded carbons are not simultaneously labeled, the dipolar coupling between connected nuclei is reduced, which results in narrower line widths with respect to uniformly labeled samples.

3.2.3
Assignment and Structure Determination

The prerequisite of structural (and relationship to functional) studies by NMR is the assignment of resonance peaks – that is, the identification of which resonance peak corresponds specifically to which nuclear site. This can be an arduous project in itself. For protein structure determinations with solid-state NMR the assignment strategy is similar to that used in solution NMR, usually from a combination of multidimensional spectra. Both, 2-D and 3-D spectra are optimized to produce crosspeaks between resonances which are in proximity to each other, for example carbonyl and Cα of an amino acid. Based on characteristic chemical shifts, and by extending the connectivity of atoms, the peaks can be assigned to specific amino acid types (e.g., valine Cα). The assessment of spectra with intra-residue and inter-residue correlation – the protein "backbone walk" – leads to site-specific assignments (e.g., Val55 Cα).

In general, the resolution of solid-state NMR spectra is poorer than that of solution NMR, and the reasons for this are discussed below. Until a few years ago, due to limited spectral resolution, uniformly or extensively labeled proteins could not be assigned by using solid-state NMR methods. However, recent technical developments in solid-state NMR – notably with regards to sample preparation, pulse sequences, and hardware development – have resulted in spectral resolution that are now sufficiently narrow for assignments. The feasibility of assignments was first demonstrated for the uniformly $^{13}C,^{15}N$-enriched bovine pancreatic trypsin inhibitor (BPTI) in 2000 [47], since which time the details of an increasing number of complete uniformly or extensively $^{13}C,^{15}N$-enriched soluble peptide and protein assignments have been reported [47–49, 60–75]. As the sequence-specific assignment of integral membrane proteins by solid-state NMR remains a difficult challenge, very few partial assignments of extensively $^{13}C,^{15}N$-enriched integral membrane proteins have been reported, although details have been published of Light-harvesting complex 2 [76–78] and Light-harvesting complex 1 [79], KcsA potassium ion channel from *Streptomyces lividans* [50, 80, 81], sensory rhodopsin II from *Natronomonas pharaonis* (NpSRII) [82], outer-membrane protein G (OmpG) [46, 83], and the *c* subunit of *E. coli* ATP synthase [84]. Promising line widths and identification of amino acid types were reported for the DGK [45] of *E. coli*.

By labeling only specific sites, the spectra become much simplified. With only a few sites labeled, distance measurements are feasible with high accuracy, and this represents an excellent means of determining the conformations of receptor-bound ligands. If a sufficient number of structural constraints is measured, the 3-D structure of a protein or ligand can be determined. In summary, solid-state NMR offers structural information analogous to that for solution NMR, but extends the range of applicability to membrane and insoluble systems.

3.2.4
NMR Techniques: Solution- versus Solid-State NMR

One of the major differences between solution- and solid-state samples is the motion of the molecules. In solution, the molecules tumble rapidly, whereas in the solid state the molecules are much more restricted in motion. Hence, the different methodologies of solution- and solid-state NMR are based on these differences. However, it should be noted that, between a classical solution (isotropic liquid) and a solid sample, there exist intermediate, partially motion restricted phases (i.e., anisotropic liquids) where both solution- and solid-state NMR techniques may be applicable.

3.2.4.1 Isotropic Liquids

Most solutions are isotropic – that is, the molecules tumble rapidly such that almost all of the orientation-dependent anisotropic interactions (i.e., dipolar couplings) are averaged out, and this results in narrow resonances at the isotropic chemical shift. Large proteins and other macromolecules (molecular weight 40 kDa) do not tumble fast enough to average out the orientation dependence, and hence this leads to broad lines and poor spectral resolution. This is a particularly serious drawback for membrane proteins in solution. Because the detergent micelles used to solubilize the membrane proteins add extra weight, a major obstacle of membrane protein structure determination by solution NMR is the large size of the membrane protein–micelle complex. Nevertheless, recent developments in solution NMR methodology have allowed for the study of large protein complexes, including some membrane proteins (for a review, see Ref. [30]).

3.2.4.2 Anisotropic Liquids

In anisotropic liquids (liquid crystals), the molecules align along an axis, which is usually parallel or perpendicular to the external magnetic field. There is still significant motion, but as the probability of different orientations is not equal the anisotropic interactions are not averaged out. In weakly aligned samples, measurements of residual dipolar couplings (RDCs) can be used to confine the bond orientations relative to a common alignment frame and to provide global structural information [85–89]. One of the most common approaches to align protein–micelle complexes has been the strain-induced alignment in polyacryalmide gel [90–93]. The application of either mechanical stretching or compression modifies the geometry of the pores in the gel, and the loss of the pores' spherical symmetry induces a slight preference in the alignment of the protein–micelles complexes. The benefit of this method is that the gel matrix is stable in the presence of lipids over a wide range of temperatures, ionic strengths, and pH values [94].

The most common techniques of weak alignment (bacteriophages and bicelles) for water-soluble proteins have not been practical in most cases for membrane proteins in solution NMR studies, because the membrane proteins bind too tightly and the system attains a high degree of order [30]. However, membrane proteins in magnetically aligned bicelles are becoming increasingly common in solid-state NMR studies. Opella and coworkers have aligned small membrane proteins in detergent micelles [67, 95, 96], with the alignment being induced by lanthanide binding at an engineered site. Recently, OmpA (a β-barrel membrane protein domain) in bicelles was aligned by a magnetic field [97]. The bicelle approach was also recently shown to produce better-resolved spectra than glass plate-aligned samples [98]. Ramamoorthy and colleagues investigated the interaction of an antidepressant molecule, desipramine, with the membrane in bicelles [99]. Although the structure of the bicelle remains the subject of debate, the results of these recent studies have confirmed that bicelles can be used as model membranes in solid- state NMR studies of structure, dynamics, and interaction of membrane-associated peptides and proteins. Hence, they also have great potential in membrane protein–ligand interaction studies.

3.2.4.3 **Solids**

In solid samples the molecular rotation is very much restricted and the anisotropic interactions are not averaged, which leads to significant line broadening and thus to low spectral resolution and sensitivity. In contrast, anisotropic terms report on the local electronic environment, and hence contain rich structural information. In a powdered static solid sample, the molecules occupy all possible orientations, and this anisotropy results in "powder shape" spectra. Chemical shift depends on the orientation of the molecule relative to the magnetic field. The broad powder spectrum arises from the overlap of individual narrow signals corresponding to the different orientation of the molecule (Fig. 3.1). In order to overcome this low resolution and sensitivity in the solid state, two major methods have been developed to obtain high-resolution spectra: (i) uniaxial orientation of the sample; and (ii) "magic angle spinning" (MAS). Alternatively, MAS can be applied to oriented samples, at which point it is known as "magic angle-oriented sample spinning" (MAOSS). Details of the various solid-state NMR methods are summarized schematically in Fig. 3.2.

3.2.4.3.1 **Magic Angle Spinning (MAS)** Magic angle spinning NMR was first introduced in 1958 as a technique to reduce the linewidths of solids [100–102], but since then it has become the most widely applied method in solid-state NMR. MAS involves rapid mechanical rotation of the sample at the "magic angle" (54.7°) relative to the external magnetic field. But what is "magic" about the "magic angle"? The broad lines in the spectra of static solid samples are due mainly to the orientation-dependent dipolar couplings between the nuclei (spins). At 54.7°, the orientation dependence $(3\cos^2\theta - 1)$ equals zero, where θ is the angle between external magnetic field and the inter-nuclear vector of a spin pair of the sample. When $(3\cos^2\theta - 1) = 0$, the dipolar coupling between the two spins will also be reduced to zero.

The solid-state sample is carefully packed into a rotor (which is generally made from zirconium oxide) that is capable of withstanding the strong forces associated with rapid spinning. Within the rotor, the sample is restricted (by spacers made

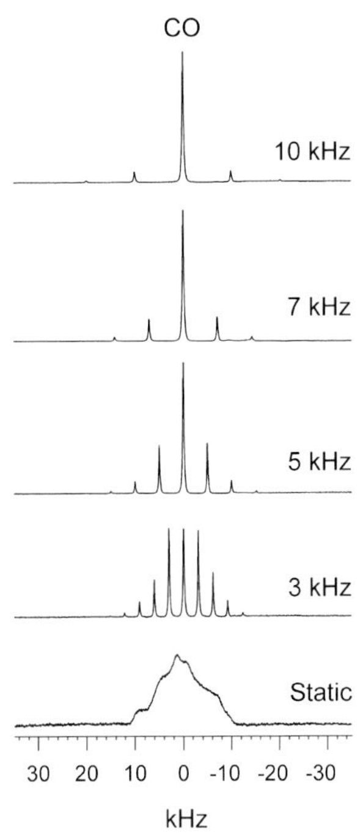

Fig. 3.1 The effect of magic angle spinning (MAS) is demonstrated by solid glycine ^{13}CO NMR spectra. Small molecules (<40 kDa) tumble rapidly in solution, which averages the orientationally dependent anisotropic interactions, such as dipolar and chemical couplings, and results in narrow resonances; this is the basis for solution-state NMR. In a powdered static solid sample, the molecules occupy all possible orientations, and this anisotropy results in homogeneously broadened lines, also called the "powder shape" spectrum (bottom spectrum). Resolution and sensitivity of solid-state spectra are improved by MAS; this involves rapid mechanical rotation of the sample at the "magic angle" (54.7°) relative to the applied magnetic field. When the spinning frequency is higher than the magnitude of weak spin interactions (i.e., dipolar couplings), the orientation dependence is removed and narrow spectral lines are obtained. MAS spectra contain a center band for each resonance at the isotropic shift, surrounded by side-bands spaced by the spinning frequency. Higher spinning frequencies result in fewer, lower-intensity sidebands, and therefore increase the center band intensity. Since the anisotropic interactions are averaged in a controlled fashion by MAS rather than random motion, the spectra still contain valuable information regarding these interactions. Typical MAS frequencies are 5 to 20 kHz. These 1-D ^{13}C cross-polarization spectra were collected on a Varian Infinity Plus 500 spectrometer (11.7 T) at 5 °C in a 4-mm HX probe.

from Teflon or Kel-F) to an area having the most homogeneous field (Fig. 3.3A). The rotor is spun around its longitudinal axis at the "magic angle" with respect to the applied field in specialized probe heads (Fig. 3.3B). The spin regulator hardware regulates the drive and bearing air to achieve the desired spinning frequency. In general, frequencies are between 5 and 20 kHz, within a precision of 5 Hz, although spinning frequencies of up to 70 kHz have been achieved in certain cases [103].

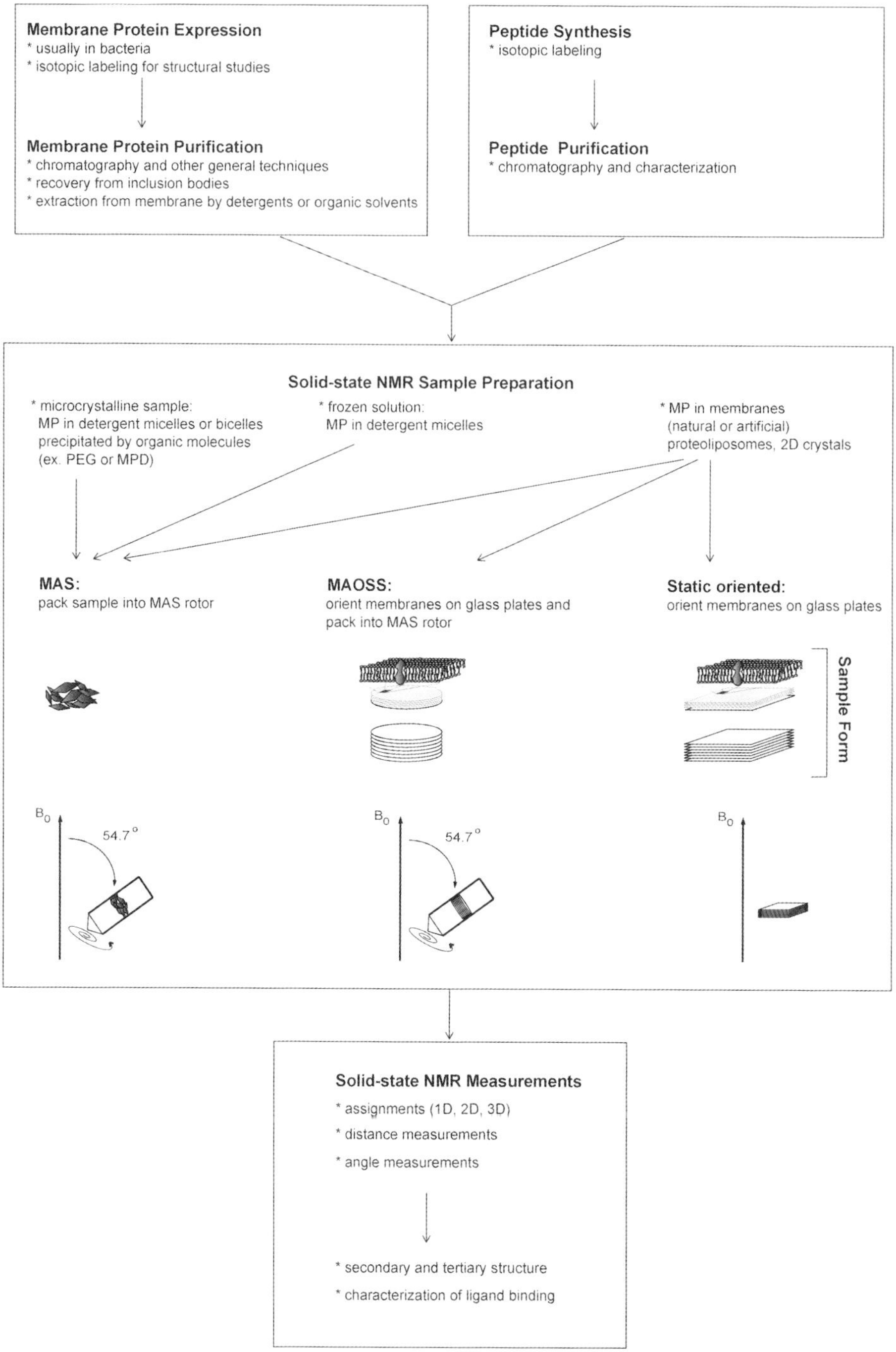

Membrane Protein Expression
* usually in bacteria
* isotopic labeling for structural studies

Membrane Protein Purification
* chromatography and other general techniques
* recovery from inclusion bodies
* extraction from membrane by detergents or organic solvents

Peptide Synthesis
* isotopic labeling

Peptide Purification
* chromatography and characterization

Solid-state NMR Sample Preparation

* microcrystalline sample: MP in detergent micelles or bicelles precipitated by organic molecules (ex. PEG or MPD)
* frozen solution: MP in detergent micelles
* MP in membranes (natural or artificial) proteoliposomes, 2D crystals

MAS:
pack sample into MAS rotor

MAOSS:
orient membranes on glass plates and pack into MAS rotor

Static oriented:
orient membranes on glass plates

Sample Form

B_0 54.7°

B_0 54.7°

B_0

Solid-state NMR Measurements
* assignments (1D, 2D, 3D)
* distance measurements
* angle measurements

* secondary and tertiary structure
* characterization of ligand binding

Fig. 3.2 An outline of membrane protein structure determination and ligand binding measurements by solid-state NMR. The first step is expression and purification of the peptide or membrane protein (MP). Isotope enrichment is required for most NMR experiments. The solid-state NMR sample can be prepared in various ways: the MP in detergents can be frozen or can be precipitated (which yields microcrystals); or it can be reconstituted into membranes to produce proteoliposomes or 2-D crystals. Depending on the sample preparation, various types of solid-state NMR technique can be applied to the sample, as indicated by the arrows. Microcrystalline, frozen, and membrane-reconstituted samples can be studied by "magic angle spinning" (MAS) experiments, where the rotor is rotated at high frequency about the magic angle (54.7°) relative to the magnetic field (B_0). MPs reconstituted in membranes can be oriented on glass plates and stacked for "magic angle oriented sample spinning" (MAOSS) or static oriented experiments. Assignments and structural (distance and angle) constraints are derived by solid-state NMR measurements, which lead to characterization of ligand binding and/or structural information about the protein. For more detail, refer to the text.

a)

b)

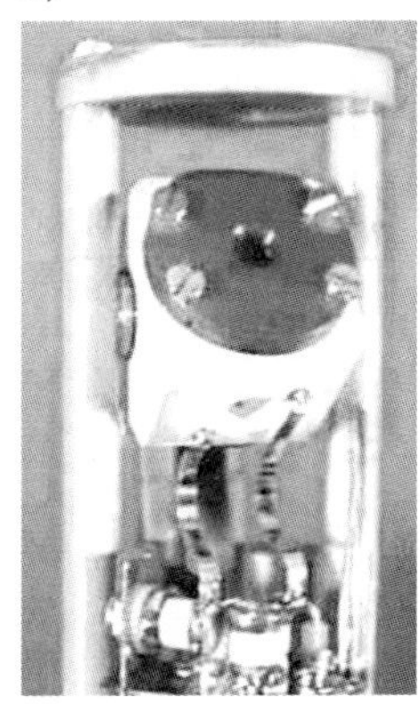

Fig. 3.3 Magic angle spinning (MAS) rotor (A) and probe spinning module (Stator) (B). High-resolution solid-state NMR relies upon the ability to rotate the sample rapidly about the "magic angle" (54.7°) relative to the magnetic field. MAS solid-state NMR samples are loaded into zirconium rotors with various diameters. In panel (A), a Varian rotor of 3.2 mm outer diameter is shown in the middle. The packing elements are displayed on the right: from top to bottom: drive-tip, spacer, sample space, end-cap. A British one pound coin is shown (left) for size reference. (B) The rotor is mechanically rotated in the probe-spinning module by blowing high-pressure compressed air (or nitrogen) at the drive-tip, while the rotor floats on a bearing air. The maximum rotation frequency for a 3.2-mm standard wall Varian rotor is currently 25 kHz. The spacer and end-cap seal the rotor and hold the sample in place. This rotor can hold up to 15 μl of sample. For protein and other hydrated samples, the use of additional seals (e.g., a rubber O-ring) is advised to prevent dehydration of the sample during the experiments.

When the spinning frequency (ω_r) is greater than the magnitude (in Hz) of weak spin interactions (e.g., dipolar couplings), then the orientational dependence is removed and narrow spectral lines are obtained. Under MAS conditions, the powder spectrum will break up into an isotropic resonance and spinning side-bands. MAS spectra contain a center band for each resonance at the isotropic shift, surrounded by side-bands spaced by the spinning frequency (ω_r) (see Fig. 3.1). Higher spinning frequencies result in fewer, lower intensity sidebands and therefore the center band intensity is increased. As the anisotropic interactions are averaged in a controlled fashion by MAS rather than random motion, the spectra still contain valuable information regarding these interactions.

Despite the technical difficulties associated with MAS, it has become the most widely used method for obtaining high-resolution spectra, based on several advantages. For example, MAS can be applied to randomly oriented molecules and allows the detection of isotropic chemical shifts; hence, well-established resonance assignment strategies of solution NMR can be implemented. In recent years, the correlation of chemical shift with secondary structure has been studied extensively, and a large statistical database has been developed for solution NMR [104, 105]. Recent advances in protein assignment by solid-state NMR initiated the comparison of solid shifts to solution shifts and the evaluation of secondary shifts (the chemical shift deviation from random coil values) for solid samples. However, the statistical data available are limited, as very few proteins have been assigned; nonetheless it was reported that, for these proteins, the chemical shifts agreed very well for the vast majority of amino acids in proteins assigned to both NMR methods [39, 48, 65]. As with liquid-state NMR [104, 105], the $^{13}C\alpha$ secondary shifts were found to be positive for α-helical structures, and negative for β-sheet secondary structures [39]. Thus, in principle, chemical shift statistics accumulated by solution NMR [105, 106] can be used to identify protein structural motifs in the solid state, as has long been practiced in solution NMR [107].

3.2.4.3.2 **Oriented Samples** For oriented samples the molecules are aligned (oriented) along a common axis. Integral membrane proteins can be oriented in membranes by layering the membranes onto glass slides. The orientational restraints of the molecule relative to the external magnetic field can then be determined from dipolar couplings and chemical shift interactions.

The alignment of membrane proteins on glass plates is a difficult procedure which usually requires the optimization of various experimental parameters (for a review, see Ref. [11]). Two main approaches have been identified for aligning lipid bilayers on glass plates:

- The lipids are dissolved in organic solvents and transferred onto glass plates. The organic solvents are then evaporated under vacuum, and the dry lipid bilayers rehydrated by water vapor from a humidity-controlled atmosphere. The glass slides are stacked and placed into a chamber where the humidity is controlled by a saturated salt solution (e.g., ammonium or potassium sulfate). The samples are then left to equilibrate and form oriented bilayers.
- The unilamellar vesicle solution is placed onto the glass plates, and the excess water is slowly evaporated in a humidity-controlled chamber. This method has the advantage that the protein in the bilayer never becomes completely dehydrated.

Other factors which affect the success of aligning membrane proteins are the composition of the lipid bilayer, the protein : lipid ratio, and the hydration level. Aligned lipid bilayers are usually mixtures of the most abundant phospholipids in the cell membrane, such as phosphatidylcholine (PC) and phosphatidylserine (PS) and/or phosphatidylethanolamine (PE). The membrane composition must be optimized for the protein of choice. For reasons of sensitivity, the typical molar ratio of proteins to lipids is between 1 : 100 and 1 : 200, although much higher protein densities have been used, albeit with a risk of aggregation and/or denaturation. The glass plates are sealed to prevent dehydration before being placed into the magnet for NMR measurements. If care is taken the samples may remain stable for some weeks.

The first step in the solid-state NMR experiments is to confirm that the lipid bilayer is in fact aligned from a 1-D ^{31}P or ^{2}H spectrum (if a small amount of deuterated lipids is present). Protein alignment can be checked with a 1-D ^{15}N spectrum; unaligned proteins exhibit a powder spectrum, while aligned samples have 2- to 5-ppm linewidths (for resolved peaks) between the regions 150 and 225 ppm. Two-dimensional spectroscopy can be used to measure the ^{1}H–^{15}N dipolar couplings, from which the orientation of secondary elements of a helical peptide or protein can be derived. The most commonly applied technique for the uniform ^{15}N-labeling of proteins is the so-called PISEMA experiment (*p*olarization *i*nversion *s*pin *e*xchange at the *m*agic *a*ngle) [108, 109]. Helical proteins display a characteristic "wheel-like" pattern, and spectral assignments of the uniformly labeled sample are confirmed from comparisons with spectra of selective ^{15}N-labeled samples.

The secondary structure and topology of membrane proteins can be modeled based on the resonance patterns in the PISEMA spectra. PISA (*p*olarity *i*ndex *s*lant *a*ngle) wheels [110, 111] and dipolar waves [112, 113] are two methods which have been developed for the interpretation and analysis of PISEMA spectra, although they may also be applied to interpret results from weakly aligned solution-state spectra (e.g., Ref. [114]). Structure calculations include the combination of experimental constraints from individual residues and the well-established covalent geometry of proteins, such as bond lengths, dihedral angles, and the planarity of peptide linkages.

The main disadvantage of this method is the intricate sample preparation (protein alignment). Although, in theory, the assignment of a uniformly ^{15}N-labeled peptide peaks in the "PISA-wheel" is possible based on one known peak and the wheel pattern (i.e., one selectively labeled sample), in practice the preparation of several selectively labeled and/or mutated protein samples is necessary because of the effects that molecular motion has on the helical wheel patterns [115]. The resolution in these experiments is often limited, especially for large membrane proteins where the number of resonances results in severe spectral overlaps. The resolution is also dependent upon the degree of orientation of the membrane sample [116].

Although these new techniques are still under development, much information on orientation has been published for several peptides and membrane proteins. For example, structural information (e.g., tilt angle, protein topology, structure) of various single transmembrane helices was determined from oriented solid-state

NMR data, including the antimicrobial peptide gramicidin A [117, 118], the M2 channel-lining segments from the proton channel of influenza [119–121] and nicotinic acetylcholine and NMDA receptors [122], the 3-D structure of Vpu (the channel-forming trans-membrane domain of virus protein “u”) from HIV-1 [123], the coat protein in fd [124] and in Pf1 bacteriophage [125], and sarcolipin (SLN) [126]. The analysis of oriented spectra of membrane proteins with multiple trans-membrane helices is much more complex, and spectral overlap may limit the resolution. Selective labeling techniques have also been used to simplify the spectra of oriented bacteriorhodopsin, an integral seven-transmembrane helical protein [127]. Likewise, the structure and orientation of some alpha-helices of ^{15}N-Met-labeled bacteriorhodopsin have been determined [127], while the dynamics of ^{15}N-Gly-labeled bacteriorhodopsin has been studied in oriented membranes [128].

3.2.4.3.3 **Magic Angle-Oriented Sample Spinning** Magic angle-oriented sample spinning (MAOSS) is an experimental approach that was proposed by Glaubitz and Watts [129], and which merges the above-described techniques: namely, it combines aligned samples with magic-angle spinning experiments. MAOSS also combines the merits of both methods; for example, it can be used to obtain orientational information and also has the benefit of high resolution of magic-angle spinning.

Sample preparation is analogous to that of the above-described static oriented samples. The intrinsic membrane proteins in membranes are layered and oriented uniaxially on round glass plates, which are then stacked and packed into an MAS rotor so that the membrane normal is parallel to the rotor axis. The lipids rotate within the membrane bilayer at the “magic angle”, which averages out strong dipolar ^{1}H–^{1}H couplings [129]. In addition, the rotor is spun at the “magic angle” at low frequency (typically 200–4000 Hz) to average out orientation defects, but not the chemical shift anisotropy. The result is a well-resolved spectrum of isotropic peaks surrounded by sharp, spinning sidebands for each labeled residue. The sideband patterns have strong orientation dependence from which the orientation constraints can be derived. The applicability of the method to membrane proteins has been demonstrated for a number of systems, including rhodopsin [130], bacteriorhodopsin in purple membranes [131, 132], and M13 coat protein [133]. As a variation of this technique, the membrane bilayers can be oriented on polymer sheets which are then rolled up and fitted into the MAS rotor.

The main difficulty of the method is the intricate sample preparation of oriented lipid bilayers/membrane proteins. Nonetheless, MAOSS represents a promising technique for obtaining unique information for structure determination, and particularly for intrinsic membrane systems.

3.3 Examples: Receptor–Ligand Studies by Solid-State NMR

Too often, only limited structural information is available regarding the target membrane protein in a receptor–ligand study. Yet, well-tailored solid-state NMR experiments can still reveal much information about the ligand and ligand–protein interactions.

- Ligand binding: a labeled ligand can be titrated into the membrane containing the receptor target
- Ligand binding to the active site of protein: by binding competitive studies
- Ligand conformation: based on distance measurements and/or chemical shift analysis for free and bound ligand
- Structural changes in the membrane protein associated by ligand binding: the membrane protein is also isotopically enriched, and chemical shifts are compared for free and ligand-bound membrane protein
- Distance measurements between the membrane protein and its ligand
- Observation of ligand mobility in binding site: from deuterium NMR experiments.

The findings of solid-state NMR measurements of receptor–ligand interactions are often utilized in modeling studies [134, 135].

3.3.1 Transport Proteins

3.3.1.1 LacS

LacS is a lactose transport protein of *Streptococcus thermophilus* which folds into 12 transmembrane helices. The K_d of [1-^{13}C]-D-galactase was measured by titrating it into its binding site of LacS [136], which illustrates how solid-state NMR can be utilized to determine the binding isotherm for weakly binding (K_d ~mM) ligands (including binding sites) by using spin labels.

3.3.2 G-Protein-Coupled Receptors and Related Proteins

Currently, G-protein-coupled receptors (GPCRs) are the most targeted class of proteins in terms of therapeutic benefit. Although, together with ion channels they contribute more than 50% of known drug targets [2], only 5% of GPCRs have yet been exploited due to a lack of structural information. One rational design approach based on GPCR structural information has shown great promise in the development of novel and/or more selective drugs for the diverse GPCR protein family [4]. In order to design selective ligands for receptors, it is essential first to realize, in atomic detail, exactly how the ligand interacts with the binding site of its target receptor. With seven transmembrane helices and more than 300 residues, GPCRs indeed represent a major target for structural studies, which in turn has led to extensive investigations world-wide into this class of drug target.

GPCRs are integral membrane proteins that function as receptors in diverse stimulus–response pathways and share a common structural motif of seven putative transmembrane α-helical domains. The binding of an extracellular signal (ligand) to the receptor triggers a cascade of intracellular responses. A wide variety of specific "ligands" is known to exist, ranging from photons, Ca^{2+} ions, and small organic molecules to complex polypeptide hormones. GPCRs have very diverse physiological functions, and affect virtually all aspects of cellular function. In fact, almost half of all GPCRs function as sensory receptors, for example as olfactory receptors when binding odorants and pheromones, or as visual receptors when binding retinals to opsins to make vision possible. The remainder of the GPCRs play important roles in intercellular communication between the cells of the immune system, in behavioral and mood regulation, and also in regulating both the sympathetic and parasympathetic nervous systems. In spite of their importance, very few high-resolution structural data are available for GPCRs, with only the crystal structure of rhodopsin having been solved [137]. Based on an homology with rhodopsin, a variety of major structural features of GPCRs can be predicted [134, 138], including the arrangement of the seven-transmembrane (7TM) spanning helices and ligand-binding sites. Bacteriorhodopsin has also served as another popular model for GPCRs.

3.3.2.1 **Bacteriorhodopsin, Rhodopsin, and Sensory Rhodopsin (NpSRII)**

The most extensively studied protein–ligand complexes using solid-state NMR are those of bacteriorhodopsin/rhodopsin and retinal. For both bacteriorhodopsin and rhodopsin, the retinal ligand (in this case as a prosthetic group) is attached covalently to a lysine residue via a Schiff-base linkage. Both proteins serve as popular models for other GPCRs.

Bacteriorhodopsin is a proton pump which is expressed in Archaebacteria and carries out photosynthesis. Although bacteriorhodopsin is not coupled to a G-protein, because of its structural similarity it has long been a model for GPCRs. The first distance measurements in a membrane protein using solid-state NMR were made for retinal in bacteriorhodopsin [139]. Distance measurements between two labeled spins, the C-8 and C-18 carbons of retinal, showed retinal to be in the 6-s-*trans* configuration. In subsequent studies, solid-state NMR spectroscopy provided detailed information about the local structural changes of retinal during the photocycle [140, 141]. Initially, retinal was deuterated at selected methyl sites and incorporated into bacteriorhodopsin. Oriented membrane films were prepared on glass plates, and the orientation and conformation within the bacteriorhodopsin was established at two stages of the photocycle, in dark-adapted [140] and in the M-state [141], by measuring the orientation vectors by deuterium NMR in aligned samples. In another study, the helix-tilt angles of bacteriorhodopsin relative to the membrane normal were determined in oriented samples [127] (Fig. 3.4).

In rhodopsin, the G-protein-binding pathway is activated by photons. An incoming photon prompts the photoisomerization of 11-*cis*-retinylidene to all-*trans*-retinylidene, which leads to a conformational change of the protein (from meta I to meta II) and resulting in a proton transfer from the ligand to a glutamate

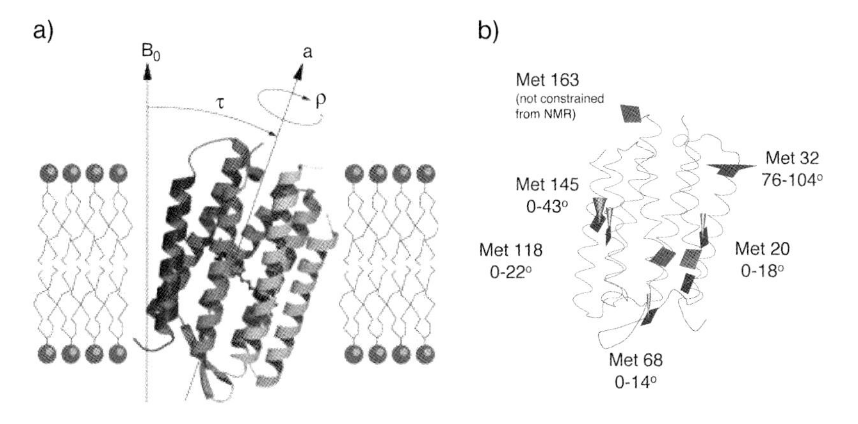

Oriented NMR **Magic Angle Oriented Sample Spinning**

Fig. 3.4 Solid-state NMR techniques: Oriented samples and magic angle oriented sample spinning (MAOSS). The different types of solid-state NMR method were summarized in Fig. 3.3. Here is illustrated the application of oriented NMR (A) and of MAOSS (B) of an integral seven-transmembrane protein, bacteriorhodopsin (bR) of *H. salinarium*; this is a well-studied membrane protein that has been used for solid-state NMR method development. For both experiments the nine Met residues of bR were selectively ^{15}N-enriched, and the purple membranes were aligned onto glass plates. (A) The tilt angle of two helices of an integral seven-transmembrane protein, bR was determined by oriented solid-state NMR [127]. The tilt angle of helix A was determined as 18–22°, while the tilt of helix B has been estimated as less than 5° from the membrane normal. (B) The orientation constraints of ^{15}N-Met-labeled bR were obtained using the MAOSS solid-state NMR technique [132]. The orientations of five Met residues were determined with respect to the membrane normal. MAOSS combines the advantages of magic angle spinning and oriented samples: it has improved sensitivity and the capacity to measure orientational constraints. (The 1C3W PDB structure defines the backbone.)

(Glu113) side chain. Several solid-state NMR studies have investigated the structure of retinal in within the binding pocket of bovine rhodopsin, which led to the resolution of the complete structural details [130, 142–147], including distance measurements, angles and dynamics, much of this information being acquired before the crystal structure of the ground state [137].

Uniform or extensive isotope labeling of a protein can be advantageous to obtain assignments for numerous residues. In a recent study, 98 residues (73%) of the sensory rhodopsin II (NpSRII) from *Natronomonas pharaonis* (an *Archaebacterium*) were assigned [148]. The protein was uniformly ^{13}C- and ^{15}N-enriched except for four common amino acids (V, L, F, Y), which relieved spectral congestion. Similar to bacteriorhodopsin and rhodopsin, NpSRII is a seven-helix transmembrane protein which has the retinal bound to a lysine residue. The feasibility of the assignments of NpSRII can lay an important foundation for future experiments of GPCRs and other large membrane proteins.

3.3.2.2 Human H_1 Receptor

In a recent report, the human H_1 GPCR was expressed in insect cells and the binding of ^{13}C,^{15}N-labeled histamine was followed in a solid-state NMR study [149]. ^{13}C and ^{15}N chemical shifts of the bound and unbound ligand were compared, and the difference in chemical shift implied ligand binding to H_1. The large increase in linewidths (4–8 ppm for bound ligand) indicated heterogeneity of the sample which was attributed to the distribution of two conformational substrates: the receptor containing either a monocationic or a dicationic histamine. Although these two states could exist in dynamic equilibrium under physiological conditions, under the experimental conditions at 203 K the equilibrium between the inactive and active substrates is likely to be frozen. The two substrates could indicate a mechanism similar to the two protonation states of rhodopsin, meta I and meta II.

3.3.2.3 Neurotensin Receptor

NTS1, one of the three neurotensin receptors identified in mammals, belongs to the GPCR family. As NTS1 is involved in many diseases, the elucidation of protein–ligand interactions may potentially be useful for developing drugs to treat pain, eating disorders, stress, schizophrenia, Parkinson's disease, Alzheimer's disease, and cancer. NTS1 becomes activated by neurotensin (NT), a tridecapeptide, but the last six amino acids (NT_{8-13}) are sufficient for biological activity. NT acts as a neuromodulator in the central nervous system and as a local hormone at the periphery. NTS1 (424 amino acids), originally isolated from rat hypothalamus, was cloned and can be expressed in *E. coli* as a fully functional GPCR. NT_{8-13} bound to the NTS1 was observed using solid-state NMR methods [150, 151].

3.3.3 Ion Channels

3.3.3.1 Nicotinic Acetylcholine Receptor

The nicotinic acetylcholine receptors (nAChRs) are involved in synaptic transmission in the nervous system. As this is a ligand-gated ion channel, two molecules of acetylcholine bind to the receptor in order to mediate synaptic transmission. The nAChRs are well-known pharmaceutical targets for the treatment of numerous neurological diseases, such as schizophrenia, Alzheimer's disease and attention deficit/hyperactivity disorder. Although a 4 Å resolution structure is available from electron diffraction studies [152], details of the neurotransmitter binding site are not known. NAChRs isolated from the electric organ of *Torpedo* can be prepared with high density, and this serves as a good model for the human neuromuscular junction receptor. The muscarinic nAChR is a large complex of five glycosylated subunits (total molecular mass 280 kDa) which surround the water-filled, ion-conducting channel. Acetylcholine binding induces opening of the ion channel which permits Na^+ flow. In solid-state NMR studies, isotopically labeled (either ^2H or ^{13}C) acetylcholine, or the acetylcholine analogue bromoacetylcholine, was bound to the functional nAChR, which reported both on the conformation and the dynamics of the agonist in the binding site [153–155]. A major dilemma concerning the mechanism of acetylcholine binding was resolved by using solid-state NMR experiments capable of selectively detecting bound acetylcholine. These results indicated that binding was mediated through interaction of the quaternary ammonium moiety, $N^+(CH_3)_3$, of acetylcholine with the π-bonded aromatic protein residue side chains, the so-called cation–π interaction.

3.3.3.2 K⁺ Ion Channel, KcsA

Potassium channels are integral membrane proteins present in both prokaryotes and eukaryotes. Two remarkable features of these channels are the very high selectivity towards K^+ ions and, at the same time, the rate of conductivity which approaches the diffusion limit. As the K^+ pore region is highly conserved, it is very likely that all K^+ channels (both ligand- and voltage-gated) from different species essentially have the same gating mechanism and pore structure [156–158]. All K^+ channels contain a critical, highly conserved amino acid "signature sequence" [159, 160], which serves as the selectivity filter of the channels (T75-G79). The best-studied K^+ ion channel is KcsA, isolated from *Streptomyces lividans*, the 3-D structure of which was determined using X-ray crystallography by MacKinnon and colleagues [156, 161]. KcsA functions as a homotetramer (70.4 kDa); the monomer is composed of two trans-membrane helices connected by an extracellular loop, including the pore helix. MAS solid-state NMR studies were conducted by Baldus and colleagues [80] on the ligand-binding properties of the chimeric KcsA-Kv1.3 [162], where Kv1.3 is a human voltage-gated K^+ channel with high sensitivity to scorpion toxin kaliotoxin (KTX) and a potential pharmacological target for the treatment of a number of diseases, including diabetes [163]. The purified, uniformly ^{13}C,^{15}N-enriched KcsA-Kv1.3 was reconstructed into proteoliposomes, and spectra were recorded in the presence and the absence of KTX. The solid-state NMR studies [80] revealed that high-affinity binding of the KTX leads to significant structural changes in both the ligand and the receptor. Based on chemical shift analysis, the structure of the membrane-spanning helices is largely maintained independent of the toxin binding. However, the residues is the channel region (KTX binding region) showed large chemical shift changes, indicating changes in both binding and conformation (see also Fig. 3.5.). These findings suggest the presence of a structural model of toxin binding more deeply into the selectivity filter than did previous models.

3.3.4 P-type ATPases

P-type ATPases (P-ATPases) are a class of ATPases which transport cations across a plasma membrane, coupled to the phosphorylation and dephosphorylation of a conserved aspartate residue. The most important P-ATPases are the H^+/K^+-ATPase, Na^+/K^+-ATPase, and Ca^{2+}-ATPase. There are two high-resolution structures available, namely the Ca^{2+}-ATPase X-ray structure [164, 165], and the nucleotide-binding N-domain of rat α1 Na^+/K^+-ATPase (~20 kDa) solution NMR structure [166]. Little

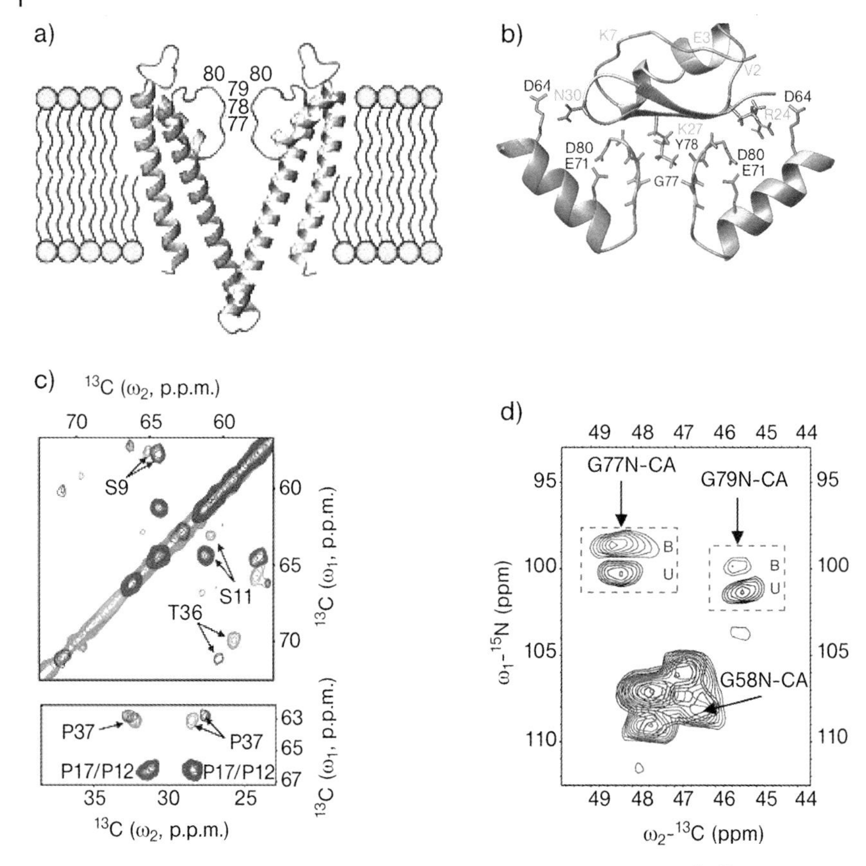

Fig. 3.5 Solid-state NMR ligand-binding study of the KcsA K^+ ion channel. Magic angle spinning (MAS) solid-state NMR techniques were used to probe the interaction of the chimeric K^+ ion channel KcsA-Kv1.3 with its ligand, the scorpion toxin kaliotoxin (KTX) [80]. KcsA-Kv1.3 is a tetramer of four identical subunits, each over 150 amino acids (A, two subunits shown). The K^+ selectivity filter is highlighted by residue numbers. The proposed structural model of the ligand interaction is shown in (B): KTX (light gray labels) and KcsA-Kv1.3 residues (black labels). Significant chemical shift changes were observed for KcsA-Kv1.3 in the region I62-D80 upon KTX binding. (C) The ^{13}C-^{13}C homonuclear spectrum enlarged region for the ligand; light gray and dark gray represent bound and unbound ligand, respectively. (D) Selected spectral region of a 2-D ^{15}N-^{13}C solid-state NMR spectrum of KcsA-Kv1.3, illustrating the chemical shift perturbation upon ligand binding for residues in the pore region (G77 and G79), but not in the transmembrane helices (G58), where the ligand bound and unbound peaks of G77 and G79 are labeled B and U, respectively. For additional details, refer to the text. (Illustrations adapted from Ref. [80].)

structural information is available about the ligand (drug) binding mechanism of these three classes of ATPases, this having been the target of several solid-state NMR studies.

The gastric H^+/K^+-ATPase, also known as the "gastric proton pump", is a P-type ATPase which secretes acid in the stomach. It is a member of the P-type E_1/E_2

ATPase family, and is composed of α and β subunits of molecular mass approximately 114 kDa and 34 kDa, respectively. As with many membrane proteins, only a low-resolution structure (18 Å) is available from electron microscopy studies [167]. Inhibition of the gastric proton pump reduces the secretion of gastric acid, thus decreasing the acidity of the stomach. Gastric proton pump inhibitors are prescribed when the acidity of the stomach is to be reduced, for example in the occurrence of gastric and duodenal ulcers, gastroesophageal reflux disease (GERD or acid-reflux), and to increase the efficacy of antibiotic treatment for *Helicobacter pylori*.

Solid-state NMR ligand-binding studies [168] were carried out on porcine gastric membranes enriched in the gastric pump (30–35% of total membrane protein). The ligands used in this study were non-covalent, K^+-competitive inhibitors, members of the SCH28080 series, imidazo[1,2-α]pyridine derivatives. Several SCH28080 analogues were synthesized which had half-maximal inhibitory concentration (IC_{50}) values in the sub-10 µM range. IC_{50} is defined in pharmacological research as the concentration of an inhibitor that is required for 50% inhibition of its target, in this case the K^+-stimulated ATPase activity. Titration of the deuterium-labeled inhibitor into H^+/K^+-ATPase membranes confirmed the inhibitor binding to the protein, and K^+ competition deuterium NMR measurements verified that the inhibitor was positioned within the active site. Conformation of the bound ligand was deduced based on ^{13}C–^{19}F distance measurements. Combined with findings of site-directed mutagenesis studies, the partially bowed inhibitor was modeled into the H^+/K^+-ATPase binding site [135, 168].

The Na^+/K^+-ATPase not only hydrolyzes ATP to drive the pumping of Na^+ and K^+ across the plasma membrane, but also acts as a receptor for digitalis compounds. These include cardiac glycosides occur mainly in plants (e.g., *Digitalis lanata*) from which the drug names have been derived. The physiological effect of these naturally occurring drugs has been known at least since 1500 BC, our ancestors having used them as emetics, diuretics, cardiovascular agents and as arrow poisons. These cardiac glycosides inhibit Na^+/K^+-ATPase, causing an increased force of myocardial contraction. Today, although they are used extensively to treat congestive heart failure and atrial fibrillation and flutter, their toxicity remains an unresolved issue.

In a solid-state NMR study [169], the structure of ouabain, the parent compound of the cardiac glycoside family, was resolved in the binding site of Na^+/K^+-ATPase. Ouabain, like other cardiac glycosides, is composed of two structural features, namely a sugar (rhamnose) and a rigid steroid moiety, which are connected through a single flexible ether link. Na^+/K^+-ATPase-enriched membranes were prepared from porcine kidney, and several active ouabain analogues synthesized that caused inhibition to various extents, albeit in the nanomolar to micromolar range (IC_{50} value). The ouabain derivatives were labeled at selected sites with NMR active isotopes (^{13}C, ^{19}F, and ^{2}H). ^{13}C–^{19}F distance measurements facilitated determination of the binding conformation of ouabain, while ^{2}H measurements provided insight into the motion of the bound ligand. The distance measurements between the sugar and steroid moieties (0.9 ± 0.05 nm) restricted the conformation to 90° relative to each other. It was considered most likely that the steroid moiety was interacting and being restrained by the protein surface, whereas the highly mobile rhamnose sugar extended into the solution.

3.3.5 Membrane Protein Soluble Alternatives

As investigations with membrane proteins can be very intricate, on occasion a shorter, water-soluble derivative of the protein may be derived for ligand-binding studies. For example, the interaction of a truncated, soluble form of the anti-apoptotic Bcl-xL with its ligand has been studied using solid-state NMR. The protein was selective for ^{13}C-labeling of the methyl groups of Leu, Val, and Ile (Cδ1), where the binding of a drug-like, organic molecule was easily monitored by the chemical shift of resonances in the Bcl-xl. This approach may also be extended to membrane proteins (especially to receptors), not only to characterize ligand interactions but also to assist in drug screening for pharmaceutical research. In another study, the labeling of unique amino acid pairs in the sequence provided insight into the binding mechanism of a substrate, *N*-palmitoylglycine, to the soluble cytochrome P450 BM-3 from *Bacillus megaterium* [170].

References

1 http://www.rcsb.org/pdb/holdings.html.
2 Russell, R. B. and Eggleston, D. S., New roles for structure in biology and drug discovery – Foreword, *Nat. Struct. Biol.*, *7*, 928 (2000).
3 Terstappen, G. C. and Reggiani, A., *In silico* research in drug discovery, *Trends Pharmacol. Sci.*, *22*, 23 (2001).
4 Klabunde, T. and Hessler, G., Drug design strategies for targeting G-protein-coupled receptors, *Chembiochem*, *3*, 929 (2002).
5 McDermott, A. E., Structural and dynamic studies of proteins by solid-state NMR spectroscopy: rapid movement forward, *Curr. Opin. Struct. Biol.*, *14*, 554 (2004).
6 Watts, A., Straus, S. K., Grage, S., Kamihira, M., Lam, Y.-H., and Xhao, Z., *Membrane protein structure determination using solid state NMR*, Vol. 278, Humana Press, New Jersey (2004).
7 Watts, A., Solid-state NMR in drug design and discovery for membrane-embedded targets, *Nat. Rev. Drug Discov.*, *4*, 555 (2005).
8 Hong, M., Oligomeric structure, dynamics, and orientation of membrane proteins from solid-state NMR, *Structure*, *14*, 1731 (2006).
9 Baldus, M., Molecular interactions investigated by multi-dimensional solid-state NMR, *Curr. Opin. Struct. Biol.*, *16*, 618 (2006).
10 Bockmann, A., Structural and dynamic studies of proteins by high-resolution solid-state NMR, *C. R. Chimie*, *9*, 381 (2006).
11 De Angelis, A. A., Jones, D. H., Grant, C. V., Park, S. H., Mesleh, M. F., and Opella, S. J., NMR experiments on aligned samples of membrane proteins, in *Nuclear Magnetic Resonance Of Biological Macromolecules, Part C*, Vol. 394, Elsevier Academic Press Inc., San Diego, pp. 350 (2005).
12 Chill, J. H., Louis, J. M., Miller, C., and Bax, A., NMR study of the tetrameric KcsA potassium channel in detergent micelles, *Protein Sci.*, *15*, 684 (2006).
13 Hwang, P. M., Choy, W. Y., Lo, E. I., Chen, L., Forman-Kay, J. D., Raetz, C. R. H., Prive, G. G., Bishop, R. E., and Kay,

L. E., Solution structure and dynamics of the outer membrane enzyme PagP by NMR, *Proc. Natl. Acad. Sci. USA, 99,* 13560 (2002).

14 Hwang, P. M., Bishop, R. E., and Kay, L. E., The integral membrane enzyme PagP alternates between two dynamically distinct states, *Proc. Natl. Acad. Sci. USA, 101,* 9618 (2004).

15 Hwang, P. M., and Kay, L. E., Solution structure and dynamics of integral membrane proteins by NMR: A case study involving the enzyme PagP, in *Nuclear Magnetic Resonance Of Biological Macromolecules, Part C,* Vol. 394, pp. 335 (2005).

16 Arora, A., Abildgaard, F., Bushweller, J. H., and Tamm, L. K., Structure of outer membrane protein A transmembrane domain by NMR spectroscopy, *Nat. Struct. Biol., 8,* 334 (2001).

17 Tamm, L. K., Abildgaard, F., Arora, A., Blad, H., and Bushweller, J. H., Structure, dynamics and function of the outer membrane protein A (OmpA) and influenza hemagglutinin fusion domain in detergent micelles by solution NMR, *FEBS Lett., 555,* 139 (2003).

18 Fernandez, C., Adeishvili, K., and Wuthrich, K., Transverse relaxation-optimized NMR spectroscopy with the outer membrane protein OmpX in dihexanoyl phosphatidylcholine micelles, *Proc. Natl. Acad. Sci. USA, 98,* 2358 (2001).

19 Fernandez, C., and Wuthrich, K., NMR solution structure determination of membrane proteins reconstituted in detergent micelles, *FEBS Lett., 555,* 144 (2003).

20 Rastogi, V. K., and Girvin, M. E., Structural changes linked to proton translocation by subunit c of the ATP synthase, *Nature, 402,* 263 (1999).

21 Ahn, H. C., Juranic, N., Macura, S., and Markley, J. L., Three-dimensional structure of the water-insoluble protein crambin in dodecylphosphocholine micelles and its minimal solvent-exposed surface, *J. Am. Chem. Soc., 128,* 4398 (2006).

22 Nussberger, S., Dorr, K., Wang, D. N., and Kuhlbrandt, W., Lipid-protein interactions in crystals of plant light-harvesting complex, *J. Mol. Biol., 234,* 347 (1993).

23 Fyfe, P. K., Hughes, A. V., Heathcote, P., and Jones, M. R., Proteins, chlorophylls and lipids: X-ray analysis of a three-way relationship, *Trends Plant Sci., 10,* 275 (2005).

24 van den Brink-van der Laan, E., Chupin, V., Killian, J. A., and de Kruijff, B., Stability of KcsA tetramer depends on membrane lateral pressure, *Biochemistry, 43,* 4240 (2004).

25 Sternberg, B., Lhostis, C., Whiteway, C. A., and Watts, A., The essential role of specific *Halobacterium halobium* polar lipids in 2D-array formation of bacteriorhodopsin, *Biochim. Biophys. Acta, 1108,* 21 (1992).

26 Sternberg, B., Watts, A., and Cejka, Z., Lipid-induced modulation of the protein packing in 2-dimensional crystals of bacteriorhodopsin, *J. Struct. Biol., 110,* 196 (1993).

27 Sabra, M. C., Uitdehaag, J. C. M., and Watts, A., General model for lipid-mediated two-dimensional array formation of membrane proteins: Application to bacteriorhodopsin, *Biophys. J., 75,* 1180 (1998).

28 Gil, T., Ipsen, J. H., Mouritsen, O. G., Sabra, M. C., Sperotto, M. M., and Zuckermann, M. J., Theoretical analysis of protein organization in lipid membranes, *Biochim. Biophys. Acta Rev. Biomembr., 1376,* 245 (1998).

29 Tian, C. L., Karra, M. D., Ellis, C. D., Jacob, J., Oxenoid, K., Sonnichsen, F., and Sanders, C. R., Membrane protein preparation for TROSY NMR screening, in *Nuclear Magnetic Resonance Of Biological Macromolecules, Part C,* Vol. 394, Elsevier Academic Press Inc., San Diego, pp. 321 (2005).

30 Sanders, C. R., and Sonnichsen, F., Solution NMR of membrane proteins: practice and challenges, *Magnet. Res. Chem., 44,* S24 (2006).

31 Petkova, A. T., Ishii, Y., Balbach, J. J., Antzutkin, O. N., Leapman, R. D., Delaglio, F., and Tycko, R., A structural model for Alzheimer's beta-amyloid fibrils based on experimental constraints from solid state NMR, *Proc. Natl. Acad. Sci. USA, 99*, 16742 (2002).

32 Tycko, R., Progress towards a molecular-level structural understanding of amyloid fibrils, *Curr. Opin. Struct. Biol., 14*, 96 (2004).

33 Jaroniec, C. P., MacPhee, C. E., Bajaj, V. S., McMahon, M. T., Dobson, C. M., and Griffin, R. G., High-resolution molecular structure of a peptide in an amyloid fibril determined by magic angle spinning NMR spectroscopy, *Proc. Natl. Acad. Sci. USA, 101*, 711 (2004).

34 Ireland, P. S., Olson, L. W., and Brown, T. L., Spin echo double resonance detection of deuterium quadrupole resonance transitions in pentacarbonylmanganese-d, *J. Am. Chem. Soc., 97*, 3548 (1975).

35 Langer, B., Schnell, L., Spiess, H. W., and Grimmer, A. R., Temperature calibration under ultrafast MAS conditions, *J. Magnet. Reson., 138*, 182 (1999).

36 Martin, R. W., Paulson, E. K., and Zilm, K. W., Design of a triple resonance magic angle sample spinning probe for high field solid state nuclear magnetic resonance, *Rev. Sci. Instrum., 74*, 3045 (2003).

37 Hong, M., Solid-state NMR determination of C-13 alpha chemical shift anisotropies for the identification of protein secondary structure, *J. Am. Chem. Soc., 122*, 3762 (2000).

38 Huster, D., Yamaguchi, S., and Hong, M., Efficient beta-sheet identification in proteins by solid-state NMR spectroscopy, *J. Am. Chem. Soc., 122*, 11320 (2000).

39 Luca, S., Filippov, D. V., van Boom, J. H., Oschkinat, H., de Groot, H. J. M., and Baldus, M., Secondary chemical shifts in immobilized peptides and proteins: A qualitative basis for structure refinement under Magic Angle Spinning, *J. Biomol. NMR, 20*, 325 (2001).

40 Pauli, J., van Rossum, B., Forster, H., de Groot, H. J. M., and Oschkinat, H., Sample optimization and identification of signal patterns of amino acid side chains in 2D RFDR spectra of the alpha-spectrin SH3 domain, *J. Magnet. Reson., 143*, 411 (2000).

41 Jakeman, D. L., Mitchell, D. J., Shuttleworth, W. A., and Evans, J. N. S., Effects of sample preparation conditions on biomolecular solid-state NMR lineshapes, *J. Biomol. NMR, 12*, 417 (1998).

42 Martin, R. W., and Zilm, K. W., Preparation of protein nanocrystals and their characterization by solid state NMR, *J. Magnet. Reson., 165*, 162 (2003).

43 Gregory, R. B., Gangoda, M., Gilpin, R. K., and Su, W., The influence of hydration on the conformation of lysozyme studied by solid-state C-13-NMR spectroscopy, *Biopolymers, 33*, 513 (1993).

44 Gregory, R. B., Gangoda, M., Gilpin, R. K., and Su, W., The influence of hydration on the conformation of bovine serum-albumin studied by solid-state C-13-NMR spectroscopy, *Biopolymers, 33*, 1871 (1993).

45 Lorch, M., Fahem, S., Kaiser, C., Weber, I., Mason, A. J., Bowie, J. U., and Glaubitz, C., How to prepare membrane proteins for solid-state NMR: a case study on the alpha-helical integral membrane protein diaglycerol kinase from *E. coli*, *ChemBioChem, 9*, 1693 (2005).

46 Hiller, M., Krabben, L., Vinothumar, K. R., Castellani, F., van Rossum, B. J., Kuhlbrandt, W., and Oschkinat, H., Solid-state magic-angle spinning NMR of outer membrane protein G from *Escherichia coli*, *ChemBioChem, 6*, 1679 (2005).

47 McDermott, A., Polenova, T., Bockmann, A., Zilm, K. W., Paulsen, E. K., Martin, R. W., and Montelione, G. T., Partial NMR assignments for uniformly (C 13, N-15)-enriched BPTI in the solid state, *J. Biomol. NMR, 16*, 209 (2000).

48 Igumenova, T., McDermott, A. E., Zilm, K. W., Martin, R. W., Paulson, E. K., and Wand, J. A., Assignments of carbon NMR resonances for microcrystalline ubiquitin, *J. Am. Chem. Soc., 126*, 6720 (2004).

49 Marulanda, D., Tasayco, M. L., McDermott, A., Cataldi, M., Arriaran, V., and Polenova, T., Magic angle spinning solid-state NMR spectroscopy for structural studies of protein interfaces. Resonance assignments of differentially enriched Escherichia coli thioredoxin reassembled by fragment complementation, *J. Am. Chem. Soc.*, *126*, 16608 (2004).

50 Varga, K., *Membrane Protein Secondary Structure and Spectral Assignments by Solid State NMR: S. lividans KcsA Potassium Channel and E. coli ATP Synthase Subunit c.* PhD Thesis, Columbia University, New York (2005).

51 Marley, J., Lu, M., and Bracken, C., A method for efficient isotopic labeling of recombinant proteins, *J. Biomol. NMR*, *20*, 71 (2001).

52 Jansson, M., Li, Y. C., Jendeberg, L., anderson, S., Montelione, G. T., and Nilsson, B., High-level production of uniformly N-15- and C-13-enriched fusion proteins in *Escherichia coli*, *J. Biomol. NMR*, *7*, 131 (1996).

53 Cai, M. L., Huang, Y., Sakaguchi, K., Clore, G. M., Gronenborn, A. M., and Craigie, R., An efficient and cost-effective isotope labeling protocol for proteins expressed in *Escherichia coli*, *J. Biomol. NMR*, *11*, 97 (1998).

54 Staunton, D., Schlinkert, R., Zanetti, G., Colebrook, S. A., and Campbell, L. D., Cell-free expression and selective isotope labelling in protein NMR, *Magnet. Reson. Chemistry*, *44*, S2 (2006).

55 Koglin, A., Klarnmt, C., Trbovic, N., Schwarz, D., Schneider, B., Schafer, B., Lohr, F., Bernhard, F., and Dotsch, V., Combination of cell-free expression and NMR spectroscopy as a new approach for structural investigation of membrane proteins, *Magnet. Reson. Chemistry*, *44*, S17 (2006).

56 Muralidharan, V., and Muir, T. W., Protein ligation: an enabling technology for the biophysical analysis of proteins, *Nature Methods*, *3*, 429 (2006).

57 Lian, L. Y., and Middleton, D. A., Labelling approaches for protein structural studies by solution-state and solid-state NMR, *Prog. Nuclear Magnet. Reson. Spectrosc.*, *39*, 171 (2001).

58 Hong, M., and Jakes, K., Selective and extensive C-13 labeling of a membrane protein for solid-state NMR investigations, *J. Biomol. NMR*, *14*, 71 (1999).

59 LeMaster, D. M., and Kushlan, D. M., Dynamical mapping of E-coli thioredoxin via C-13 NMR relaxation analysis, *J. Am. Chem. Soc.*, *118*, 9255 (1996).

60 Straus, S. K., Bremi, T., and Ernst, R. R., Side-chain conformation and dynamics in a solid peptide: CP-MAS NMR study of valine rotamers and methyl-group relaxation in fully C-13-labelled antamanide, *J. Biomol. NMR*, *10*, 119 (1997).

61 Detken, A., Hardy, E. H., Ernst, M., Kainosho, M., Kawakami, T., Aimoto, S., and Meier, B. H., Methods for sequential resonance assignment in solid, uniformly C-13, N-15 labelled peptides: Quantification and application to antamanide, *J. Biomol. NMR*, *20*, 203 (2001).

62 Hong, M., and Griffin, R. G., Resonance assignments for solid peptides by dipolar-mediated C-13/N-15 correlation solid-state NMR, *J. Am. Chem. Soc.*, *120*, 7113 (1998).

63 Rienstra, C. M., Tucker-Kellogg, L., Jaroniec, C. P., Hohwy, M., Reif, B., McMahon, M. T., Tidor, B., Lozano-Perez, T., and Griffin, R. G., De novo determination of peptide structure with solid-state magic-angle spinning NMR spectroscopy, *Proc. Natl. Acad. Sci. USA*, *99*, 10260 (2002).

64 Pauli, J., Baldus, M., van Rossum, B., de Groot, H., and Oschkinat, H., Backbone and side-chain C-13 and N-15 signal assignments of the alpha-spectrin SH3 domain by magic angle spinning solid-state NMR at 17.6 tesla, *Chembiochem*, *2*, 272 (2001).

65 van Rossum, B. J., Castellani, F., Rehbein, K., Pauli, J., and Oschkinat, H., Assignment of the nonexchanging protons of the alpha-spectrin SH3 domain by two- and three-dimensional H-1-C-13 solid-state magic-angle spinning NMR and comparison of solution and solid-state proton chemical shifts, *ChemBioChem*, *2*, 906 (2001).

66 van Rossum, B. J., Castellani, F., Pauli, J., Rehbein, K., Hollander, J., de Groot, H. J. M., and Oschkinat, H., Assignment of amide proton signals by combined evaluation of HN, NN and HNCA MAS-NMR correlation spectra, *J. Biomol. NMR, 25,* 217 (2003).

67 Castellani, F., van Rossum, B., Diehl, A., Schubert, M., Rehbein, K., and Oschkinat, H., Structure of a protein determined by solid-state magic-angle-spinning NMR spectroscopy, *Nature, 420,* 98 (2002).

68 Castellani, F., van Rossum, B. J., Diehl, A., Rehbein, K., and Oschkinat, H., Determination of solid-state NMR structures of proteins by means of three-dimensional N-15-C-13-C-13 dipolar correlation spectroscopy and chemical shift analysis, *Biochemistry, 42,* 11476 (2003).

69 Hong, M., Resonance assignment of C-13/N-15 labeled solid proteins by two- and three-dimensional magic-angle-spinning NMR, *J. Biomol. NMR, 15,* 1 (1999).

70 Igumenova, T., Wand, J. A., and McDermott, A., Assignment of the backbone resonances for microcrystalline ubiquitin, *J. Am. Chem. Soc., 126,* 5323 (2004).

71 Zech, S. G., Wand, A. J., and McDermott, A. E., Protein structure determination by high-resolution solid-state NMR spectroscopy: Application to microcrystalline ubiquitin, *J. Am. Chem. Soc., 127,* 8618 (2005).

72 Seidel, K., Etzkorn, M., Heise, H., Becker, S., and Baldus, M., High-Resolution Solid-State NMR Studies on Uniformly [13C, 15N]-Labeled Ubiquitin, *ChemBioChem, 9,* 1638 (2005).

73 Bockmann, A., Lange, A., Galinier, A., Luca, S., Giraud, N., Juy, M., Heise, H., Montserret, R., Penin, F., and Baldus, M., Solid state NMR sequential resonance assignments and conformational analysis of the 2×10.4 kDa dimeric form of the *Bacillus subtilis* protein Crh, *J. Biomol. NMR, 27,* 323 (2003).

74 Fujiwara, T., Todokoro, Y., Yanagishita, H., Tawarayama, M., Kohno, T., Wakamatsu, K., and Akutsu, H., Signal assignments and chemical-shift structural analysis of uniformly C-13, N-15-labeled peptide, mastoparan-X, by multidimensional solid-state NMR under magic-angle spinning, *J. Biomol. NMR, 28,* 311 (2004).

75 Franks, W. T., Zhou, D. H., Wylie, B. J., Money, B. G., Graesser, D. T., Frericks, H. L., Sahota, G., and Rienstra, C. M., Magic-angle spinning solid-state NMR spectroscopy of the beta 1 immunoglobulin binding domain of protein G (GB1): N-15 and C-13 chemical shift assignments and conformational analysis, *J. Am. Chem. Soc., 127,* 12291 (2005).

76 Egorova-Zachernyuk, T. A., Hollander, J., Fraser, N., Gast, P., Hoff, A. J., CogDell, R., de Groot, H. J. M., and Baldus, M., Heteronuclear 2D-correlations in a uniformly C-13, N-15 labeled membrane-protein complex at ultra-high magnetic fields, *J. Biomol. NMR, 19,* 243 (2001).

77 van Gammeren, A. J., Buda, F., Hulsbergen, F. B., Kiihne, S., Hollander, J. G., Egorova-Zachernyuk, T. A., Fraser, N. J., Cogdell, R. J., and De Groot, H. J. M., Selective chemical shift assignment of B800 and B850 bacteriochlorophylls in uniformly [C-13,N-15]-labeled light-harvesting complexes by solid-state NMR spectroscopy at ultra-high magnetic field, *J. Am. Chem. Soc., 127,* 3213 (2005).

78 van Gammeren, A. J., Hulsbergen, F. B., Hollander, J. G., and de Groot, H. J. M., Residual backbone and side-chain 13C and 15N resonance assignments of the intrinsic transmembrane light-harvesting 2 protein complex by solid-state Magic Angle Spinning NMR spectroscopy, *J. Biomol. NMR, 31,* 279 (2005).

79 Huang, L., *Solid state NMR spectral properties and partial sequential assignment of a membrane protein, the Light Harvesting Complex 1 (LH1).* PhD Thesis, Columbia University, New York (2005).

80 Lange, A., Giller, K., Hornig, S., Martin-Eauclaire, M.-F., Pongs, O., Becker, S., and Baldus, M., Toxin-induced conformational changes in a potassium

channel revealed by solid-state NMR, *Nature, 440,* 959 (2006).

81 Lange, A., Giller, K., Pongs, O., Becker, S., and Baldus, M., Two-dimensional solid-state NMR applied to a chimeric potassium channel, *J. Receptors Signal Transduct., 26,* 379 (2006).

82 Etzkorn, M., Martell, S., Andronesi, O. C., Seidel, K., Engelhard, M., and Baldus, M., Secondary structure, dynamics, and topology of a seven-helix receptor in native membranes, studied by solid-state NMR spectroscopy, *Angew. Chem. Int. Ed., 9999,* NA (2006).

83 Jehle, S., Hiller, M., Rehbein, K., Diehl, A., Oschkinat, H., and van Rossum, B. J., Spectral editing: selection of methyl groups in multidimensional solid-state magic-angle spinning NMR, *J. Biomol. NMR, 36,* 169 (2006).

84 Kobayashi, M., Matsuki, Y., Yumen, I., Fujiwara, T., and Akutsu, H., Signal assignment and secondary structure analysis of a uniformly [^{13}C, ^{15}N]-labeled membrane protein, H$^+$-ATP synthase subunit *c*, by magic-angle spinning solid-state NMR, *J. Biomol. NMR 36,* 279 (2006).

85 Bax, A., Kontaxis, G., and Tjandra, N., Dipolar couplings in macromolecular structure determination, *Methods Enzymol., 339,* 127 (2001).

86 Bax, A., Weak alignment offers new NMR opportunities to study protein structure and dynamics, *Protein Sci., 12,* 1 (2003).

87 de Alba, E., and Tjandra, N., NMR dipolar couplings for the structure determination of biopolymers in solution, *Prog. NMR Spectrosc., 40,* 175 (2002).

88 Gronenborn, A. M., The importance of being ordered: Improving NMR structures using residual dipolar couplings, *C. R. Biol., 325,* 957 (2002).

89 Lee, S., Mesleh, M. F., and Opella, S. J., Structure and dynamics of a membrane protein in micelles from three solution NMR experiments, *J. Biomol. NMR, 26,* 327 (2003).

90 Chou, J. J., Gaemers, S., Howder, B., Louis, J. M., and Bax, A., A simple apparatus for generating stretched polyacrylamide gels, yielding uniform alignment of proteins and detergent micelles, *J. Biomol. NMR, 21,* 377–382. (2001).

91 Jones, D. H., and Opella, S. J., Weak alignment of membrane proteins in stressed polyacrylamide gels, *J. Magnet. Reson., 171,* 258–269 (2004).

92 Meier, S., Haussinger, D., and Grzesiek, S., Charged acrylamide copolymer gels as media for weak alignment, *J. Biomol. NMR, 24,* 351–356 (2002).

93 Sass, H. J., Musco, G., Stahl, S. J., Wingfield, P. T., and Grzesiek, S., Solution NMR of proteins within polyacrylamide gels: Diffusional properties and residual alignment by mechanical stress or embedding of oriented purple membranes, *J. Biomol. NMR,* 303–309 (2000).

94 Tycko, R., Blanco, F. J., and Ishii, Y., Alignment of biopolymers in strained gels: A new way to create detectable dipole-dipole couplings in high-resolution biomolecular NMR, *J. Am. Chem. Soc., 122,* 9340 (2000).

95 Ma, C., and Opella, S. J., Lanthanide ions bind specifically to an added "EF-hand" and orient a membrane protein in micelles for solution NMR spectroscopy, *J. Magnet. Reson., 146,* 381 (2000).

96 Veglia, G., and Opella, S. J., Lanthanide ion binding to adventitious sites aligns membrane proteins in micelles for solution NMR spectroscopy, *J. Am. Chem. Soc., 122,* 11733 (2000).

97 Triba, M. N., Zoonens, M., Popot, J. L., Devaux, P. F., and Warschawski, D. E., Reconstitution and alignment by a magnetic field of a beta-barrel membrane protein in bicelles, *Eur. Biophys. J. Biophys. Lett., 35,* 268 (2006).

98 De Angelis, A. A., Nevzorov, A. A., Park, S. H., Howell, S. C., Mrse, A. A., and Opella, S. J., High-resolution NMR spectroscopy of membrane proteins in aligned bicelles, *J. Am. Chem. Soc., 126,* 15340 (2004).

99 Dvinskikh, S. V., Durr, U. H. N., Yamamoto, K., and Ramamoorthy, A., High-resolution 2D NMR spectroscopy of bicelles to measure the membrane interaction of ligands, *J. Am. Chem. Soc.*, *129*, 794 (2007).

100 Andrew, E. R., Bradbury, A., and Eades, R. G., Nuclear magnetic resonance spectra from a crystal rotated at high speed, *Nature*, *4650*, 1659 (1958).

101 Andrew, E. R., Bradbury, A., and Eades, R. G., Removal of dipolar broadening of nuclear magnetic resonance spectra of solids by specimen rotation, *Nature*, *183*, 1802 (1959).

102 Lowe, I. J., Free induction decay of rotating solids, *Phys. Rev. Lett.*, *2*, 285 (1959).

103 Samoson, A., Tuherm, T., Past, J., Reinhold, A., Anupold, T., and Heinmaa, I., New horizons for magic-angle spinning NMR, *Top. Curr. Chem.*, *246*, 15 (2005).

104 Spera, S., and Bax, A., Empirical correlation between protein backbone conformation and C-alpha and C-beta C-13 nuclear-magnetic-resonance chemical-shifts, *J. Am. Chem. Soc.*, *113*, 5490 (1991).

105 Wishart, D. S., and Sykes, B. D., *Chemical-shifts as a tool for structure determination, in Nuclear Magnetic Resonance, Part C*, Vol. 239, Academic Press Inc, San Diego, pp. 363 (1994).

106 Zhang, H. Y., Neal, S., and Wishart, D. S., RefDB: A database of uniformly referenced protein chemical shifts, *J. Biomol. NMR*, *25*, 173 (2003).

107 Cornilescu, G., Delaglio, F., and Bax, A., Protein backbone angle restraints from searching a database for chemical shift and sequence homology, *J. Biomol. NMR*, *13*, 289 (1999).

108 Wu, C. H., Ramamoorthy, A., and Opella, S. J., High-resolution heteronuclear dipolar solid-state NMR-spectroscopy, *J. Magnet. Reson. Ser. A*, *109*, 270 (1994).

109 Ramamoorthy, A., Wu, C. H., and Opella, S. J., Experimental aspects of multidimensional solid-state NMR correlation spectroscopy, *J. Magnet. Reson.*, *140*, 131 (1999).

110 Marassi, F. M., and Opella, S. J., A solid-state NMR index of helical membrane protein structure and topology, *J. Magnet. Reson.*, *144*, 150 (2000).

111 Wang, J., Denny, J., Tian, C., Kim, S., Mo, Y., Kovacs, F., Song, Z., Nishimura, K., Gan, Z., Fu, R., Quine, J. R., and Cross, T. A., Imaging membrane protein helical wheels, *J. Magnet. Reson.*, *144*, 162 (2000).

112 Mesleh, M. F., Lee, S., Veglia, G., Thiriot, D. S., Marassi, F. M., and Opella, S. J., Dipolar waves map the structure and topology of helices in membrane proteins, *J. Am. Chem. Soc.*, *125*, 8928 (2003).

113 Mesleh, M. F., and Opella, S. J., Dipolar waves as NMR maps of helices in proteins, *J. Magnet. Reson.*, *163*, 288 (2003).

114 Howell, S. C., Mesleh, M. F., and Opella, S. J., NMR structure determination of a membrane protein with two transmembrane helices in micelles: MerF of the bacterial mercury detoxification system, *Biochemistry*, *44*, 5196 (2005).

115 Straus, S. K., Scott, W. R. P., and Watts, A., Assessing the effects of time and spatial averaging in N-15 chemical shift/N-15-H-1 dipolar correlation solid state NMR experiments, *J. Biomol. NMR*, *26*, 283 (2003).

116 Vosegaard, T., and Nielsen, N. C., Towards high-resolution solid-state NMR on large uniformly ^{15}N- and [^{13}C, ^{15}N]-labeled membrane proteins in oriented lipid bilayers, *J. Biomol. NMR*, *22*, 225 (2002).

117 Ketchem, R. R., Hu, W., and Cross, T. A., High-resolution conformation of gramicidin A in a lipid bilayer by solid-state NMR, *Science*, *261*, 1457 (1993).

118 Ketchem, R. R., Lee, K. C., Huo, S., and Cross, T. A., Macromolecular structural elucidation with solid-state NMR-derived orientational constraints, *J. Biomol. NMR*, *8*, 1 (1996).

119 Wang, J. F., Kovacs, F., and Cross, T. A., Structure of the transmembrane region of the M2 protein H^+ channel, *Protein Sci.*, *10*, 2241 (2001).

120 Tian, C., Tobler, K., Lamb, R. A., Pinto, L. H., and Cross, T. A., Expression and initial structural insights from solid state NMR of the M2 proton channel from influenza A virus, *Biochemistry, 41,* 11294 (2002).

121 Nishimura, K., Kim, S. G., Zhang, L., and Cross, T. A., The closed state of a H^+ channel helical bundle combining precise orientational and distance restraints from solid state NMR – 1, *Biochemistry, 41,* 13170 (2002).

122 Opella, S. J., Marassi, F. M., Gesell, J. J., Valente, A. P., Kim, Y., Oblatt-Montal, M., and Montal, M., Structures of the M2 channel-lining segments from nicotinic acetylcholine and NMDA receptors by NMR spectroscopy, *Nat. Struct. Biol., 6,* 374 (1999).

123 Park, S. H., Mrse, A. A., Nevzorov, A. A., Mesleh, M. F., Oblatt-Montal, M., Montal, M., and Opella, S. J., Three-dimensional structure of the channel-forming trans-membrane domain of virus protein "u" (Vpu) from HIV-1, *J. Mol. Biol., 333,* 409 (2003).

124 Zeri, A. C., Mesleh, M. F., Nevzorov, A. A., and Opella, S. J., Structure of the coat protein in fd filamentous bacteriophage particles determined by solid state NMR spectroscopy, *Proc. Natl. Acad. Sci. USA, 100,* 6458 (2003).

125 Thiriot, D. S., Nevzorov, A. A., Zagyanskiy, L., Wu, C. H., and Opella, S. J., Structure of the coat protein in Pf1 bacteriophage determined by solid-state NMR spectroscopy, *J. Molec. Biol., 341,* 869 (2004).

126 Buffy, J. J., Traaseth, N. J., Mascioni, A., Gor'kov, P. L., Chekmenev, E. Y., Brey, W. W., and Veglia, G., Two-dimensional solid-state NMR reveals two topologies of sarcolipin in oriented lipid bilayers, *Biochemistry, 45,* 10939 (2006).

127 Kamihira, M., Vosegaard, T., Mason, A. J., Straus, S. K., Nielsen, N. C., and Watts, A., Structural and orientational constraints of bacteriorhodopsin in purple membranes determined by oriented-sample solid-state NMR spectroscopy, *J. Struct. Biol., 149,* 7 (2005).

128 Kamihira, M., and Watts, A., Functionally relevant coupled dynamic profile of bacteriorhodopsin and lipids in purple membranes, *Biochemistry, 45,* 4304 (2006).

129 Glaubitz, C., and Watts, A., Magic angle-oriented sample spinning (MAOSS): A new approach toward biomembrane studies, *J. Magnet. Reson., 130,* 305 (1998).

130 Gröbner, G., Burnett, I. J., Glaubitz, C., Chol, G., Mason, A. J., and Watts, A., Observation of light-induced structural changes of retinal within rhodopsin, *Nature, 405,* 810 (2000).

131 Glaubitz, C., Burnett, I. J., Gröbner, G., Mason, A. J., and Watts, A., Deuterium-MAS NMR spectroscopy on oriented membrane proteins: applications to photointermediates, *J. Am. Chem. Soc., 121,* 5787 (1999).

132 Mason, A. J., Grage, S. L., Straus, S. K., Glaubitz, C., and Watts, A., Identifying anisotropic constraints in multiply labeled bacteriorhodopsin by ^{15}N MAOSS NMR: A general approach to structural studies of membrane proteins, *Biophys. J., 86,* 1610 (2004).

133 Glaubitz, C., Gröbner, G., and Watts, A., Structural and orientational information of the membrane embedded M13 coat protein by 13C-MAS NMR spectroscopy, *Biochim. Biophys. Acta, 1463,* 151 (2000).

134 Becker, O. M., Shacham, S., Marantz, Y., and Noiman, S., Modeling the 3D structure of GPCRs: Advances and application to drug discovery, *Curr. Opin. Drug Discov. Dev., 6,* 353 (2003).

135 Kim, C. G., Watts, J. A., and Watts, A., Ligand docking in the gastric H+/K+-ATPase: Homology modeling of reversible inhibitor binding sites, *J. Med. Chem., 48,* 7145 (2005).

136 Spooner, P. J. R., Veenhoff, L. M., Watts, A., and Poolman, B., Structural information on a membrane transport protein from nuclear magnetic resonance spectroscopy using sequence-selective nitroxide labeling, *Biochemistry, 38,* 9634 (1999).

137 Palczewski, K., Kumasaka, T., Hori, T., Behnke, C. A., Motoshima, H., Fox, B. A., Le Trong, I., Teller, D. C., Okada, T., Stenkamp, R. E., Yamamoto, M., and Miyano, M., Crystal structure of rhodopsin: A G protein-coupled receptor, *Science, 277,* 687 (2000).

138 Stenkamp, R. E., Teller, D. C., and Palczewski, K., Rhodopsin: A Structural Primer for G-Protein Coupled Receptors, *Archiv. der Pharmazie, 338,* 209 (2005).

139 Creuzet, F., McDermott, A., Gebhard, R., Vanderhoef, K., Spijkerassink, M. B., Herzfeld, J., Lugtenburg, J., Levitt, M. H., and Griffin, R. G., Determination of membrane-protein structure by rotational resonance NMR – Bacteriorhodopsin, *Science, 251,* 783 (1991).

140 Ulrich, A. S., Heyn, M. P., and Watts, A., Structure determination of the cyclohexene ring of retinal in bacteriorhodopsin by solid-state deuterium NMR, *Biochemistry, 31,* 10390 (1992).

141 Ulrich, A. S., Wallat, I., Heyn, M. P., and Watts, A., Re-orientation of retinal in the M-photointermediate of bacteriorhodopsin, *Nat. Struct. Biol., 2,* 190 (1995).

142 Spooner, P. J. R., Sharples, J. M., Verhoeven, M. A., Lugtenburg, J., Glaubitz, C., and Watts, A., Relative orientation between the beta-ionone ring and the polyene chain for the chromophore of rhodopsin in native membranes, *Biochemistry, 41,* 7549 (2002).

143 Spooner, P. J. R., Sharples, J. M., Goodall, S. C., Seedorf, H., Verhoeven, M. A., Lugtenburg, J., Bovee-Geurts, P. H. M., DeGrip, W. J., and Watts, A., Conformational similarities in the beta-ionone ring region of the rhodopsin chromophore in its ground state and after photoactivation to the metarhodopsin-I intermediate, *Biochemistry, 42,* 13371 (2003).

144 Feng, X., Verdegem, P. J. E., EDen, M., Sandstrom, D., Lee, Y. K., Bovee-Geurts, P. H., de Grip, W. J., Lugtenburg, J., de Groot, H. J., and Levitt, M. H., Determination of a molecular torsional angle in the metarhodopsin-I photointermediate of rhodopsin by double-quantum solid-state NMR, *J. Biomol. NMR, 16,* 1 (2000).

145 Creemers, A. F., Kiihne, S., Bovee-Geurts, P. H., DeGrip, W. J., Lugtenburg, J., and de Groot, H. J., (1)H and (13)C MAS NMR evidence for pronounced ligand-protein interactions involving the ionone ring of the retinylidene chromophore in rhodopsin, *Proc. Natl. Acad. Sci. USA, 99,* 9101 (2002).

146 Salgado, G. F. J., Struts, A. V., Tanaka, K., Fujioka, N., Nakanishi, K., and Brown, M. F., Deuterium NMR structure of retinal in the ground state of rhodopsin, *Biochemistry, 43,* 12819 (2004).

147 Carravetta, M., Zhao, X., Johannessen, O. G., Lai, W. C., Verhoeven, M. A., Bovee-Geurts, P. H. M., Verdegem, P. J. E., Kiihne, S., Luthman, H., de Groot, H. J. M., deGrip, W. J., Lugtenburg, J., and Levitt, M. H., Protein-induced bonding perturbation of the rhodopsin chromophore detected by double-quantum solid-state NMR, *J. Am. Chem. Soc., 126,* 3948 (2004).

148 Etzkorn, M., Martell, S., Andronesi, O., Seidel, K., Engelhard, M., and Baldus, M., Secondary structure, dynamics, and topology of a seven-helix receptor in native membranes, studied by solid-state NMR spectroscopy, *Angew. Chem. Int. Ed., 46,* 459 (2007).

149 Ratnala, V. R. P., Kiihne, S. R., Buda, F., Leurs, R., deGroot, H. J. M., and DeGrip, W. J., Solid-State NMR evidence for a protonation switch in the binding pocket of the H1 receptor upon binding of the agonist histamine, *J. Am. Chem. Soc., 129,* 867 (2007).

150 Williamson, P. T. F., Bains, S., Chung, C., Cooke, R., and Watts, A., Probing the environment of neurotensin whilst bound to the neurotensin receptor by solid state NMR, *FEBS Lett., 518,* 111 (2002).

151 Luca, S., White, J. F., Sohal, A. K., Filippov, D. V., van Boom, J. H., Grisshammer, R., and Baldus, M., The conformation of neurotensin bound to its G protein-coupled receptor, *Proc. Natl. Acad. Sci. USA, 100,* 10706 (2003).

152 Miyazawa, A., Fujiyoshi, Y., and Unwin, N., Structure and gating mechanism of the acetylcholine receptor pore, *Nature, 423,* 949 (2003).

153 Williamson, P. T. F., Grobner, G., Spooner, P. J. R., Miller, K. W., and Watts, A., Probing the agonist binding pocket in the nicotinic acetylcholine receptor: A high-resolution solid-state NMR approach, *Biochemistry, 37,* 10854 (1998).

154 Williamson, P. T. F., Watts, J. A., Addona, G. H., Miller, K. W., and Watts, A., Dynamics and orientation of N+(CD3)(3)-bromoacetylcholine bound to its binding site on the nicotinic acetylcholine receptor, *Proc. Natl. Acad. Sci. USA, 98,* 2346 (2001).

155 Williamson, P. T. F., Verhoeven, A., Miller, K. W., Watts, A., and Meier, B. H., Structural studies of acetylcholine bound to the nicotinic acetylcholine receptor, *Biophys. J., 84,* 278A (2003).

156 Doyle, D. A., Cabral, J. M., Pfuetzner, R. A., Kuo, A. L., Gulbis, J. M., Cohen, S. L., Chait, B. T., and MacKinnon, R., The structure of the potassium channel: Molecular basis of K+ conduction and selectivity, *Science, 280,* 69 (1998).

157 Long, S. B., Campbell, E. B., and MacKinnon, R., Crystal structure of a mammalian voltage-dependent Shaker family K+ channel, *Science, 309,* 897 (2005).

158 Jiang, Y. X., Lee, A., Chen, J. Y., Cadene, M., Chait, B. T., and MacKinnon, R., The open pore conformation of potassium channels, *Nature, 417,* 523 (2002).

159 Heginbotham, L., Abramson, T., and Mackinnon, R., A functional connection between the pores of distantly related ion channels as revealed by mutant K+ channels, *Science, 258,* 1152 (1992).

160 Heginbotham, L., Lu, Z., Abramson, T., and Mackinnon, R., Mutations in the K+ channel signature sequence, *Biophys. J., 66,* 1061 (1994).

161 Zhou, Y. F., Morais-Cabral, J. H., Kaufman, A., and MacKinnon, R., Chemistry of ion coordination and hydration revealed by a K+ channel-Fab complex at 2.0 angstrom resolution, *Nature, 414,* 43 (2001).

162 Legros, C., Pollmann, V., Knaus, H. G., Farrell, A. M., Darbon, H., Bougis, P. E., Martin-Eauclaire, M. F., and Pongs, O., Generating a high affinity scorpion toxin receptor in KcsA-Kv1.3 chimeric potassium channels, *J. Biol. Chem., 275,* 16918 (2000).

163 Xu, J. C., Wang, P. L., Li, Y. Y., Li, G. Y., Kaczmarek, L. K., Wu, Y. L., Koni, P. A., Flavell, R. A., and Desir, G. V., The voltage-gated potassium channel Kv1.3 regulates peripheral insulin sensitivity, *Proc. Natl. Acad. Sci. USA, 101,* 3112 (2004).

164 Toyoshima, C., Nakasako, M., Nomura, H., and Ogawa, H., Crystal structure of the calcium pump of sarcoplasmic reticulum at 2.6 angstrom resolution, *Nature, 405,* 647 (2000).

165 Toyoshima, C., and Nomura, H., Structural changes in the calcium pump accompanying the dissociation of calcium, *Nature, 418,* 605 (2002).

166 Hilge, M., Siegal, G., Vuister, G. W., Guntert, P., Gloor, S. M., and Abrahams, J. P., ATP-induced conformational changes of the nucleotide-binding domain of Na,K-ATPase, *Nat. Struct. Biol., 10,* 468 (2003).

167 Xian, Y. J., and Hebert, H., Three-dimensional structure of the porcine gastric H,K-ATPase from negatively stained crystals, *J. Struct. Biol., 118,* 169 (1997).

168 Watts, J. A., Watts, A., and Middleton, D. A., A model of reversible inhibitors in gastric H^+/K^+-ATPase binding site determined by rotational echo double resonance NMR, *J. Biol. Chem., 276,* 43197 (2001).

169 Middleton, D. A., Rankin, S., Esmann, M., and Watts, A., Structural insights into the binding of cardiac glycosides to the digitalis receptor revealed by solid-state NMR, *Proc. Natl. Acad. Sci. USA, 97,* 13602 (2000).

170 Jovanovic, T., and McDermott, A. E., Observation of ligand binding to cytochrome P450-BM-3 by means of solid-state NMR spectroscopy, *J. Am. Chem. Soc., 127,* 13816 (2005).

Part III
Molecular Interaction and Large Assemblies

4
Analytical Ultracentrifugation: Membrane Protein Assemblies in the Presence of Detergent

Christine Ebel, Jesper V. Møller and Marc le Maire

4.1
Introduction

Analytical ultracentrifugation (AUC) is the classical method for determining the molecular mass and size of proteins. Despite the fact that, at present, precise molecular masses must be accessed through extensive sequencing information or mass spectroscopy, analytical ultracentrifugation still plays a leading role, as it has the ability to combine particle separation and analysis in a powerful technique for the determination – in a rigorous, thermodynamic manner – of the distribution, size, mass, composition, and interactions of even complex macromolecular assemblies in solution.

Perhaps, today, nowhere is the use of AUC of more importance than in the case of membrane proteins when attempting to correlate organization with regulatory aspects, and to define the role that quaternary structure (monomeric/oligomeric organization) plays in basic protein function. These areas include the translocation of substrates through transport proteins (see e.g., Refs. [1–6]), the phosphorylation of membrane receptors ([7]), and protein–protein interactions, for example in GTP-regulated membrane proteins (for a review, see Ref. [8]).

In contrast to methods such as crosslinking, native gel electrophoresis, and fluorescence depolarization, unambiguous information on the aggregation state can be acquired by using AUC [9, 10]. Whilst small-angle laser light-scattering is also rigorously founded, it must be coupled to chromatography for best effect [11]. The main problem encountered is that both light scattering and AUC require the use of solubilizing detergents to transfer integral membrane proteins from their membranous environment to a micellar phase (with or without the retention of some membranous lipid). Moreover, the effect of this transfer must be taken into account in terms of both structural and functional properties [12].

The aim of this chapter is to highlight the current potential and perspectives of analytical ultracentrifugation in the detergent-based solubilization of membrane proteins. Following a brief description of the present-day instrumentation and typical experiments, the macromolecular parameters that determine

Biophysical Analysis of Membrane Proteins. Investigating Structure and Function. Edited by Eva Pebay-Peyroula

ISBN: 978-3-527-31677-9

sedimentation will be outlined. The theory of macromolecule transport will be then be briefly described, before some general principles of modern data analysis are detailed. For clarity, selected applications and the perspectives of the method are discussed throughout the chapter.

Historical references relating to the use of analytical ultracentrifugation for water-soluble proteins can be found in Refs. [13–15], whilst equivalent information for membrane proteins, where Tanford and colleagues were the pioneers [16], is also available. It should be noted that many general biophysics textbooks will contain good descriptions of AUC concepts [17–20], while specific reviews for the AUC of membrane proteins are also available [21–25].

Following a decline during the 1970s, when the classical Beckman Model E ultracentrifuge was outphased, the commercialization of modern, easy-to-use instrumentation, incorporating numerical data acquisition and absorption optics (Beckman-Coulter model XLA), followed by the model XLI with interference optics as an additional feature [26], and recently also fluorescence detection [27, 28], has led to a renaissance of the technique during the 1990s. Consequently, a new generation of AUC enthusiasts has developed novel strategies for the design of experiments or for data analysis including hydrodynamics modeling, thereby allowing highly efficient, relatively easy-to-use and freely available procedures. On a routine basis the AUC scientific community shares its knowledge, general information and software at its forum "Reversible Associations in Structural and Molecular Biology" (RASMB [29]). Further evidence of the present-day vitality of the field is available not only in recent reference books [30, 31] but also in a selection of tutorial reviews [32–36].

4.2
Instrumentation and the Principle of Typical Experiments

Macromolecules subjected to a centrifugal field (of up to 300 000 g in commercially available instruments) will be redistributed in solution, and the resulting sedimentation profiles may be measured by using absorbance and interference optics:

- Absorbance can be related to concentration by knowing the extinction coefficient. Although most proteins absorb at 280 nm according to their content of aromatic amino acids, many detergents and lipids do not absorb at this wavelength.
- The interference optics generates interference fringes, which measure radial changes in the index of refraction of the solution. This technique is sensitive to concentration variations of all components in particular protein, lipid and detergent.

Fringe shifts, ΔJ, are related to the concentration change, c, via the refractive index increment $(\partial n/\partial c)$ and the laser wavelength λ:

$$\Delta J = [(\partial n/\partial c)/\lambda]\, lc. \tag{1}$$

Typical values of $(\partial n/\partial c)$ are 0.186 mL g^{-1} for proteins, whereas values between 0.1 and 0.15 mL g^{-1} are reported for detergents [37].

A schematic representation of two very normal types of experiment, sedimentation velocity (SV) and sedimentation equilibrium (SE), is presented in Fig. 4.1. In sedimentation velocity, transport of the macromolecule is dominated by the sedimentation process, whereas in sedimentation equilibrium the effect of the centrifugal force is counterbalanced by the diffusion process, which results in a distribution that is dependent on the ratio of the sedimentation coefficient and the diffusion coefficient. A variety of other experimental devices allow specific geometries and experiments to be employed. For example, special centerpieces in the sample compartment at the start of the experiment allow the superposition of two layers of solution, where one layer contains the macromolecule and the other layer the solvent. Such an approach allow examination of the sedimentation of a band [38–40] or the determination of the diffusion coefficient from boundary spreading at low rotor speeds [17].

As an alternative, if an XLA or XLI analytical ultracentrifuge is not available, then a preparative ultracentrifuge may be used to determine molar masses by sedimentation equilibrium [41, 42]. This approach requires an analysis of the protein distribution after the run in a density-stabilized sample column [43]. Although density gradient ultracentrifugation to study protein–detergent complexes, micelles, lipoproteins, or vesicles in the analytical ultracentrifuge are beyond the scope of this chapter, the interested reader is welcome to consult various examples elsewhere [44–47].

4.3 General Theoretical Background

4.3.1 Equation of the Transport

During an analytical ultracentrifugation experiment the sedimenting/floating material (see Fig. 4.1) undergoes zone broadening as the result of diffusion of the macromolecular constituent back into the depleted part of the solvent column. The net movement (flux) at any given position r (radial distance) as the result of these opposing forces is given by:

$$J = s\omega^2 rc - D(\partial c/\partial r)\iota \tag{2}$$

In this equation, the sedimentation coefficient s is defined as the velocity of the particle per unit of centrifugal field: $s = v/\omega^2 r$. The unit of s is the second, but it is generally expressed in Svedberg units (S): $1\,\text{S} = 10^{-13}\,\text{s}$. The diffusion coefficient, D, is related to the Brownian motion of the particles in solution, and expresses the

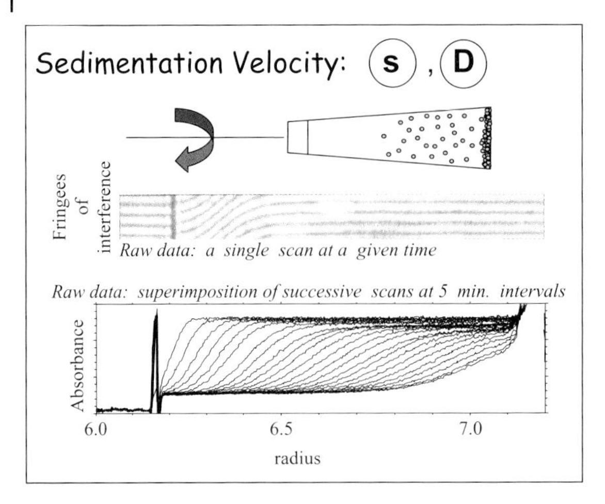

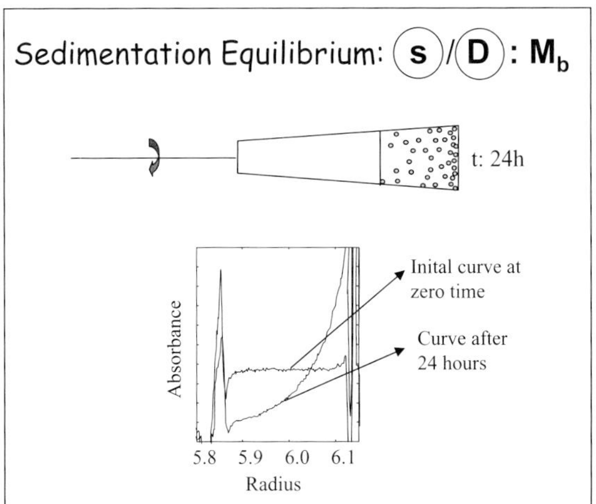

Fig. 4.1 Schematic representation of sedimentation velocity and sedimentation equilibrium experiments. (Modified from Ref. [33].) Sedimentation velocity uses a long column (typically 1 cm long, 400 μL and 100 μL of sample for 12- and 3-mm path length centerpieces, respectively) and a high rotor speed. The material sediments and the formed boundary moves. Sedimentation profiles are obtained as a function of time, and measured by interference or absorbance. They provide insight into the sedimentation and diffusion coefficients, s and D. The transport is dominated by the sedimentation process. Equilibrium sedimentation uses a short column (typically 3 mm long) and a rather low rotor speed. One equilibrium profile is obtained after typically 24 h and is related to the ratio s/D (i.e., the buoyancy molar mass). Users often use three equilibrium profiles obtained at three different rotor speeds to characterize their system.

transport resulting from a gradient of concentration. For $J = 0$, this equation can be used to derive the distribution of the sedimenting or floating species at equilibrium (see Section 4.5). During the sedimentation, for $J \neq 0$, the effect of having a sector-shaped cell must be taken into account when expressing the change in concentration as a function of time, $(\partial c/\partial t)$, that occurs when the macromolecule passes through the different layers of the sample:

$$(\partial c/\partial t) = -(1/r)\partial/\partial r[r(cs\omega^2 r - D\partial c/\partial r)] + q \quad (3)$$

This is the *Lamm equation* (also known as the "transport equation") which expresses rigorously, and in general terms, the spatial and temporal evolution of c, the concentration in the ultracentrifuged sample of each species as a function of s and D. In this equation, q defines the quantity, by unit of time and volume, of the appearance or disappearance of solute from chemical reactions or interactions. In case of interacting systems, c and q for the various species are linked. Equation (3) and its derivatives can be reformulated in terms of kinetic and equilibrium association constants. In the case of the sedimentation equilibrium condition, corresponding to the absence of net flux at each radial position, the concentration gradient is governed by the ratio of the sedimentation and diffusion coefficients. Apart from this situation, exact solutions of the Lamm equation during the approach to equilibrium are highly complex, and computer programs are required to obtain numerical solutions to fit the experimental data.

4.3.2 The Macromolecular Parameters: R_S, M_b, M, and $\bar{v}$

For each individual non-interacting species, the sedimentation and diffusion coefficients are related to two macromolecular parameters: the Stokes radius or hydrodynamic radius, R_S, and the buoyant molar mass, M_b, which is the molar mass of the particle, M, minus the mass of solvent corresponding to the volume occupied by a mole of particle. With the solvent density denoted as ρ_0 and the partial specific volume of the particle as $\bar{v}$, the relation of M_b and M is given by:

$$M_b = M(1 - \rho_0\bar{v}). \quad (4)$$

Furthermore, for a non-interacting macromolecule, the Stokes–Einstein law relates D and R_S.

$$D = RT/N_A 6\pi\eta R_S \quad (5)$$

$$f = 6\pi\eta R_S \quad (6)$$

where R is the gas constant, T is the absolute temperature, N_A is Avogadro's number, and η is the solvent viscosity. R_S is the radius of a hypothetical sphere that would have the same hydrodynamic behavior (same D) as the diluted particle, and f is the frictional coefficient. For comparative purposes, D is often expressed as $D_{20,w}$, corrected for temperature and solvent viscosity to the solvent conditions of water at 20 °C (T_{20} = 293.45 K; $\eta_{20,w}$ = 1.002 cp):

$$D_{20,w} = D(T_{20}/T)(\eta/\eta_{20,w}). \tag{7}$$

4.3.3 The Svedberg Equation

For a non-interacting macromolecule, the Svedberg equation relates s to R_S and M_b or, alternatively, to R_S, M_b and $\bar{v}$:

$$s = M_b/(N_A 6\pi\eta R_S) = M(1-\rho_0\bar{v})/(N_A 6\pi\eta R_S). \tag{8}$$

The sedimentation coefficient generally is expressed as $s_{20,w}$, after correction for solvent density and viscosity to the solvent conditions of water at 20 °C ($\rho_{20,w}$=0.99832 g mL^{-1}):

$$s_{20,w} = s\{(1-\rho_{20,w}\bar{v})/(1-\rho\bar{v})\}(\eta/\eta_{20,w}) \tag{9}$$

4.3.3.1 Mean values of M_b and s

In the classical Svedberg procedure, M_b is determined on the basis of independent measurements of the sedimentation and diffusion coefficients by $s/D=M_b/RT$, whereas sedimentation equilibrium is mostly used for this purpose nowadays. For heterogeneous systems, sedimentation equilibrium measures mean (weight average) buoyant molar masses $\bar{M}_b$ at each position, and globally within the cell. Mean sedimentation coefficients $\bar{s}$ can also be measured, and in a quite easy way, as will be detailed below. They also represent weight average values. Thus, with c_i in weight units:

$$\bar{M}_b = \sum_i (c_i M_{b,i}) \Big/ \sum_i c_i \tag{10}$$

$$\bar{s} = \sum_i (c_i s_i) \Big/ \sum_i c_i \tag{11}$$

Changes in M_b as a function of sample concentration and the amount of bound constituents form the basis for the formulation of possible association schemes by modeling the values of the association numbers and constants, in conjunction with an analysis of the sedimentation coefficients for the different species present in solution. For a recent example concerning analysis of sedimentation equilibrium data for heterogeneous association of soluble proteins, see Ref. [48]. An analysis of the concentration dependencies of $\bar{M}_b$ and $\bar{s}$ is very often used to estimate dissociation constants and number of subunits in multimeric complexes of membrane proteins (e.g., Refs. [9, 49]).

4.3.4 Non-Ideality

For concentrated solutions, typically above 1 mg mL^{-1}, weak particle–particle interactions can lead to non-ideality of the solutions. For concentrations typically up to

100 mg mL^{-1}, s and D and apparent molar mass M_{app} can be described by linear approximations as

$$s = s°(1 - k_s'c + \ldots) \quad (12)$$

$$D = D°(1 + k_D c + \ldots) \quad (13)$$

$$1/M_{app} = 1/M + 2A_2 c + \ldots, \quad (14)$$

where $s°$ and $D°$ are the coefficients corresponding to the particle at infinite dilution, and A_2 is the second virial coefficient. The analysis of non-ideal solutions by sedimentation velocity experiments is described in Ref. [50]. Rosenbuch et al. have reported s-value dependencies on centrifugal field – possibly related to detergent compressibility – for two detergent-solubilized membrane proteins [23].

4.4 Membrane Proteins: Measurement of R_S, M_b, M, and $\bar{v}$

4.4.1 Composition and Molar Mass

For a protein of molar mass M_P, considering the binding of detergent and lipids in terms of gram per gram protein to be δ_D and δ_L, the molar mass of the protein–detergent complex is:

$$M^* = M_P(1 + \delta_D + \delta_L) \quad (15)$$

The binding of other molecules (chlorophyll, plastoquinone, etc.) can be treated in the same way [51, 52]. Detergent binding was found to be related to the hydrophobic area of the protein [53]. The amount of bound detergent varies widely depending on the type of protein and, to a lesser extent, on the type of detergent. As an example, bacteriorhodopsin was measured to bind 2.4 g octaethylene glycol monododecyl ether ($C_{12}E_8$) g^{-1} protein and 4.1 g dodecyl-β-D-maltoside (DDM) g^{-1} protein, while Ca^{2+}-ATPase, with its large cytosolic domains, binds only 0.5 g $C_{12}E_8$ g^{-1} protein and 0.73 g DDM g^{-1} protein [53]. The amount of lipid depends on the solubilization conditions, as was shown for example for EmrE, where the amounts of bound lipid were found as 0.13 and 1.28 g g^{-1} protein in two different protocols of extraction and purification [2]. Detergent binding can be estimated with the aid of radioactively labeled detergent, if the detergent-solubilized sample is subjected to gel chromatography in order to isolate it from other components, such as mixed detergent/lipid micelles and contaminating proteins present in the sample, before analytical ultracentrifugation. The detergent binding can then be estimated from the increase in detergent concentration associated with the relevant protein peak. Column chromatography also allows the determination of any lipid retained by the protein. However, it should be realized that often lipid is only gradually stripped from the protein during chromatography, a circumstance which may lead to a considerable rise in detergent concentration above the baseline, and overlapping with a much smaller protein peak. In order to achieve an effective separation, the method also assumes that the size of the solubilized protein is sufficiently different from that of the mixed lipid–detergent micelles. Therefore, it is sometimes advisable to remove any lipid material before the gel chromatography stage, for example by using ion-exchange chromatography or by two successive stages of size-exclusion chromatography (SEC). (For procedures of detergent-binding measurements, see Ref. [53].) Lipid binding measurements can also be performed on the column fractions, for example by determining phosphate levels for phospholipids, or by thin-layer chromatography (TLC) [53, 54]. As an alternative, the (detergent + lipid)-content can be estimated from the ratio of the number of fringes and absorbance at 280 nm of the detergent-solubilized protein boundary in sedimentation velocity experiments, under the assumption that the protein boundary does not overlap with that of the detergent micelle (see Sections 4.6.5 and 4.6.6.1).

4.4.2 Values of $\bar{v}$

Partial specific volumes are thermodynamically precisely defined as the volume occupied by a weight unit of component in solution – that is, as the inverse of component density. With respect to detergents, the partial specific volume may be slightly different below and above the critical micelle concentration (CMC) [16]. $\bar{v}$ can be measured with a precise density-meter (Anton PAAR) from the change in density related to the variation in the concentration of the dry component (the solvent composition must remain unchanged). The density increment $(\partial\rho/\partial c)$ is then given by:

$$(\partial\rho/\partial c) = 1 - \rho_0\bar{v} \tag{16}$$

$\bar{v}$ can also be estimated from the variation of the buoyant molar mass estimated with sedimentation equilibrium or sedimentation velocity performed in H_2O and D_2O solvents (see Sections 4.5.5 and 4.6.6.3), but in this case the slight increase of M_b related to hydrogen–deuterium exchange must be considered [55–57].

In practice, the partial specific volume $\bar{v}^*$ for a protein–detergent–lipid complex is most often estimated on the basis of the partial specific volumes of the protein- ($\bar{v}_P$), detergent-($\bar{v}_D$) and lipid ($\bar{v}_L$) moieties, respectively, according to:

$$\bar{v}^* = (\bar{v}_P + \delta_D\bar{v}_D + \delta_L\bar{v}_L)/(1 + \delta_D + \delta_L) \tag{17}$$

where δ_D and δ_L denote the amount of bound detergent and lipid, respectively, in terms of grams per gram protein. The contribution of water is generally not expressed, being infinitesimally small in dilute aqueous buffer (see also Sections 4.4.3 and 4.5.4). Values of the partial specific volumes for proteins, detergents, and lipids can in general be estimated from chemical composition [58]. For proteins, a typical value is $0.74\,mL\,g^{-1}$, and it can be calculated from the amino-acid composition, using the program Sednterp [59]. For detergent micelles, $\bar{v}_D$ can be below or above 1, leading to positive and negative buoyancy terms in water for $(1-\rho_0\bar{v}_D)$, and thus to a tendency for either sedimentation or flotation of the protein complex upon centrifugation in water. The partial specific volumes of detergents depend on the type and bulk of the polar head and hydrocarbon chain length. In Section 4.5.4, we consider examples of detergents with the same density as that of the medium, leading to $(1-\rho_0\bar{v}_D)=0$, and no effect on buoyancy; reported values of $\bar{v}_D$ for *n*-octyl-β-D-glucoside (OG) and DDM are in the range 0.81 to $0.86\,mL\,g^{-1}$, whilst for LAPAO and dodecyldimethyl-*N*-amineoxide (DDAO) they are in the range 1.07 to $1.13\,mL\,g^{-1}$. Hence, with the latter detergents the membrane proteins tend to float during centrifugation. In contrast, deoxycholate ($\bar{v}_D = 0.778\,mL\,g^{-1}$) has a density close to that of proteins, which is an advantage in structural analysis by low-angle angle X-ray scattering [60]. For detergent and bile salts, a number of values relevant for analytical ultracentrifugation have been compiled [12, 23, 24, 61]. For lipids, $\bar{v}_L$-values of 0.981 and $0.93\,mL\,g^{-1}$ are reported for egg yolk phosphatidylcholine and phosphatidylserine [16, 21, 62]. An average value of $1.02\,mL\,g^{-1}$ is considered appropriate for the lipid moieties of lipoproteins [58].

4.4.3
Buoyant Mass for Detergent-Solubilized Membrane Proteins, M_b*

M_b* can be written as a function of molar mass and partial specific volume of the protein–detergent–lipid complex, $\bar{v}$*, as calculated from Eq. (17):

$$M_b{*} = M{*}(1-\rho_0\bar{v}{*}) \qquad (18)$$

From multi-component thermodynamics [24, 63, 64], M_b* can be expressed rigorously as a function of the molar mass of the protein M_P:

$$M_b{*} = M_P(\partial\rho/\partial c_P)_\mu \qquad (19)$$

where $(\partial\rho/\partial c_P)_\mu$ indicates the change in the solution density related to the concentration of protein (not to that of the complex), and also reflects the interactions of the protein with all components of the solution. M_b* is measured by reference to a protein-free dialysate, in which all components except protein have the same chemical potential as in the protein solution. Density measurements under such conditions are extremely difficult to perform, due to the large size of the detergent micelles and the very long times required for equilibration. Knowledge is also required of the absolute concentration in protein. Such measurements have been made, however, with proteins that are large when compared to the micelle, or following the use of affinity columns where the detergent concentration can be equilibrated, for example in the case of His-tagged proteins [2, 24, 65]. In these cases, the two sets of experimental values, with $(\partial\rho/\partial c_P)_\mu$ obtained from density measurements, and M_b* from analytical ultracentrifugation, can be combined to derive the molar mass M_P of the protein within the protein–detergent complexes, without any hypothesis concerning the detergent and lipid binding.

Another approach is to calculate the density increment $(\partial\rho/\partial c_P)_\mu$ from the protein–detergent–lipid composition.

$$(\partial\rho/\partial c_P)_\mu = (1+\delta_D+\delta_L) - \rho_0(\bar{v}_P+\delta_D\bar{v}_D+\delta_L\bar{v}_L) \quad (20)$$

In this case, $M_b{}^*$ can be written as the sum of separate contributions of the protein-, detergent-, and lipid moieties to the buoyant molar mass:

$$M_b{}^* = M_P(1-\rho_0\bar{v}_P)+\delta_D M_P(1-\rho_0\bar{v}_D)+\delta_L M_P(1-\rho_0\bar{v}_L) \quad (21)$$

Thus, any component i whose buoyant term $(1-\rho_0\bar{v}_i)$ is null does not contribute to the buoyant molar mass of the complex. This is the case for some detergents in water, such as C_8E_4 ($\bar{v}_D = 0.997\,\mathrm{mL\,g^{-1}}$). Lipids, with a density close to water, contribute in a limited way also. Ordinarily, the hydration term $\delta_W(1-\rho_0\bar{v}_W)$ for bound water does not affect the buoyant molar mass of particle in dilute aqueous buffer, but this is not the case in a sucrose gradient where this term can contribute significantly [16]. Solvent density variation will be discussed further in Section 4.5.4.

Note that an operational partial specific volume ϕ', which relates $M_{b\,complex}$ to M_P in the given experimental condition, is often used to replace by one term all the partial specific volumes given in Eq. (21) [16]:

$$M_b{}^* = M_P(1-\rho_0\phi') \quad (22)$$

4.4.4
Stokes Radius, Frictional Ratio

For homogenous samples, R_S can be obtained from boundary spreading in sedimentation velocity experiments (see below) or extracted from measurements of M_b and s. In principle, dynamic light-scattering experiments could also provide R_S values but, because of the presence of the detergent micelle of a similar size as the protein–detergent complex, the determination will be hazardous for detergent-solubilized proteins. Calibrated SEC can provide estimates of R_S, provided that there is an absence of interactions with the matrix or dissociation of an oligomeric state on the column. However, with some detergents, such as $C_{12}E_8$ and deoxycholate, overestimation of the size has been noted [66].

R_S is often compared to a minimum radius value, R_{min}, calculated for a sphere having the anhydrous volume of the particle $(M\bar{v}/N_A)$. For the specific case of membrane protein associated with detergent and lipids:

$$R_{min} = [(3M_P(\bar{v}_P+\delta_D\bar{v}_D+\delta_L\bar{v}_L)/(4\pi N_A)]^{1/3} \quad (23)$$

The frictional ratio f/f_{min} links the actual values of R_S to the minimum values R_{min}:

$$R_S/R_{min} = f/f_{min} \quad (24)$$

It should be noted that water hydration contributes to the hydrodynamic dimensions. Equation (23) does not consider bound water (typically 0.3 g g^{-1}), but values of f/f_{min} include the effect of hydration, in addition to surface rugosity and deviations from spherical symmetry. Experimentally, the values of $f/f_{min} \approx 1.2$–1.3 that characterize compact and moderately hydrated proteins are also found for some solubilized membrane proteins. However, there are examples of f/f_{min} with values as low as 1.18, such as for the membrane proteins FhuA and BmrA, where bound DDM indicates a very compact structure of these protein–detergent complexes [54, 67]. However, high values have been found in a number of other cases: typically, $f/f_{min} = 1.43$ for the reaction center in $C_{12}E_8$, or 1.6 for the Ca^{2+}-ATPase in deoxycholate (for a list of f/f_{min} values see Table 4 in Ref. [22]). It should be noted that high values of f/f_{min} indicate, in principle, strongly anisotropic shapes: thus, for non-hydrated and mathematically defined ellipsoids with axial ratios of 4 and 10, the values of f/f_{min} increase from a factor of 1.17 to 1.5 as compared to a sphere of the same mass [18]. Hydrodynamic bead modeling represents a general method to model arbitrary shapes for particles represented as a rigid assembly of non-overlapping spherical elements of any size [68, 69]. In all cases, interpretation of f/f_{min} in terms of shape requires the evaluation of the error on its determination.

4.4.5 The Example of the Membrane Protein BmrA

In order to demonstrate what can be achieved by the analytical ultracentrifugation of detergent-solubilized membrane proteins, as performed in practice, herein are presented data acquired with a bacterial membrane protein, BmrA from *Bacillus subtilis*, a member of the multidrug resistance (MDR) ABC transporter family. Although biophysical evidence for a dimeric (and possibly tetrameric) organization of this half ABC transporter (which contains only six putative transmembrane segments) has been obtained after reconstitution in a lipid bilayer environment, the precise oligomeric order in a detergent-solubilized state has not been addressed until recently by analytical ultracentrifugation [54]. Here, BmrA was purified by affinity chromatography after heterologous expression in *E. coli* membranes and solubilization with retention of high ATPase activity by 0.05% DDM. Subsequently, SEC was used to estimate the Stokes radius of the complex (5.6 nm). Furthermore, DDM binding (1.5 g g^{-1} protein) was measured using radiolabeled DDM, and the phospholipid content (0.07 g g^{-1} protein) by TLC analysis of BmrA. Both, the SEC elution profile and sedimentation velocity at 44 000 rpm (~140 000 g) in the ultracentrifuge indicated the essential homogeneity of the DDM-solubilized membrane protein, for which a sedimentation coefficient ($s_{20,w}$) of 8.9S was obtained. In combination, these data provide a buoyant molar mass of 47 kDa. Sedimentation equilibrium profiles, obtained at 7000 rpm (3500 g) and three different protein concentrations over a period of four days indicated the absence of reversible and irreversible aggregation, and provided a virtually identical value of 50 kDa, corresponding to molar masses for the complex of 334 kDa and for the

Table 4.1 BmrA structural parameters as deduced from size-exclusion chromatography (SEC), detergent and phospholipids assay and analytical ultracentrifugation.

		0.05% (or 0.01%) DDM	*0.02% $C_{12}E_8$ Main species*
SEC data (1a)	R_S (nm)	5.6 ± 0.4	~5.6[a]
SEC data (1b)	δ_D (g ± g^{-1})	1.5 ± 0.6	1.5[a]
Phospholipids assay (2)	δ_L (g ± g^{-1})	0.07 ± 0.02	0.07[a]
Sedimentation velocity data (3)	$s_{20,w}$ (S)	8.9 ± 0.3	7 ± 0.2
Sedimentation equilibrium data (4)	M_b* (kDa)	50 ± 4.5	
Protein sequence (5)	M_P (kDa)	66.254	66.254
Combining (1a and 3)	M_b* (kDa)	46.9 ± 1.9	29.7 ± 1.4
Combining (1a, 1b, 2 and 3)	M* (kDa)	314 ± 15	420 ± 20
Combining (1a, 1b, 2 and 3)	M_P (kDa)	110 ± 20	146 ± 20
Combining (1b, 2 and 4)	M* (kDa)	334 ± 30	
Combining (1b, 2 and 4)	M_P (kDa)	117 ± 20	
Combining (1b, 2 and 5 as a dimer)	M* (kDa)	340 ± 80	340 ± 80
Combining (1a, 1b, 2 and 5 as a dimer)	f/f_{min}	1.18 ± 0.08	1.14 ± 0.08

Modified from Ref. [54].

(a) Values are assumed, based on the values obtained with DDM.

protein within the complex of 117 kDa. The data in Table 4.1 show how the various measurements obtained by analytical ultracentrifugation can be combined to provide a consistent picture of the size and mass of the detergent-solubilized protein, leading to the unequivocal conclusion of a dimeric organization of BmrA in the complex.

For the $C_{12}E_8$-solubilized protein, in addition to the dimer, identified from the value of $s_{20,w}$ for the main species (Table 4.1), both the SEC profile and the sedimentation boundary formed in sedimentation velocity experiments indicated the formation of unspecific aggregates with higher $s_{20,w}$ values than corresponding to the dimer. The maximum activity was threefold lower than in the presence of DDM. In conclusion, the data implicate the dimer of BmrA as the *minimal* functional unit of BmrA compatible with activity and probably also with transport competence, while self-association in $C_{12}E_8$ is the consequence of inactivation and secondary, irreversible aggregation. It should be noted, however, that the limitation of analytical ultracentrifugation is that, due to the different environments presented to BmrA by a lipid membrane and a detergent, the data do not definitively rule out the possibility that, in the native membrane, BmrA could exist in a higher supramolecular state than a dimer. In conclusion, this example illustrates the interest of analytical ultracentrifugation both as an independent method and its coupling with other methods such as SEC (see also Ref. [67]). It also demonstrates the benefits of data collection in several detergents for an optimal physicochemical characterization.

4.5 Sedimentation Equilibrium Data Analysis

4.5.1 Equation of Sedimentation Equilibrium and Comments on the Experimental Set-Up

At sedimentation equilibrium, the signal in the ultracentrifuge $a(r)$ is a linear combination of the signals from the concentrations of each individual species:

$$a(r) = \varepsilon l c(r) = c_0 \varepsilon l \exp[\omega_2 M_b / 2RT](r^2 - r_0^2) \tag{25}$$

where c_0 is the concentration at an arbitrary reference radial position r_0 (often chosen to be close to the meniscus). The attainment of sedimentation equilibrium allows us to obtain in a thermodynamically rigorous way the buoyant molar masses of macromolecules in solution. For the more general case of heterogeneous and/or interacting dilute systems, the equilibrium sedimentation condition can be written as the sum of the exponential terms of each species, as in Eq. (25). The interpretation of such expressions must be analyzed within the context of explicit models.

Of course, analysis of the data requires that a true equilibrium state has been obtained – which excludes unstable systems leading to irreversible aggregation or proteolysis, for example, on the time scale of the experiments (typically one or more days). The analysis assumes the measured signal to be proportional to the concentration of the macromolecular species. Large absorbances must therefore be avoided to ensure the linearity in the response of the optics; furthermore, unstable small solutes that can absorb light in a time-dependent manner must also be avoided. A detailed data analysis is usually not trivial, as sedimentation equilibrium profiles have a limited content of information with respect to heterogeneity, because they must be decomposed in terms of contributions from exponentials corresponding to the different species present in the sample. The equilibration time is heavily dependent on sample volume, and may require a couple of days for a medium-sized column. However, for a sufficiently stable system it is advantageous to use as large a column as possible, as this leads to a significant increase in the information content of the data [36].

For chemical equilibria, there are additional conditions linking the buoyant molar masses and the concentrations at reference r_0 for the various species. For the example of a monomer–dimer equilibrium with an association constant K_a, the conditions are: $M_{dimer} = 2M_{monomer}$ and $c_{0,dimer} = K_a c_{0,\,monomer}{}^2$. Species in equilibrium or mixtures or non-interacting species (of same molar masses) cannot be distinguished if a single equilibrium condition experiment is performed. As an example, a monomer–dimer equilibrium and a non-interacting mixture of monomers and dimers cannot be distinguished from a single sedimentation profile: in both cases, they give rise to two exponential terms in Eq. (25). However, they can be distinguished by using a set of experiments performed at different loading concentrations, which is why a typical protocol for sedimentation equilibrium experiments uses three such concentrations.

4.5.2 Simulation of Sedimentation Equilibrium for a Mixture of Particles

The theoretical sedimentation profiles of DDM micelles and various detergent–protein complexes with different molar masses and detergent binding are illustrated in Fig. 4.2. Under the conditions of the simulation, smaller particles (e.g., detergent monomer) would be almost homogeneously distributed, a behavior that would often be analyzed as a constant baseline contribution. On the other hand, larger particles (such as very aggregated material) would be essentially pelleted. Thus, the analysis of a sedimentation equilibrium experiment can "miss" both small and large particles. Consequently, typical protocols would use three angular velocities, in order to "reveal" the different species in solution: the larger particles at lower angular velocity, and small particles at a larger angular velocity. In general, on completion of the experiment the angular velocity would be increased to a maximum for a couple of hours in order to estimate experimentally the level of the baseline.

In the centrifuged samples, detergent micelles are present together with detergent-solubilized membrane proteins, and will redistribute in a similar fashion as protein complexes, depending upon the buoyant properties of the detergent.

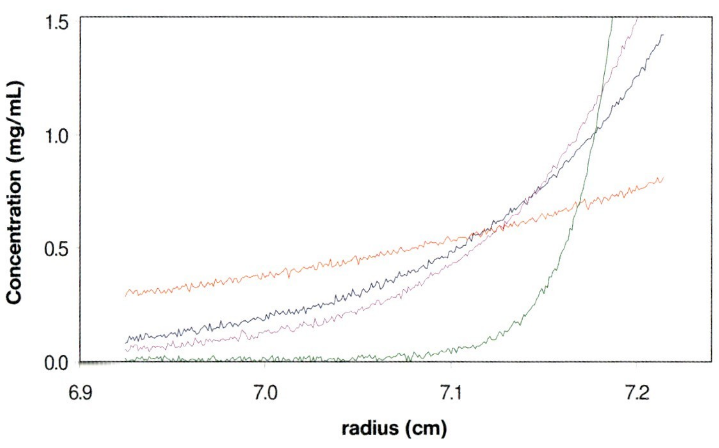

Fig. 4.2 Sedimentation equilibrium for four types of particle. Simulation was carried out at 10000 rpm in water at 20 °C after 26 h for four types of particle, each with a loading concentration of 0.5 mg mL^{-1}. The curves correspond to a DDM micelle of 60000 g mol^{-1} (red) and three detergent–protein complexes of respectively 150000 g mol^{-1} (blue) with δ_D of 4 g g^{-1} (i.e., 120000 g mol^{-1} detergent associated to 30000 g mol^{-1} protein), 180000 g mol^{-1} (pink) with δ_D of 1 g g^{-1} (i.e., 90000 g mol^{-1} detergent associated to 90000 g mol^{-1} protein) and 600000 g mol^{-1} (green), with also a δ_D of 1 g g^{-1}. The details of the particles are listed in Table 4.2.

Because of the poor control with the exact detergent concentration in the different sample compartments (in the reference solvent as well as in the dilution series), and because of the intrinsic complex behavior of detergents (equilibrium between monomer and micelle), interference optics are not used for sedimentation equilibrium experiments. The analysis of solubilized proteins is greatly simplified when their contribution is masked. Most detergents do not absorb at 280 nm and so will not be detected at this wavelength with absorbance optics (unless they contain absorbing impurities which, unfortunately, is often the case). In the opposite situation (e.g., when using Triton X-100), one way to avoid the problem is to follow the protein via a prosthetic group; for example, bacteriorhodopsin can be followed by the absorption of retinal at 555 nm [43].

The detergent–protein complexes will sediment according to their buoyant molar masses, which include the detergent contribution. In the example given in Fig. 4.2, the complex of 30 000 g mol^{-1} detergent associated with 120 000 g mol^{-1} protein and the complex of 90 000 g mol^{-1} detergent associated with 90 000 g mol^{-1} protein have buoyant molar masses of 30 200 and 40 300 g mol^{-1}, respectively, and are hardly distinguishable one from another. In other words, it would be difficult to distinguish a monomer with a large amount of bound detergent from a trimer of the same protein where hydrophobic association of the transmembrane segments might give rise to the binding of a much smaller amount of detergent. Nevertheless, for a mixture of these two species, the unequal repartition of different particles as a function of the radial position, with a larger proportion of the heaviest species being present at larger radii, can be made apparent in the careful analysis of the experimental data. Finally, the large species of 600 000 g mol^{-1} is under-represented in the conditions of Fig. 4.2, as the angular velocity is too large for a proper analysis.

4.5.3
Analysis of Data

Methods of data analysis for sedimentation equilibrium can be divided into two general forms. The first form is model-independent, and is based on the measurement of $\partial \ln c/\partial r^2$, using for example the program SEGAL [70]. At each radial position, $\partial \ln c/\partial r^2$ gives an estimate of the mean buoyant molar mass, which increases with c and r in the case of heterogeneous and/or interacting systems. Traditionally, before the advent of computer facilities, this was the way to determine buoyant molar masses by sedimentation equilibrium. Although this approach is highly versatile, the applications are limited by the quality of the data acquired: hence, this type of analysis is often made as a first step before a more complex analysis, or when homogeneous samples are used.

The second (and most commonly used) form of data analysis is performed within the framework of a model (or of different models of increasing complexity that are successively applied), such as a single ideal species, reversible and irreversible association, and mixtures of species that can either float or sediment. The analysis is made globally on a number of different files obtained at different concentrations, and if necessary with different optics to cover the largest concentration range and velocities. A careful inspection of deviations between modeled and experimental data (residuals) allows the acceptation/rejection of models, or discrimination to be made between them. Different packages exist (e.g., see the RASMB web site [29]; these are also reviewed in Ref. [35]). For example, SEDPHAT [71] allows a global analysis of profiles to be obtained with different samples and experimental conditions, the selection of parameters that are either fixed or fitted, mass conservation (which is useful in the case of heterogeneous or associating systems), and systematic noise evaluation as well as a combined analysis of sedimentation velocity with sedimentation equilibrium profiles.

4.5.4 Matching of Surfactant and Solvent Densities

Even when there is no reliable estimate of $(\partial\rho/\partial c_P)_\mu$ or detergent binding available [Eqs. (19) and (20)], it is still possible to measure M_P and to obtain other interesting data if the solution conditions are well chosen, either in terms of detergent type or solution density.

In dilute aqueous buffers, C_8E_4($\bar{v}_D = 0.997\,\mathrm{mL\,g^{-1}}$) has the same density as the solvent; that is, the term $(1-\rho_0\bar{v}_D)$ is null in Eq. (21). Thus, the detergent micelles neither sediment nor float, and their concentration – apart from the result of the interaction with the macromolecules – does not change within the cell in the ultracentrifuge. In addition, the detergent does not contribute to the buoyant molar mass of the particle, the behavior of which – despite not knowing the value of δ_D – can be interpreted in terms of protein association state and dissociation constant. A less-expensive detergent octyl-POE (which has a heterogeneous polyethylene content; $\bar{v}_D = 0.99\,\mathrm{mL\,g^{-1}}$) is often used in preference to the less-denaturing C_8E_5; however, it should be noted that in general these short-chain detergents are rather deleterious towards the enzymatic activity of especially the larger-sized membrane proteins [72]. However, for such cases $C_{12}E_8$ with a density of $0.973\,\mathrm{cm^3\,g^{-1}}$ may be used [12] (see also Section 4.4.2). The contribution of lipids, if any, is often neglected, but should be discussed for each specific case. Under these conditions, sedimentation equilibrium allows the characterization of the protein-association state and equilibrium association constant [25, 73].

For the same purpose, detergents with $\bar{v}_D$ between 0.9 and $1\,\mathrm{mL\,g^{-1}}$ can be matched effectively in H_2O/D_2O mixtures; for example, C_{14}-sulfobetain and dodecylphosphocholine (DPC) can be density-matched at densities between 1.0 and $1.11\,\mathrm{g\,mL^{-1}}$ [25, 74]. The effect of deuterium exchange must be evaluated in particular for compounds with partial specific values close to 1 [56, 57]. D_2O^{18} can, in principle, be used to extend the range to $\bar{v}_D$ above $0.82\,\mathrm{mL\,g^{-1}}$ and ρ_0 up to $1.21\,\mathrm{g\,mL^{-1}}$, although commercial samples may be very expensive and also be of very poor (in particular optical) quality. The present authors recently used D_2O^{18} for the characterization of new amphiphilic polymers designed to stabilize membrane proteins [57, 75], and also to determine the association state of bacteriorhodopsin within the complex [76].

More generally, masking the contribution of a component by changing the density of the solvent, can be achieved by adding a co-solute, such as sucrose, nicodenz, or salts. However, it must be realized, as mentioned previously, that this will change the buoyancy terms of bound water and lipids – an effect which may be far from negligible when the solvent density is different from that of water. One way of overcoming the hydration term is to work at a solvent density which corresponds to the experimentally determined matched density of the detergent micelle [77, 78]. This type of approximation is likely to be justifiable for membrane proteins that are entirely embedded in membranes with no cytoplasmic or extra-cytoplasmic regions, but it is not feasible for proteins with a large extramembranous regions, such as the Ca^{2+}-ATPase.

4.5.5
Determining the Association States and Dissociation Constant in the Presence of Non-Density-Matched Detergent

Sedimentation equilibrium profiles obtained in different H_2O/D_2O mixtures can be used in contrast variation experiments for the determination of M^* and $\bar{v}^*$. A global analysis of the data provides a new and interesting perspective, with the potential to obtain precise results even in the absence of density matching of the detergent [79]. Alternatively (see Section 4.4.3 and references therein), the measurement of $(\partial\rho/\partial c_P)_\mu$ allows determination of the molar mass of the protein, and thus the association states and dissociation constants from sedimentation equilibrium experiments, even when the detergent is not density-matched by the solvent. The experimental determination of $(\partial\rho/\partial c_P)_\mu$ should be facilitated by the expanding progress in overexpressing tagged proteins purified in large amounts through tag-affinity chromatography. Beyond these new possibilities, a note of caution must be placed on the possible dependency of the association constant in the presence of varying amounts of detergent, as considered in the following section.

4.5.6
Dependency of Association Constants on Detergent Concentration

It is well known that, for the classical detergents used to study membrane proteins, detergent monomers are in equilibrium with micelles and with solubilized membrane proteins or hydrophobic peptides. The thermodynamic treatment of such an equilibrium was first derived by Tanford [80]. Initially, it was considered that each membrane protein would bind the equivalent of a single micelle of detergent, but measurements performed with different detergents and several proteins showed this not to be the case [53], except perhaps for small hydrophobic peptides [12]. The current view is that the detergent forms a monolayer on the hydrophobic surface of the protein. Membrane proteins have a natural tendency to autoassociate as a function of protein concentration in detergent solution, and sometimes also in their native membrane. This tendency is obviously increased at lower detergent concentrations, near or below the CMC of the detergent, whereas high detergent concentrations and low protein concentrations favor monomer formation (see Refs. [12, 81]). This tendency is also strongly dependent on the type of detergent; for example, long-chain detergents (C_{16}–C_{18}) or the presence of residual lipids favor dimer and oligomer formation, so that it is always difficult to infer the state of association in the native membrane from the state of association in detergent solution (see the discussion in Ref. [8] and the example of BmrA protein presented in Section 4.5.4 [54]).

Different approaches have been followed to rationalize the effect of detergent concentrations and types on monomer/dimer/higher oligomer equilibria from the measurement of association constants (or ΔG and other thermodynamic constants) obtained using analytical ultracentrifugation or other techniques (e.g., using Wyman formalism [81]; or volume considerations that may be relevant in the lipidic phase of a membrane [82, 83]). The example [81] of human plasma paraoxonase, an enzyme which requires the use of detergent to isolate it from high-density lipoprotein (HDL), is particularly illustrative of the potential of analytical ultracentrifugation. Here, sedimentation velocity measurements showed one or two overlapping $c(s)$ peaks, with mean s-values that were increased when the loading protein concentration was raised at a given detergent concentration, or when the detergent concentration was reduced at a given protein concentration. An analysis indicated an association–dissociation process for the protein (in this case a monomer/dimer equilibrium) that was modulated by both the detergent and protein concentrations. The detergent used was $C_{12}E_8$, which is almost density-matched (see Section 4.4.4), and this allowed the determination from sedimentation equilibrium experiments of the monomer/dimer equilibrium constant for different samples. Decreasing the detergent concentration shifted the equilibrium towards the formation of dimers. This shift was linked to a relative decrease in detergent binding, which changed from 100 to 50 molecules per monomer in the dimeric complex. A similar decrease in detergent binding was also found by oligomeric association of sarcoplasmic Ca^{2+}-ATPase in $C_{12}E_8$ [53, 84, 85].

4.6 Sedimentation Velocity Data Analysis

The potential of sedimentation velocity has increased dramatically during the past few years, in relation to the increased possibility of data treatment. During the 1990s, both time-derivative methods [86] (which eliminated systematic noise from the data) and the Van Holde Weischet method [87] (qualifying sample heterogeneity) utilized these new opportunities for the global fitting of sedimentation profiles obtained at different times.

4.6.1 Numerical Solutions of the Lamm Equation

During the past few years, further progress has been made by the use of simulated data. Following the pioneering studies of Claverie [88–90], numerical solutions of

$\chi(s, D, r, t)$ of the Lamm equation [Eq. (3)] can now be obtained in routine manner. Three program packages have been developed that use numerical solutions of this equation for the analysis of sedimentation velocity (and sedimentation equilibrium) data: by Peter Schuck with SEDFIT/SEDPHAT [71]; by Boris Demeler with ULTRASCAN [91], and Walter Stafford with SEDANAL [92], with the latter system using differences between scans. These procedures allow the simulation of sedimentation velocity profiles corresponding to any $s/D/q$ condition. These simulated profiles are used for data analysis; moreover, the programs have opened the possibility of describing, in a global analysis, the behavior of complex macromolecular systems. In general, they consider the entire sedimentation process without restriction such that data acquisition is made on a large time scale (typically overnight) in order to obtain the maximum information available for data treatment. Of these programs, SEDFIT and SEDPHAT (see below) are particularly versatile.

4.6.2 Analysis in Terms of Non-Interacting Species: Principle

The analysis of sedimentation velocity profiles can be made by considering a limited number of species (typically one to three non-interacting species; see Section 6.4 for how a model-independent $c(s)$ analysis can help to establish the validity of the hypothesis). In that case, the experimental sedimentation velocity profiles are analyzed as a linear combination of the Lamm equation:

$$a(r, t) = \sum \chi(s, D, r, t) \tag{26}$$

This allows the determination, for each species, of s and D. This analysis can include a compound that is floating in addition to one that is sedimenting. Then, by using the Svedberg equation [Eq. (8)], the buoyant molar mass, M_b and R_S can be obtained and interpreted in terms of particle composition (see Section 4.4). The limitation of this approach is that the underlying hypothesis – which concerns the number of species considered in the analysis as well as the assumption that the particles are not interacting – must be strictly correct. If this is not the case – for example, if we fail to take into account a minor heterogeneity – then the D and related R_S, M_b and M-values obtained will be notably wrong.

4.6.3 Analysis in Terms of Non-Interacting Species: Applications to Detergent and the Membrane Protein EmrE

In solutions of detergent-solubilized protein(s), detergent is exchanged between the monomer and micellar forms of the detergent and complex(es). For membrane proteins, the validity of analysis of sedimentation velocity data in terms of s and D thus needs to be tested as the species are not truly non-interacting. Sedimentation velocity measurements on detergent solutions of DDM and $C_{12}E_8$ have led to aggregation numbers of 130 ± 4 and 118 ± 8, respectively, which are compatible with literature data obtained by sedimentation equilibrium or other biophysical techniques [93]. For EmrE in 2% DDM, M_b* from s and D differs by 8% from that derived by s and R_S, and by 1% from that calculated from the composition [3]. Thus, a very good check on the validity of the sedimentation velocity analysis of detergent-solubilized membrane proteins was obtained in these experiments.

4.6.4
c(*s*) Analysis: Principle

The analysis of sedimentation velocity data in terms of a continuous distribution of sedimentation coefficients $c(s)$, as proposed by Peter Schuck in 2000 [94], constitutes a major advance for the analysis of sedimentation velocity experiments, which is extremely powerful and is widely used [95, 96]. The sample is described as an assembly of non-interacting particles; for example, by $N=200$ particles, defined by their minimum and maximum s-values. These particles are assumed to have the same density and general shape factors – that is, the same $\bar{v}$ and f/f_{min} characteristic values. This allows each s-value to be mathematically linked to a unique D-value. Thus, for N particles defined by s and related $D(s)$ values, N sets of sedimentation velocity profiles are simulated, corresponding to the experimental geometry and centrifugal conditions of the sedimentation profiles under analysis. The signal $a(r, t)$ in the ultracentrifuge at position r and time t is described as a linear combination of the signals $\chi(s, D(s), r, t)$ related to each individual species.

$$a(r,t) = \int c(s)\chi(s, D(s), r, t)ds \tag{27}$$

The program calculates the best distribution of sets of simulated profiles modeling the experimental data, taking advantage of a systematic noise evaluation procedure [97]. This provides a distribution of sedimentation coefficients, named $c(s)$ (actually a signal distribution $a(s)$ would be a more exact designation, as the distribution is given in absorbance or fringe numbers). A regularization procedure is generally applied particularly for complex species distributions, which provides a more regular $c(s)$ distribution.

The $c(s)$ analysis is in general a robust method to describe samples in terms of sedimentation coefficients, as only the details of the $c(s)$ distribution are sensitive to the hypothesis. This is because the sedimentation profiles are mainly determined by the values of s, and in a minor way by the value of D. The considered D-values are most often good approximations of the actual values, which allows diffusion to be nicely deconvoluted from sedimentation. Even assemblies of particles the sizes of which vary within a large range – including small particles which diffuse extensively – are analyzed, and even particles of similar size can be resolved at quite high resolution. For example, non-interacting monomers and dimers are very easily resolved.

4.6.5
Sedimentation Velocity Simulation and $c(s)$ Analysis for a Hypothetical Sample of Membrane Proteins

To demonstrate how the $c(s)$ analysis operates, we have simulated sedimentation velocity profiles corresponding to a hypothetical mixture of four particles, each at 0.5 mg mL^{-1}, representative of detergent and the membrane proteins presented in Table 4.2, as already simulated under sedimentation equilibrium conditions in Fig. 4.2. In the upper part of Panels A and B of Fig. 4.3 are shown the sedimentation velocity profiles for absorbance optics and interference optics, respectively, simulated with reasonable absorption coefficient and $\partial n/\partial c$ values for each of the partners (see Table 4.2). The "small" detergent micelle (s = 2.9S) is only detectable by interference optics, and cannot be readily discerned by visual inspection, because it is characterized by sedimentation velocity boundaries that broaden significantly by diffusion. Because the two smaller protein–detergent complexes are similar in size and mass, the presence of the two components is hardly distinguished in the raw data. In contrast, the largest particles of 600 000 g mol^{-1} are

Table 4.2 Examples (characteristics) of four particles representative of detergent (I) and membrane proteins (II–IV).

	I	*II*	*III*	*IV*
M^* (kDa)	60	150	180	600
δ_D (g g^{-1})	–	4	1	1
M_D (kDa)	60	120	90	300
M_P (kDa)	0	30	90	300
M_b (kDa)	11.2	30.2	40.3	134.3
R_S (nm)	3.4	4.5	4.8	7.1
s (S)	2.9	5.9	7.4	16.6
c (mg mL^{-1})	0.25	0.25	0.25	0.25
A	0.00	0.06	0.15	0.15
ΔJ	0.59	0.64	0.71	0.71
$\Delta J/A$	∞	10.6	4.7	4.7

M_D and M_P represent the molar masses of the detergent and protein, respectively, within the complexes. Partial specific volumes of 0.815 (corresponding to that of DDM) and 0.74 mL g^{-1}, $E_{0.1\%}$ values of 0 for the detergent and 1 M^{-1} cm^{-1} for the proteins, $\partial n/\partial c$ values of 0.133 and 0.186 mL g^{-1}, were considered for detergent and protein partners, respectively. s-values are calculated with f/f_{min} = 1.25; ρ_0 = 0.99832 g mL^{-1}, η = 1.002 cp (i.e., typical values for globular compact particles in water at 20 °C). c is the total concentration.
A and ΔJ are calculated for an optical path length of 12 mm. The simulations of Figs. 4.2 and 4.3 were made with standard deviation for the noise of 0.01.

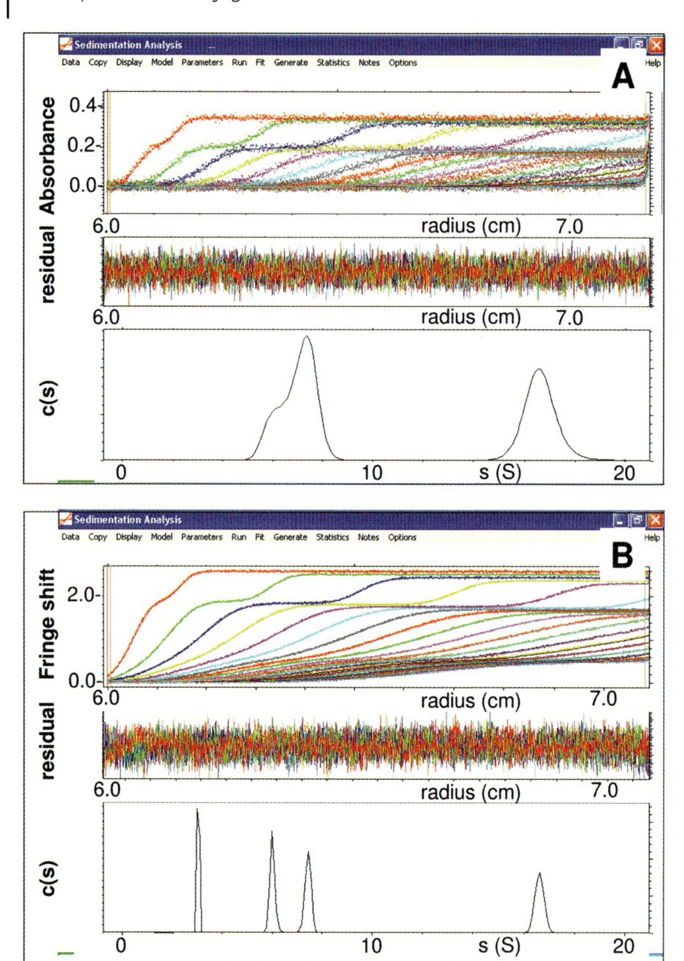

Fig. 4.3 Sedimentation velocity and $c(s)$ analysis of a mixture of four particles. Panel A shows Absorbance; Panel B shows Interference. Simulation was carried out for 20 profiles every 30 min at 42 000 rpm (~130 000g) for a mixture of the four particles described in Table 4.2. $c(s)$ analysis was made considering 200 particles of s between 0.2 and 20S, $\bar{v} = 0.77\,mL\,g^{-1}$, $f/f^{\circ} = 1.25$, and with a confidence level of 0.68.

easily differentiated from the smaller particles. The data in Fig. 4.3 illustrate the potential of sedimentation velocity for the study of multicomponent highly heterogeneous systems. The results of the $c(s)$ analysis are shown in the lower panels of Fig. 4.3. The $c(s)$ analysis – by deconvoluting even in an approximate manner the diffusion processes from sedimentation processes – resolves all of the species, whichever the optical systems of detection (except for the detergent micelles in absorbance, of course, as they are not "seen" in this case). This is particularly remarkable when considering the low signal-to-noise ratio in absorbance optics.

It should be noted, however, that by using interference optics the two peaks corresponding to $s=5.9$S and 7.4S are perfectly well resolved, which is not the case for absorbance optics. The position and area under the peaks provide values of the mean s-values and of the initial concentration of the related species, as measured in signal units. The relative proportions of the different species vary significantly in Panels A and B, corresponding to the absorbance or fringe shifts, respectively, as the detergent and protein contents are different for the different species (see Table 4.2). Thus, the data in Fig. 4.3 illustrates how the acquisition of joint absorbance and interference scans can be used to estimate surfactant binding (together with that of lipid, if present).

4.6.6 Example of Characterization of a Membrane Protein by Sedimentation Velocity

4.6.6.1 Association State of Na^+-K^+-ATPase Expressed in *Pichia pastoris* and of Sarcoplasmic Ca^{2+}-ATPase

As examples to illustrate how $c(s)$ analysis can be used to unravel membrane protein function, we shall consider experiments aimed at characterizing the functional unit of Na^+,K^+-ATPase [4] and Ca^{2+}-ATPase [84, 85]. In this study, the Na^+,K^+-transporting enzyme was heterologously expressed in yeast (*Pichia pastoris*). After purification from the yeast membranes by affinity chromatography, the protein was solubilized by DDM and fractionated into two peaks by SEC. This was achieved with an eluant containing a DDM/phosphatidylserine mixture (since the presence of this lipid together with detergent was necessary to maintain the ATPase in an active state).

Two successive fractions from SEC of the peak with the smallest size were subjected to sedimentation velocity study. By visual inspection, the sedimentation boundary appeared homogenous, but $c(s)$ analysis by absorption and interference fringes indicated that, in addition to the main component with an $s_{20,w}$-value of 10.2 ± 0.15S, small amounts of both a smaller and a larger component were present. For the main species the same sedimentation rate was obtained at three different concentrations, and thus these experiments did not provide any evidence of auto-association of the enzymatically active ATPase. Sedimentation velocity using interference optics showed a very high ratio $J/A=21$ for the smallest species, which was interpreted to represent lipid–detergent complexes possibly in combination with a small peptide. For the main species, after chromatography, $J/A=6$ indicated a total of 1.4 g of lipid and/or detergent per gram protein. SEC with radiolabeled detergent gave a ratio of 0.5 g of detergent per gram protein, and a R_S value of 6.5 nm (which would correspond to the size of a soluble globular protein of 450 kDa). For a globular compact complex with 0.5 g detergent and 0.9 g lipid (the maximum value compatible with J/A) per gram protein, it was possible to calculate R_S and $s_{20,w}$-values of 6.3 ± 0.2 nm and 10.5 ± 0.5S, respectively, for a monomer; whereas, the corresponding values for a dimer would be 7.9 ± 0.2 nm and 15.9 ± 0.9S. The experimental and calculated values thus unequivocally indicate the monomeric α,β-form as a *minimal* structural and functional unit of Na^+,K^+-ATPase [4].

There was little evidence for higher-order oligomers, except in the void volume of SEC. These may be related to the presence of Na^+,K^+-ATPase dimers or tetramers, as proposed from other studies (see e.g. Refs. [98–100]). Thus issue was also raised of whether different pools of Na^+,K^+-ATPase are present in many membrane systems, a topic which is hotly debated on the basis of various biophysical and cellular biology experiments.

In previous studies on sarcoplasmic reticulum Ca^{2+}-ATPase [84, 85], the $C_{12}E_8$-solubilized enzyme was shown, by SEC, to associate irreversibly during enzymatic inactivation. Examination of the sedimentation boundary in the analytical ultracentrifuge indicated the conversion of Ca^{2+}-ATPase from a monomeric to a dimeric state, together with heavier and rapidly sedimenting aggregates. Ca^{2+}-ATPase solubilized with retention of enzyme activity was also found to self-associate at high protein and low detergent concentrations; however, in contrast to the irreversibly aggregated and inactive Ca^{2+}-ATPase, this association was fully reversible. Given the high concentration of Ca^{2+}-ATPase in the sarcoplasmic reticulum membrane, the conclusion is almost inescapable that the transport protein is present in an oligomeric state in the membranous state, as is also experimentally indicated by an increased tendency for dimer formation in $C_{12}E_8$-lipid mixtures [12]. Yet, detailed investigation of the functional properties indicates that the monomeric form is functionally fully competent with respect to its enzymatic and transport properties [1, 101], in agreement with 3-D structural evidence for the presence of a central Ca^{2+}-conducting pathway in the monomer [102, 103].

4.6.6.2 Complex Behavior in Solution of New Amphiphilic Compounds

Although detergents are generally used to solubilize membrane proteins in aqueous buffers, they often lead to irreversible protein inactivation. This may be due to removal of the lipid phase, and in some cases of essential lipids, when the detergent concentration is much higher than the CMC (or aggregation) during inactivation, or when the detergent concentration is close to, or lower than the CMC. In order to overcome these problems, new surfactants compounds are being designed [104–107]. The $c(s)$ analysis has proved to be very powerful for the characterization of these new surfactants and the complexes they form with membrane proteins. For example, a homologue to DDM detergent with hemifluorinated chains forms large and heterogeneous assemblies (most likely cylindrical micelles), the size of which increases with concentration. This homologue also stabilizes the b_6f dimer even at high surfactant concentrations [52]. Another example is that of the amphiphilic polymer A8-35 which, despite the variable length of the chains and the random distribution of hydrophobic groups along them, was found to self-organize into well-defined assemblies [75] and to define also compact homogeneous particles with bacteriorhodopsin [76]. In the case of the Ca^{2+}-ATPase- A8-35 complexes however, some heterogeneity was noted by analytical ultracentrifugation [108].

4.6.6.3 The s_H/s_D Method

The comparison of $c(s)$ obtained in D_2O and H_2O solutions can provide information on the partial specific volume of a complex if the association state and macromolecule conformation (thus R_S) are the same in the two solvents. The accuracy, advantages, and limits of the method have been studied in detail, using model macromolecules (DNA, protein, and polysaccharide). This approach was also used for acidic polymers to describe their effective charge [57]. The values of the partial specific volumes of detergents obtained in this way are identical to those obtained from density measurements [93]. This method has the potential to determine the partial specific volume $\bar{v}^*$ of protein–detergent complexes (which is linked to the particle composition: see Section 4.4.2) within heterogeneous samples.

4.6.7 General Potentials of the *c*(*s*) Analysis *per se* as a Prelude to more Sophisticated Analysis

The $c(s)$ analysis method offers the possibility to characterize sample homogeneity or heterogeneity in a very efficient and versatile manner. For a ligand that absorbs where the protein does not, sedimentation velocity offers a particularly elegant means of characterizing the binding process (e.g., see the binding of vitamin B_{12} to the vitamin B_{12} receptor BtuB followed by sedimentation velocity at 358 nm [109]). The effects of solvent composition, aging, macromolecule concentration and so forth, are easily described in terms of a continuous size distribution, $c(s)$. Because s is related to mass and size, the $c(s)$ analysis may detect changes in the conformation or interaction between macromolecules. The s-values can be compared to theoretical values and exclude certain hypotheses (as considered in Section 4.6.6.1). The $c(s)$ analysis primarily provides information on homogeneity, number of species and their possible associations. This is essential information prior to a more detailed analysis, which, implicitly or not, is generally made in the framework of a model: for example, the analysis in terms s and D-values, and thus of M_b and R_S (Section 4.6.2), requires a limited number of non-interacting species. Recent procedures in SEDPHAT [71] allow a combined analysis with a description in terms of $c(s)$ distribution for one or two intervals of s-values, in addition to a description in terms of a limited number of discrete non-interacting species. This opens the possibility of characterizing macromolecules in terms of s and D, even in the presence of contaminants. From the concentration dependency of $c(s)$, it can be checked if one peak in the $c(s)$ analysis corresponds to one species or to different species in interaction, this being a prerequisite for a detailed analysis in terms of discrete species. It is for this reason that sedimentation velocity is performed routinely with three samples at different concentrations. Shifts in mean s-values or changes in the relative proportions of the different species in solution indicate interactions between the species. The concentration dependency of mean s-values can be analyzed with equations derived from Eq. (12) in terms of s-values for different oligomers and dissociation constants (see e.g. Ref. [9]).

Applied until now only to soluble proteins, more sophisticated programs can be used to characterize the equilibrium and dynamic properties of association–dissociation processes [48, 110]. New approaches of data analysis tend to analyze globally data from sedimentation velocity, sedimentation equilibrium and possibly other techniques related to the same macromolecular parameters (M_b and R_S) such as dynamic light scattering. They also tend to analyze data obtained at various concentrations, with different optics, in solvents of different densities, in order to describe the hydrodynamics and thermodynamics of interacting systems. For mixtures of different species having different optical signatures – as generally occur in detergents and proteins – a "multi-wavelength analysis" was recently proposed to describe sedimentation velocity data obtained with different optical detection (different wavelengths and/or interference) to the distribution of components in molar units as a function of s [111].

4.7 Analytical Ultracentrifugation and SANS/SAXS

Detergent-solubilized membrane proteins can be subjected to rather detailed structural study by using small-angle X-ray scattering (SAXS) and small-angle neutron-scattering (SANS) techniques (see also Chapter 9). Today, not only new synchrotron X-ray and high neutron flux reactors, but also new possibilities of data analysis [112] are becoming available to determine *ab initio* from such studies the low-resolution solution structures for detergent-solubilized proteins [113]. Previous examples of the use of SAXS and SANS for detergent-solubilized membrane proteins are cited in Ref. [12]. Whilst the combination of analytical ultracentrifugation and SANS/SAXS is highly desirable and informative, it should be noted that scattering techniques place stringent requirements on the monodispersity of samples, which must first be checked with analytical ultracentrifugation to test the effect of detergent (type and concentration) and protein concentration on monodispersity (as scattering intensities are weak, rather high protein concentrations of 5–10 $mg\,mL^{-1}$ are required). An early example of combined analytical ultracentrifugation and SAXS was a study on sarcoplasmic reticulum Ca^{2+}-ATPase, solubilized in monomeric form with deoxycholate, a detergent which has several distinct advantages for structural study by SAXS, namely moderate binding and a density close to that of proteins [60]. In a completely different approach, the contribution of the protein component to X-ray scattering was minimized by variation of the sucrose concentration of the medium in a study on the DDAO–rhodopsin complex [114]. The same protein was analyzed by SANS after solubilization with partially deuterated alkylpolyoxyethylene glycol detergent, so that the contribution of the detergent to scattering could be minimized by addition of D_2O to the medium [115]. In these three studies analytical ultracentrifugation was used with success prior to the SAXS/SANS analysis.

4.8 Conclusions

The movement of macromolecules in a centrifugal field forms the basis for analytical ultracentrifugation, by both sedimentation velocity and sedimentation equilibration analysis. During recent years an improved design of analytical ultracentrifuges and computational treatment of the data have led to significant advances in terms of multicomponent analysis, including the characterization of reversible associations. In this chapter, these developments have been reviewed in relation to their applications to membrane proteins, extracted from the membranous environment as detergent-solubilized complexes. Prior to ultracentrifugal analysis these complexes are often purified by affinity- and/or SEC chromatography, with the retention of modest amounts of lipid. With such preparations, consideration must be made as to how to correct for the effects of bound detergent on molecular mass and size (Stokes radius). Consideration was also given to the problems arising from the coexistence of detergent micelles of similar size, including their effect on monomer/oligomer equilibria. On the basis of selected examples, it was demonstrated how a careful analysis of functionally intact preparations can provide information on the quaternary organization that it is impossible or difficult to obtain from studies of membrane proteins in their membranous environment.

Acknowledgments

The authors are grateful to David Stroebel for his critical reading of the manuscript. C. E. and M. l. M also thank the CEA DSV for a grant to develop the study of detergent-solubilized membranes proteins.

References

1 J. V. Møller, B. Juul, M. le Maire, *Biochim. Biophys. Acta* **1996**, *1286*, 1–51.
2 P. J. Butler, I. Ubarretxena-Belandia, T. Warne, C. G. Tate, *J. Mol. Biol.* **2004**, *340*, 797–808.
3 T. L. Winstone, M. Jidenko, M. le Maire, C. Ebel, K. A. Duncalf, R. J. Turner, *Biochem. Biophys. Res. Commun.* **2005**, *327*, 437–445.
4 E. Cohen, R. Goldshleger, A. Shainskaya, D. M. Tal, C. Ebel, M. le Maire, S. J. Karlish, *J. Biol. Chem.* **2005**, *280*, 16610–16618.
5 A. Musatov, J. Ortega-Lopez, N. C. Robinson, *Biochemistry* **2000**, *39*, 12996–13004.
6 A. Musatov, N. C. Robinson, *Biochemistry* **2002**, *41*, 4371–4376.
7 J. Schlessinger, *Cell* **2002**, *110*, 669–672.
8 M. Chabre, M. le Maire, *Biochemistry* **2005**, *44*, 9395–9403.
9 R. H. Heuberger, L. M. Veenhoff, R. H. Duurkens, R. H. Friesen, B. Poolman, *J. Mol. Biol.* **2002**, *317*, 591–600.
10 L. M. Veenhoff, E. H. Heuberger, B. Poolman, *Trends Biochem. Sci.* **2002**, *27*, 242–249.
11 Y. Hayashi, H. Matsui, T. Takagi, *Methods Enzymol.* **1989**, *172*, 514–528.
12 M. le Maire, P. Champeil, J. V. Møller, *Biochim. Biophys. Acta* **2000**, *1508*, 86–111.
13 T. Svedberg, K. O. Pedersen, *The Ultracentrifuge*, Oxford University Press, UK, **1940**.
14 H. K. Schachman, *Ultracentrifugation in Biochemistry*, Academic Press, USA, **1959**.
15 H. Fujita, *Foundations of Ultracentrifugal Analysis*, Wiley, USA, **1975**.
16 C. Tanford, Y. Nozaki, J. A. Reynolds, S. Makino, *Biochemistry* **1974**, *13*, 2369–2376.

17 C. Tanford, *Physical Chemistry of Macromolecules*, Wiley, USA, **1961**.
18 C. R. Cantor, P. R. Schimmel, in: *Biophysical Chemistry*, Part 2, pp. 344–846, W. H. Freeman, USA, **1980**.
19 K. E. Van Holde, *Physical Biochemistry*, Prentice-Hall, USA, **1985**.
20 I. Serdyuk, J. Zaccai, N. Zaccai, *Methods in Molecular Biophysics: Structure, Dynamics, Function*, Cambridge University Press, UK, **2007**.
21 C. Tanford, J. A. Reynolds, *Biochim. Biophys. Acta* **1976**, *457*, 133–170.
22 J. V. Møller, J. P. Andersen, M. le Maire, in: *Progress in Protein-Lipid Interactions*, pp. 147–196, Elsevier, The Netherlands, **1986**.
23 J. P. Rosenbusch, A. Lustig, M. Grabo, M. Zulauf, M. Regenass, *Micron*, **2001**, *32*, 75–90.
24 P. J. P. Butler, C. G. Tate, in: *Modern Analytical Ultracentrifugation: Techniques and Methods*, pp. 133–151, The Royal Society of Chemistry, UK, **2006**.
25 K. G. Fleming, in: *Modern Analytical Ultracentrifugation: Techniques and Methods*, pp. 432–448, The Royal Society of Chemistry, UK, **2006**.
26 http://www.beckmancoulter.com/products/splashpage/xla/.
27 http://www.bitc.unh.edu.
28 I. K. MacGregor, A. L. Anderson, T. M. Laue, *Biophys. Chem.* **2004**, *108*, 165–185.
29 http://www.bbri.org/RASMB/rasmb.html.
30 D. J. Scott, S. E. Harding, A. J. Rowe, *Modern Analytical Ultracentrifugation: Techniques and Methods*, The Royal Society of Chemistry, UK, **2006**.
31 L. Börger, W. Mächtle, *Analytical Ultracentrifugation of Polymers and Nanoparticles*, Springer, Germany, **2006**.
32 J. Lebowitz, M. S. Lewis, P. Schuck, *Protein Sci.* **2002**, *11*, 2067–2079.
33 C. Ebel, *Prog. Colloid Polymer Sci.* **2004**, *127*, 73–82.
34 S. E. Harding, *Carbohydr. Res.* **2005**, *340*, 811–826.
35 C. Ebel, in: *Protein Structures: Methods in Protein Structure and Stability Analysis*, Nova Science, USA, **2007**.
36 A. Balbo, P. Schuck, in: *Protein-Protein Interactions: a molecular cloning manual*, 2nd edn., Chapter 14, Cold Spring Harbor Laboratory Press, USA, **2005**.
37 P. Strop, A. T. Brunger, *Protein Sci.* **2005**, *14*, 2207–2211.
38 J. Vinograd, R. Bruner, R. Kent, J. Weigle, *Acad. Sci. USA* **1963**, *49*, 902–910.
39 J. Lebowitz, in: *BeckmanTechnical Application Information Bulletins*, **1995**.
40 J. Lebowitz, M. Teale, P. W. Schuck, *Biochem. Soc. Trans.* **1998**, *26*, 745–749.
41 R. J. Pollet, B. A. Haase, M. L. Standaert, *J. Biol. Chem.* **1979**, *254*, 30–33.
42 A. P. Minton, *Anal. Biochem.* **1989**, *176*, 209–216.
43 M. Garrigos, F. Centeno, S. Deschamps, J. V. Møller, M. le Maire, *Anal. Biochem.* **1993**, *208*, 306–310.
44 V. N. Schumaker, D. L. Puppione, *Methods Enzymol.* **1986**, *128*, 155–170.
45 A. Lustig, A. Engel, M. Zulauf, *Biochim. Biophys. Acta* **1991**, *1115*, 89–95.
46 Z. Bozóky, L. Fülöp, L. Köhidai, *Eur. Biophys. J.* **2001**, *29*, 621–627.
47 C. E. MacPhee, R. Y. Chan, W. H. Sawyer, W. F. Stafford, G. J. Howlett, *J. Lipid Res.* **1997**, *38*, 1649–1659.
48 J. Dam, P. Schuck, *Biophys. J.* **2005**, *89*, 651–656.
49 E. Cabezon, P. J. Butler, M. J. Runswick, J. E. Walker, *J. Biol. Chem.* **2000**, *275*, 25460–25464.
50 A. Solovyova, P. Schuck, L. Costenaro, C. Ebel, *Biophys. J.* **2001**, *81*, 1868–1880.
51 E. Rivas, F. Reiss-Husson, M. le Maire, *Biochemistry* **1980**, *19*, 2943–2950.
52 F. Lebaupain, A. G. Salvay, B. Olivier, G. Durand, A.-S. Fabiano, N. Michel, J.-L. Popot, C. Ebel, C. Breyton, B. Pucci, *Langmuir* **2006**, *22*, 8881–8890.
53 J. V. Møller, M. le Maire, *J. Biol. Chem.* **1993**, *268*, 18659–18672.
54 S. Ravaud, M. A. Do Cao, M. Jidenko, C. Ebel, M. le Maire, J. M. Jault, A. Di Pietro, R. Haser, N. Aghajari, *Biochem. J.* **2006**, *395*, 345–353.
55 S. J. Edelstein, H. K. Schachman, *J. Biol. Chem.* **1967**, *242*, 306–311.
56 S. J. Edelstein, H. K. Schachman, *Methods Enzymol.* **1973**, *27*, 82–98.
57 Y. Gohon, G. Pavlov, P. Timmins, C. Tribet, J.-L. Popot, C. Ebel, *Anal. Biochem.* **2004**, *334*, 318–334.

58 H. Durchschlag, P. Zipper, in: *Modern Analytical Ultracentrifugation: Techniques and Methods*, The Royal Society of Chemistry, UK, **2006**.

59 http://www.jphilo.mailway.com/.

60 M. le Maire, J. V. Møller, A. Tardieu, *J. Mol. Biol.* **1981**, *150*, 273–296.

61 K.-J. Tiefenbach, H. Durchschlag, R. Jaenicke, *Prog. Colloid Polym. Sci.* **1999**, *113*, 135–141.

62 B. W. Koenig, K. Gawrisch, *Biochim. Biophys. Acta* **2005**, *1715*, 65–70.

63 H. Eisenberg, *Biological macromolecules and polyelectrolytes in solution*, Clarendon Press, UK, **1976**.

64 C. Ebel, in: *Protein Interactions – Biophysical Approaches for the Study of Complex Reversible Systems*, Springer, USA, **2007**.

65 P. J. Butler, W. Kuhlbrandt, *Acad. Sci. USA* **1988**, *85*, 3797–3801.

66 M. le Maire, L. P. Aggerbeck, C. Monteilhet, C. Monteilhet, J. P. Andersen, J. V. Møller, *Anal. Biochem.* **1986**, *154*, 525–535.

67 P. Boulanger, M. le Maire, M. Bonhivers, S. Dubois, M. Desmadril, L. Letellier, *Biochemistry* **1996**, *35*, 14216–14224.

68 J. Garcia de la Torre, V. A. Bloomfield, *Q. Rev. Biophys.* **1981**, *14*, 81–139.

69 J. Garcia de la Torre, S. Navarro, M. C. Lopez Martinez, F. G. Diaz, J. J. Lopez Cascales, *Biophys. J.* **1994**, *67*, 530–531.

70 http://www.biozentrum.unibas.ch/auc/.

71 http://www.analyticalultracentrifugation.com.

72 S. Lund, S. Orlowski, B. de Foresta, P. Champeil, M. le Maire, J. V. Møller, *J. Biol. Chem.* **1989**, *264*, 4907–4915.

73 K. G. Fleming, A. L. Ackerman, D. M. Engelman, *J. Mol. Biol.* **1997**, *272*, 266–275.

74 K. G. Fleming, C. C. Ren, A. K. Doura, M. E. Eisley, F. J. Kobus, A. M. Stanley, *Biophys. Chem.* **2004**, *108*, 43–49.

75 Y. Gohon, F. Giusti, C. Prata, D. Charvolin, P. Timmins, C. Ebel, C. Tribet, J.-L. Popot, *Langmuir* **2006**, *22*, 1281–1290.

76 Y. Gohon, T. Dahmane, R. W. H. Ruigrok, P. Schuck, D. Charvolin, F. Rappaport, P. Timmins, D. M. Engelman, C. Tribet, J.-L. Popot, C. Ebel, in preparation.

77 A. Lustig, A. Engel, G. Tsiotis, E. M. Landau, W. Baschong, *Biochim. Biophys. Acta* **2000**, *1464*, 199–206.

78 R. J. Center, P. Schuck, R. D. Leapman, L. O. Arthur, P. L. Earl, B. Moss, J. Lebowitz, *Acad. Sci. USA* **2001**, *98*, 14877–14882.

79 D. Noy, J. R. Calhoun, J. D. Lear, *Anal. Biochem.* **2003**, *320*, 185–192.

80 C. Tanford, in: *Hydrophobic Effect: Formation of Micelles and Biological Membranes*, p. 90, Krieger Publishing Company, USA, **1991**.

81 D. Josse, C. Ebel, D. Stroebel, A. Fontaine, F. Borges, A. Echalier, D. Baud, F. Renault, M. le Maire, E. Chabrieres, P. Masson, *J. Biol. Chem.* **2002**, *277*, 33386–33397.

82 K. G. Fleming, *J. Mol. Biol.* **2002**, *323*, 563–571.

83 L. E. Fisher, D. M. Engelman, J. N. Sturgis, *Biophys. J.* **2003**, *85*, 3097–3105.

84 J. P. Andersen, B. Vilsen, H. Nielsen, J. V. Møller, *Biochemistry* **1986**, *25*, 6439–6439.

85 J. V. Møller, M. le Maire, J. P. Andersen, *Methods Enzymol.* **1988**, *157*, 261–270.

86 W. F. Stafford, *Curr. Opin. Biotechnol.* **1997**, *8*, 14–24.

87 B. Demeler, H. Saber, J. C. Hansen, *Biophys. J.* **1997**, *72*, 397–407.

88 J. M. Claverie, H. Dreux, R. Cohen, *Biopolymers* **1975**, *14*, 1685–1700.

89 R. Cohen, J. M. Claverie, *Biopolymers* **1975**, *14*, 1701–1716.

90 J. M. Claverie, *Biopolymers* **1976**, *15*, 843–857.

91 http://www.ultrascan.uthscsa.edu.

92 http://www.bbri.org/aucrl.html.

93 A. G. Salvay, C. Ebel, *Prog. Colloid Polym. Sci.* **2006**, *131*, 74–82.

94 P. Schuck, *Biophys. J.* **2000**, *78*, 1606–1619.

95 P. Schuck, M. A. Perugini, N. R. Gonzales, G. J. Howlett, D. Schubert, *Biophys. J.* **2002**, *82*, 1096–1111.

96 J. Dam, P. Schuck, *Methods Enzymol.* **2004**, *384*, 185–212.

97 P. Schuck, B. Demeler, *Biophys. J.* **1999**, *76*, 2288–2296.

98 M. Laughery, M. Todd, J. H. Kaplan, *J. Biol. Chem.* **2004**, *279*, 36339–36348.

99 N. Shinji, Y. Tahara, E. Hagiwara, T. Kobayashi, K. Mimura, H. Takenaka, Y. Hayashi, *Ann. N. Y. Acad. Sci.* **2003**, *986*, 235–237.

100 K. Taniguchi, S. Kaya, K. Abe, S. Mardh, *J. Biochem. (Tokyo)* **2001**, *129*, 335–342.

101 J. P. Andersen, K. Lassen, J. V. Møller, *J. Biol. Chem.* **1985**, *260*, 371–380.

102 C. Toyoshima, G. Inesi, *Annu. Rev. Biochem.* **2004**, *73*, 269–292.

103 V. J. Møller, P. Nissen, T. L. Sorensen, M. le Maire, *Curr. Opin. Struct. Biol.* **2005**, *15*, 387–393.

104 J.-L. Popot, E. A. Berry, D. Charvolin, C. Creuzenet, C. Ebel, D. M. Engelman, M. Flotenmeyer, F. Giusti, Y. Gohon, Q. Hong, J. H. Lakey, K. Leonard, H. A. Shuman, P. Timmins, D. E. Warschawski, F. Zito, M. Zoonens, B. Pucci, C. Tribet, *Cell Mol. Life Sci.* **2003**, *60*, 1559–1574.

105 C. Breyton, C. Tribet, J. Olive, J.-P. Dubacq, J.-L. Popot, *J. Biol. Chem.* **1997**, *272*, 21892–21900.

106 C. R. Sanders, A. Kuhn Hoffmann, D. N. Gray, M. H. Keyes, C. D. Ellis, *Chembiochem* **2004**, *5*, 423–426.

107 J. I. Yeh, S. Du, A. Tortajada, J. Paulo, S. Zhang, *Biochemistry* **2005**, *44*, 16912–16919.

108 P. Champeil, T. Menguy, C. Tribet, J.-L. Popot, M. le Maire, *J. Biol. Chem.* **2000**, *275*, 18623–18637.

109 R. Taylor, J. W. Burgner, J. Clifton, W. A. Cramer, *J. Biol. Chem.* **1998**, *273*, 31113–31118.

110 W. F. Stafford, *Methods Enzymol.* **2000**, *323*, 302–325.

111 A. Balbo, K. H. Minor, C. A. Velikovsky, R. A. Mariuzza, C. B. Peterson, P. Schuck, *Acad. Sci. USA* **2005**, *102*, 81–86.

112 M. H. Koch, P. Vachette, D. I. Svergun, *Q. Rev. Biophys.* **2003**, *36*, 147–227.

113 A. Johs, M. Hammel, I. Waldner, R. P. May, I. Laggner, R. Prass, *J. Biol. Chem.* **2006**, *281*, 19732–19739.

114 C. Sardet, A. Tardieu, V. Luzzati, *J. Mol. Biol.* **1976**, *105*, 383–407.

115 H. B. Osborne, C. Sardet, M. Michel-Villaz, M. Chabre, *J. Mol. Biol.* **1978**, *123*, 177–206.

5
Probing Membrane Protein Interactions with Real-Time Biosensor Technology

Iva Navratilova, David G. Myszka and Rebecca L. Rich

5.1
Introduction

The release of the first commercially viable biosensor technology (Biacore) in 1990 by the biosensor division of Pharmacia revolutionized the study of molecular interactions. For the first time, the binding interactions of biomolecules could be studied routinely in real time and without labeling requirements. In the intervening years, biosensor technology has become a standard tool for life science and drug discovery research. Characterizing monoclonal antibody activity and screening for small molecule binders to target proteins are just two examples of the success that biosensor technology has achieved in studying soluble molecular systems. Today, improvements in biosensor technology and methodology are launching a new wave of biosensor applications centered on the study of interactions of molecules with, and embedded in, membrane surfaces. In this chapter, a brief review is provided of the fundamentals of biosensor analysis, and recent advances in the applications of biosensor technology related to membrane systems are detailed.

There are a number of commercial biosensor technologies available, or soon to be released. While many of these systems use different detection methods, the basics of the assay remain the same. Most of these systems monitor mass or refractive index changes near the sensor surface as complexes form and break down over time. By far the most abundant and popular technology is Biacore, with more than 1000 primary publications cited per year using these systems [1, 2]. Biacore instruments use a detector based on surface plasmon resonance (SPR). Regardless of the detection method, the fundamentals of the assay involve attaching one of the binding partners to the sensor surface and flowing the other partner over the surface to study their interactions (Fig. 5.1A). Complex formation and dissociation is monitored in real time to generate a binding response, as shown in Fig. 5.1B. A key advantage of biosensor technology is that the amount of complex is measured in the presence of free material. This allows binding constants to be measured without perturbing the reaction equilibrium. This is in

Biophysical Analysis of Membrane Proteins. Investigating Structure and Function. Edited by Eva Pebay-Peyroula

ISBN: 978-3-527-31677-9

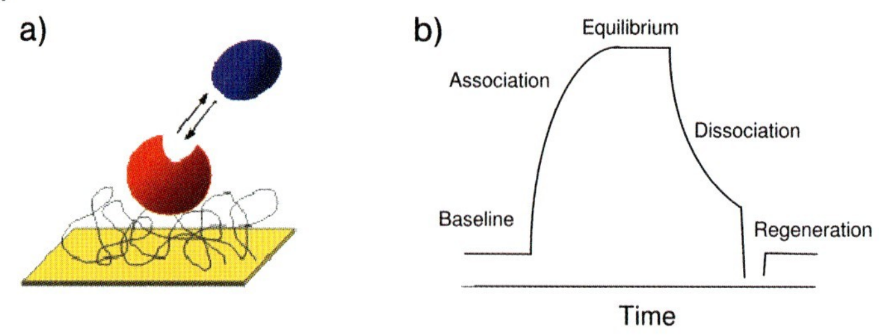

Fig. 5.1 The basics of a biosensor analysis. (A) Biosensor assay design. One reactant is immobilized on a surface and its partner is flowed across the surface. (B) The main features of biosensor response data collected in one binding cycle.

contrast to separation assays such as radioligand binding methods which require washing steps to separate reactants and products before quantitation – and also makes them less effective in determining weak interactions. Another advantage of the biosensor is that the reactants do not need to be purified. Using tagged systems, for example, it is possible to capture target proteins out of crude cell supernatants or lysates. Moreover, the binding partner in solution (typically referred to as the analyte) can even be detected in contaminated samples. How these features make biosensor technology an ideal tool to characterize membrane-associated systems will be discussed later in the chapter.

Historically, critics of biosensor technology have argued that the immobilization of a target molecule onto the sensor surface will change its binding properties. Unfortunately, what these critics do not realize is that the vast majority of experiments are not performed with the target adsorbed directly onto a flat surface. Instead, the target is suspended in solution on a non-crosslinked hydrogel (most often dextran) that coats the sensor chip surface. It has been shown by several groups that the binding constants, when properly measured, from the biosensor do in fact match those obtained from solution-based methods [1, 3–7]. The preponderance of evidence suggests that, in fact, biosensor technology can be used as a biophysical tool. It can also be argued that biosensor technology also affords a unique opportunity for studying membrane-associated systems precisely because it employs the use of a surface. As discussed below, in fact a variety of ways exist to configure biosensor assays to mimic both soluble and membrane-associated systems.

The goal of the chapter is not to burden the reader with details on how the different commercial biosensor systems work. Frankly, too many commercial instruments exist today to provide a comprehensive review of all systems, but some of the latest developments in biosensor technology are highlighted in Refs. [8–11]. In addition, several recent review articles illustrate how biosensor technology is being applied to membrane studies [12–15]. In particular, the review by Cooper provides a comprehensive overview of biosensor and complementary technologies that are applicable for studying diverse membrane proteins [12]. Building on these previously published reviews, herein is summarized the evolution of biosensor-based membrane studies, with emphasis being placed on the most recent developments. In addition, given that the focus of this book is membrane systems and not sensor technology per se, the analyses of membrane systems described here use only commercially available systems, although the techniques should be readily adaptable to non-commercial biosensor technologies. For ease of reading, this chapter is divided into four main areas of application, namely receptor extracellular domains, soluble proteins binding membrane surfaces, targets embedded in membrane surfaces, and solubilized membrane proteins.

5.2 Interactions of Extracellular Domains

Some of the earliest applications of biosensor technology involved the characterization of ligand–receptor interactions. Much information was gained from studies of the extracellular domains of membrane-associated receptors. Not only was it possible to monitor the binding of a ligand with its receptor, but the conformational freedom of the hydrogel on the biosensor surface allowed the ligand–receptor complexes to assemble in a manner similar to how they would form within a lipid membrane. This important feature of biosensor technology is exemplified by ligand-binding studies of growth hormone, erythropoietin, and interleukin-2 (IL-2).

James Wells' research group's analysis of human growth hormone (hGH) and its receptor elegantly illustrate the range of information that can be obtained from biosensor measurements. These investigators showed that hGH could crosslink two receptor domains on the sensor surface (Fig. 5.2) [16], and that the stoichiometry of the receptor–hormone complex was hGH concentration-dependent. In

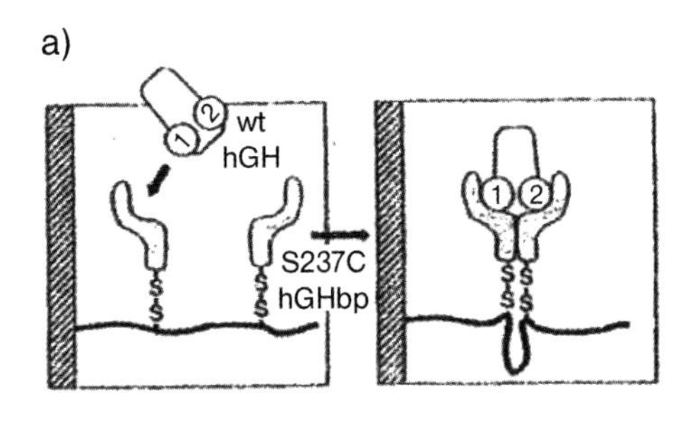

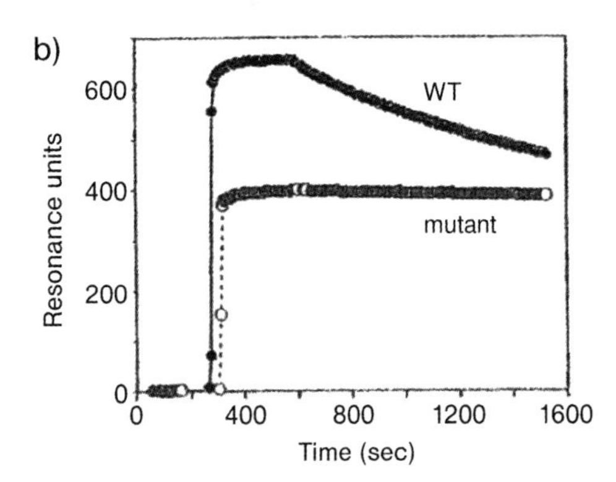

Fig. 5.2 Human growth hormone (hGH) binding to its receptor (hGHbp). (A) Cartoon of hGH binding to two receptor subunits immobilized on the sensor chip surface. (B) Responses for wild-type (WT) and mutant hGH binding to the immobilized receptor. (Reproduced from Ref. [16], with permission from Elsevier; © 1993.)

continuing their biosensor-based analyses, the Wells group probed the functional activity of the receptor's binding interface by determining how its hGH affinity was altered by single amino acid substitutions [17]. By mapping which residues in the hormone–receptor binding interface were critical for complex formation, Wells introduced the concept of binding energy "hotspots". Together, these studies showed how the biosensor could be used in progressive stages to characterize a biomolecular interaction, from the initial verification of binding partners and stoichiometry determination to the detailed mapping of the binding interface.

By using the biosensor, erythropoietin (EPO) binding to engineered Fc-constructs of its transmembrane receptor were characterized [18]. By constructing a soluble, covalently homodimerized form of the receptor (EPOR), in which extracellular receptor regions were fused to Fc domains (Fig. 5.3A), it was possible to determine the stoichiometry, kinetics, and affinity of the EPO–EPOR interaction. The EPOR-Fc chimera was reproducibly captured on Protein A-coated sensor chips, and a concentration series of EPO was flowed across the captured EPOR (Fig. 5.3B). By using this approach, the levels of both EPOR capture and EPO binding could be quantitated. Since the intensities of the responses shown in Fig. 5.3B were proportional to the molecular masses of EPOR and EPO, the stoichiometry of this interaction could be determined: each dimeric EPOR-Fc bound one EPO ligand. This stoichiometry was consistent with data from isothermal titration calorimetry and analytical ultracentrifugation measurements. In addition, the kinetics and affinity of the ligand–receptor interaction could be determined (Fig. 5.3C). The affinity, 3 pM, determined from the biosensor analysis, was consistent with affinities from cellular proliferation assays.

Taking receptor subunit assembly a step further, the biosensor chip was used as a mimic of the cell surface to demonstrate that it is possible to create mixed receptor surfaces using the α and β subunits of the IL-2 receptor [19, 20]. As shown in Fig. 5.4A, the IL-2 ligand would form a trimeric complex on the mixed receptor surfaces. Data shown in Fig. 5.4B–D show that IL-2 bound to the individual α and β monomer surfaces with rapid dissociation rates and lower affinities compared to the $\alpha\beta$ dimer surface. Another important feature of the biosensor is that it allowed order-of-addition experiments to be performed, and for the formation of a quaternary complex between α, β, and γ in the presence of IL-2 ligand to be characterized (Fig. 5.4E). These findings (summarised in Fig. 5.4f) clearly illustrated that the sensor surface itself could serve as an excellent model of the constraining properties imposed by a membrane surface: the contained volume and flexibility of the dextran layer provides the opportunity to assemble receptor complexes in a space resembling the two-dimensional plane of a membrane surface.

5.3 Interactions of Soluble Proteins with Lipid Layers

A major leap forward into membrane surface applications occurred with Biacore's introduction of the HPA and L1 sensor chips (Fig. 5.5). The HPA chip contains a

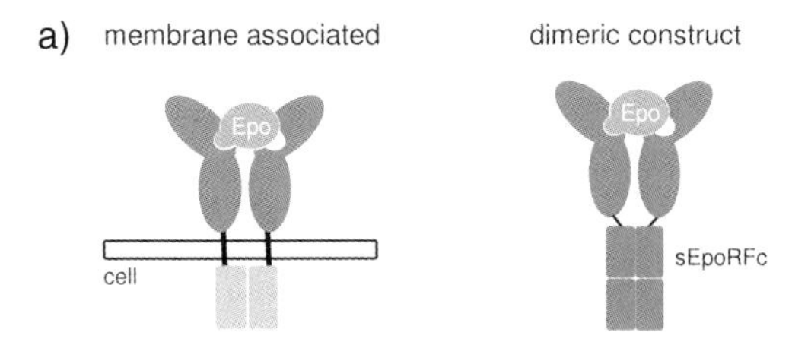

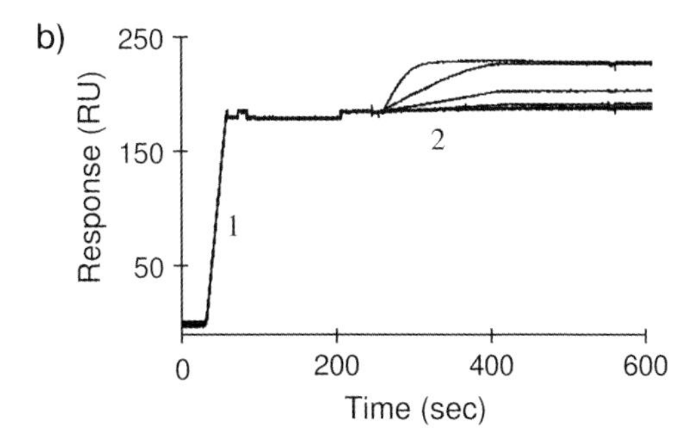

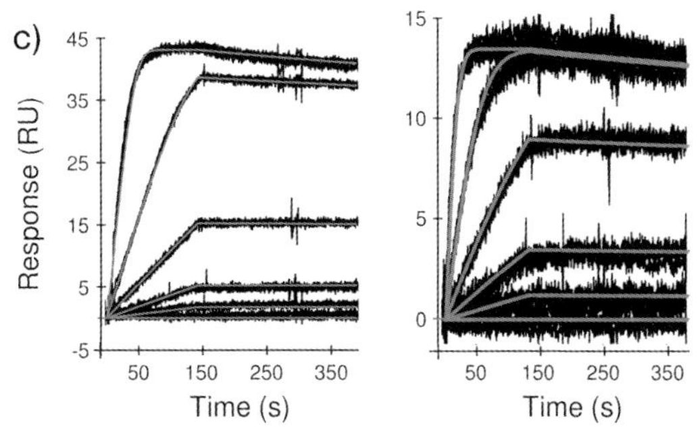

Fig. 5.3 Erythopoietin (EPO) binding to its receptor. (A) Cartoons of *in-vivo* and engineered Epo–receptor complexes. (B) Responses for EPOR-Fc capture on a Protein A chip (1) and the subsequent binding of 0–2 nM EPO (2) injected across the surface. (C) Kinetic analysis of EPO binding to EPOR-Fc captured at high (left) and low (right) densities. Experimental data (black lines) of five repeat injections of each Epo concentration were fitted globally to a partially mass transport-limited 1:1 interaction model (red lines) to obtain $k_a = 8.09 \times 10^7\,M^{-1}\,s^{-1}$, $k_d = 2.44 \times 10^{-4}\,s^{-1}$, and $K_D = 3.01$ pM. (Panels B and C reproduced from Ref. [18], with permission from Academic Press; © 2000.)

self-assembled hydrophobic monolayer to which lipids spontaneously fuse to create a lipid monolayer. Numerous groups have used the HPA chip to create monolayers impregnated with specific lipid groups to study protein binding [21–24]. The challenge with the HPA chip is that the surface is a little difficult to work with. First, the surface is extremely hydrophobic, which can encourage significant non-specific binding if it is not uniformly coated with lipids. Second, lipids adsorbed to this surface are not particularly stable. Indeed, they can be easily

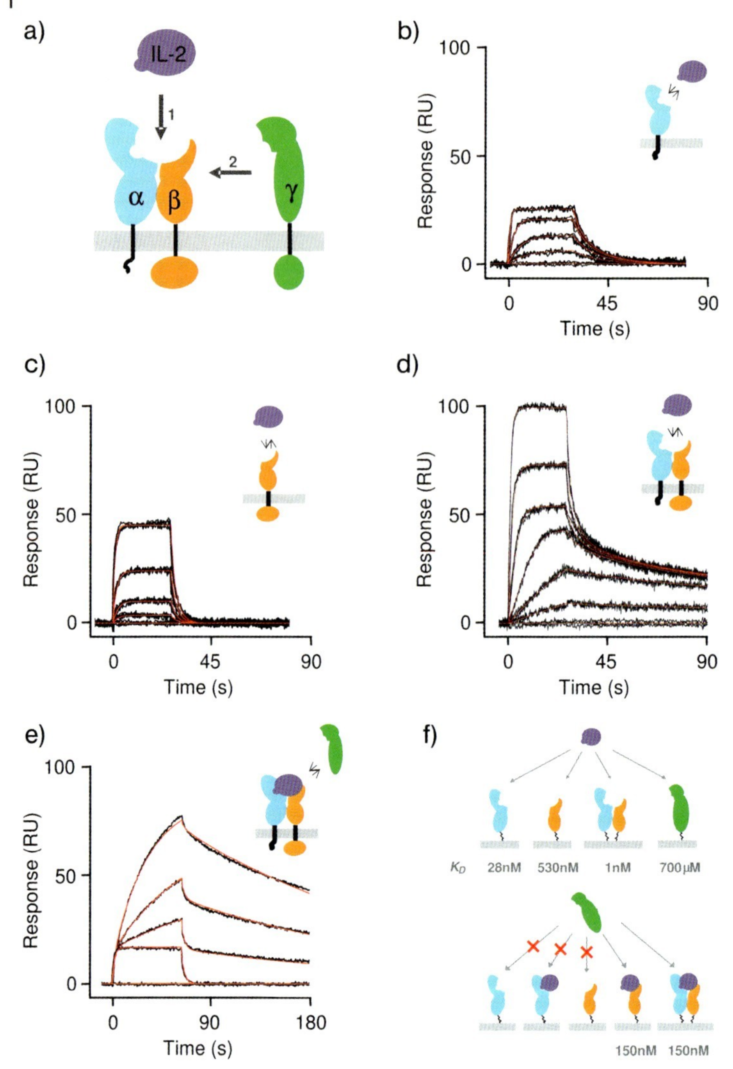

Fig. 5.4 Biosensor-based characterization of interleukin-2 (IL-2) interaction with the three IL-2 receptor (IL-2R) subunits α, β, and γ. (A) Cartoon of the two steps involved in complex assembly. (B) Kinetic analysis of IL-2 binding to immobilized IL-$2R_\alpha$ to yield $k_a=1.1\times10^7\,M^{-1}\,s^{-1}$, $k_d=0.30\,s^{-1}$, and $K_D=28\,nM$. (C) Kinetic analysis of IL-2 binding to immobilized IL-$2R_\beta$ to yield $k_a=5.8\times10^5\,M^{-1}\,s^{-1}$, $k_d=0.0.31\,s^{-1}$, and $K_D=530\,nM$. (D) Kinetic analysis of IL-2 binding to immobilized heterodimeric IL-$2R_{\alpha\beta}$ to yield $k_a=1.7\times10^7\,M^{-1}\,s^{-1}$, $k_d=0.018\,s^{-1}$, and $K_D=1.1\,nM$. (E) Kinetic analysis of IL-$2R_\gamma$ binding to immobilized IL-2/IL-$2R_{\alpha\beta}$ to yield $k_a=3.4\times10^4\,M^{-1}\,s^{-1}$, $k_d=5.1\times10^{-3}\,s^{-1}$, and $K_D=150\,nM$. (F) Summary of IL-2/IL-2R interactions examined using the biosensor. In panels (B) and (C), the binding responses (black lines) are superimposed with the fit (red lines) of a partially mass transport-limited 1:1 interaction model. In panels (D) and (E), the binding responses are superimposed with the fit of interaction models described in Refs. [19] and [20], respectively. [The data shown in panels (B)–(D) are reproduced from Ref. [19]; data in panel (E) are reproduced from Ref. [20], with permission from the Protein Society (© 1996) and the American Chemical Society (© 2002), respectively.]

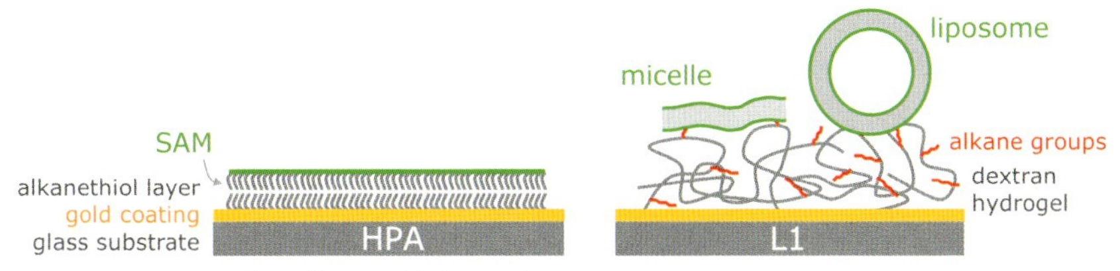

Fig. 5.5 Cartoons of a self-assembled monolayer (SAM) prepared on an HPA chip (left) and the bilayer formed by micelle deposition and intact liposomes present on an L1 chip (right).

removed from the surface with the slightest air bubble flowing through the system; this exposes patches of the hydrophobic surface underneath the lipid layer and results in increased and uncontrollable non-specific binding. By using a different approach to construct lipid layers, the L1 chip contains alkane groups immobilized to the standard hydrogel surface [25]. These surfaces can be used to create stable lipid surfaces from either micelle or liposome preparations. While micelles routinely form a continuous bilayer on the L1 chip surface, several groups have shown that vesicles and liposomes can either remain intact or can be modified to form a fused bilayer, depending on the preparation conditions [12, 25, 26].

Some of the early studies of lipid membrane surfaces involved using the biosensor to examine how drugs can partition into liposome and bilayer surfaces [27, 28]. The biosensor allowed the characterization of how strongly a given drug would interact with the lipid surface. Thus, drug binding was shown to correlate linearly with the density of lipid deposited on the chip surface, such that compounds could be ranked based on their propensity to bind to lipid membranes. From these analyses, a high correlation was found between drugs that bind with high affinity to liposome surfaces and those that lead to a physiological effect called *phospholipidosis* [28]. In addition, determining how drugs interact with lipids yields critical baseline information that is required before the binding of drugs to proteins embedded in a lipid background was investigated. Overall, the formation of liposome and bilayer surfaces was found to be both straightforward and reproducible using the L1 chip. Thus, based on experience, the present authors recommend that, for most membrane studies, the L1 chip should be utilized in preference to the HPA chip.

The L1 chip is an ideal scaffold for capturing liposomes that have been seeded with different binding components in order to assess the binding of proteins to target molecules presented in a membrane environment. This approach has been used to identify the membrane-binding region of the protein the dysfunction of which results in maturity-onset obesity in mice (Fig. 5.6A) [29]. Many other research groups have taken advantage of the robustness and reliability of the L1 chip to explore a wide range of protein–membrane interactions. Recent examples include the screening by Amon et al. of sugar- and lipid-modified T-cell antigen receptor peptides binding to model membranes correlated with the peptides'

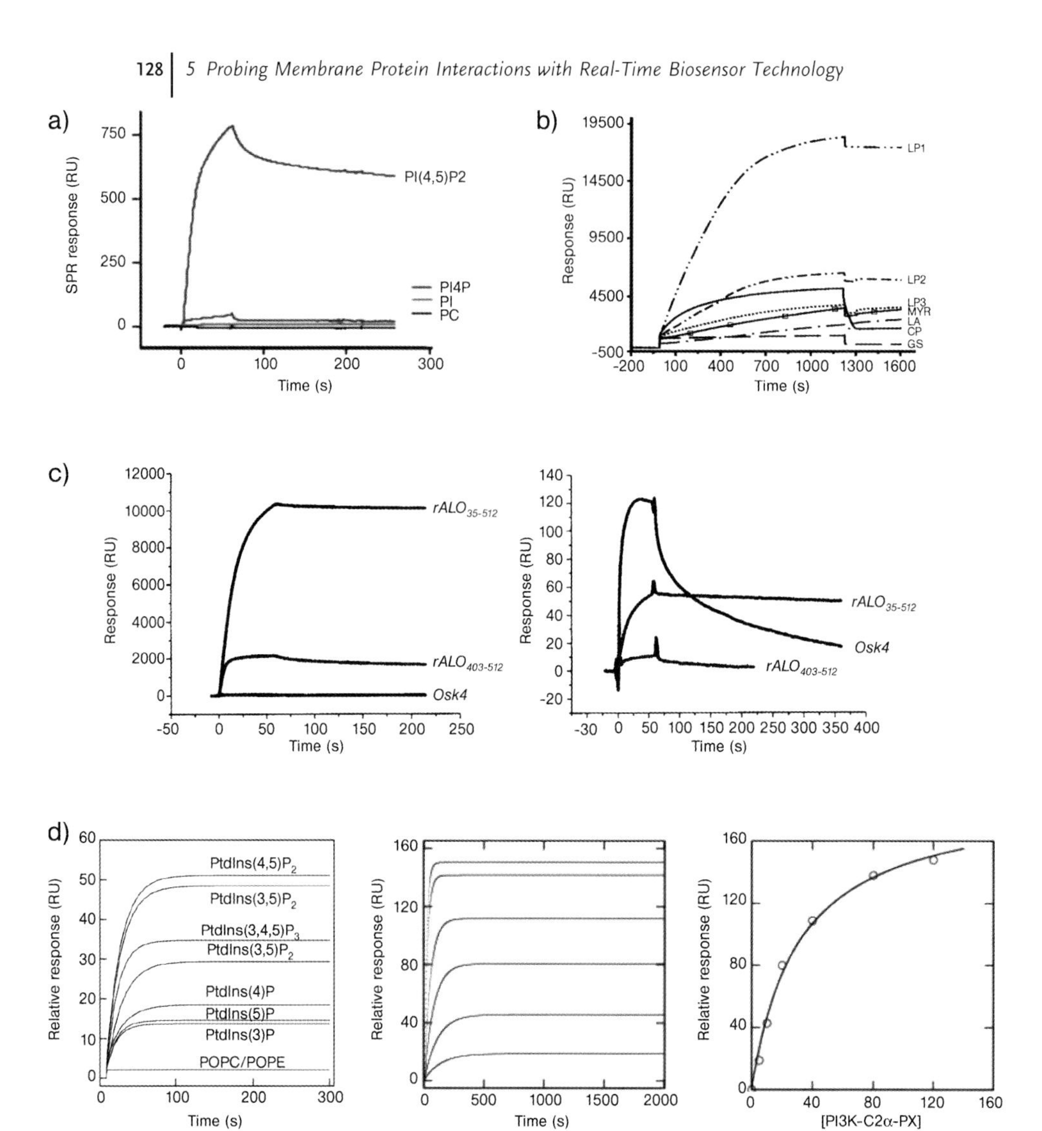

Fig. 5.6 Protein–lipid interactions. (A) Responses for tubby C-terminal domain protein binding to various phosphoinositide-containing lipid bilayers. (Reproduced from Ref. [29] with permission from the American Academy for Advancement of Sciences; © 2001.) (B) Responses for a T-cell antigen receptor transmembrane sequence-derived peptide and its sugar- and lipid-conjugates binding to a model membrane. (Reproduced from Ref. [30] with permission from Elsevier; © 2006.) (C) Responses for two anthrolysin O constructs and the yeast oxysterol binding protein Osh4 binding to model membranes containing cholesterol (left) and ergosterol (right). (Reproduced from Ref. [31], with permission from John Wiley & Sons; © 2006. (D) Left: responses for PI3K-C2αC2 domain binding to lipid layers containing a variety of phosphoinositides. Center: Responses for 2 to 60 nM PI3K-C2α PX binding to a PtdIns(4,5) P_2 lipid surface. Right: Equilibrium analysis of the PI3K-C2α PX/PtdIns(4,5)P_2 lipid interaction, which yielded $K_D = 25$ nM. (Reproduced from Ref. [32], with permission from the American Society for Biochemistry and Molecular Biology; © 2006.)

in-vivo retardation of arthritis progression (Fig. 5.6B) [30]. Likewise, when Cocklin et al. studied the binding of the *Bacillus anthracis* hemolysin anthrolysin O (a potential anthrax virulence factor) to various model membranes, cholesterol was found to be essential for this protein–membrane interaction (Fig. 5.6C) [31]. Finally, Stahelin et al. conducted a specificity and affinity analysis of Phox homology (PX) domains binding to phosphoinositide-containing membranes, and established that the PX domain-containing protein's penetration of membranes was enhanced by a single phosphoinositide (Fig. 5.6D) [32].

5.4 Interactions of Proteins Embedded in Lipid Layers

The ability to reproducibly and stably capture liposomes onto the L1 chip surface allows the next step to be taken in membrane surface applications, namely to characterize the binding of analytes to proteins embedded in the liposomes. The challenge at this point is how to acquire the lipo/protein vesicles initially, since although the biosensor is a label-free, real-time technology, it is not exceptionally sensitive. In order to detect the binding of an analyte to a lipo/protein surface, the number of binding sites on the surface must be relatively high. Typically, the amount of active sites in the lipo/protein should most likely be 1000-fold higher than would be needed for fluorescence detection assays. Moreover, as most biosensor detection systems are mass-based, the need for large amounts of target protein only increases as the size of the analyte decreases.

Several approaches are available to deal with the lipo/protein content and concentration issue. If the receptor of interest can be highly overexpressed, then one approach would be to harvest cellular membranes through blebs [33] and to capture them directly onto an L1 surface. This approach should succeed if the target analyte was relatively large such that the overall expression level required to detect binding would not have to be exceptionally high. For example, if the analyte was a monoclonal antibody, the bleb approach most likely succeed. However, in cases where expression levels are not exceptionally high, the use of alternative expression systems may prove beneficial. As an example, Doms and coworkers have taken advantage of viral packaging systems to produce viral pseudoparticles containing membrane-embedded receptors and have used these to study the binding of protein analytes [34].

5.4.1 On-Surface Reconstitution of G-Protein-Coupled Receptor

One of the most novel approaches to creating lipo/protein surfaces has been that derived by Biacore's in-house group. In 2002, Karlsson and Löfås showed that it was possible to immobilize rhodopsin on an L1 surface and then reconstruct a lipid membrane environment around the receptor (Fig. 5.7) [35]. Using a specially modified biosensor that included an internal light source, it was shown that the

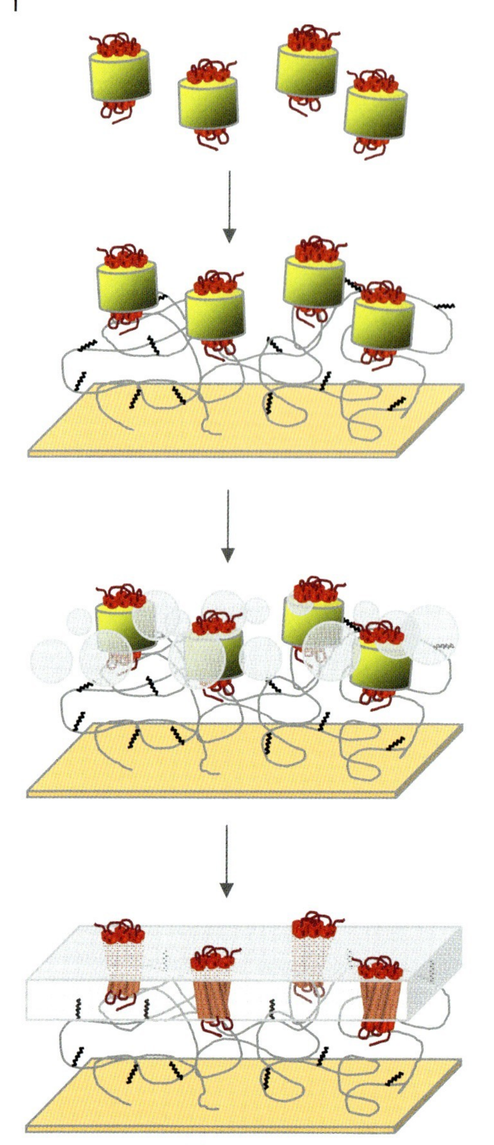

Fig. 5.7 Cartoon of the steps involved in on-surface reconstitution: rhodopsin was solubilized and immobilized on the chip surface, and a mixed micelle preparation, which fused to form the lipid layer, was flowed across the immobilized receptor.

surface-tethered rhodopsin could interact with transducin in solution. Whilst this approach was exceedingly elegant, it suffered from the requirement for a highly purified and highly active G-protein-coupled receptor (GPCR). Thus, the method could not be applied to typical membrane-associated receptors, which are often less abundant and less stable than rhodopsin.

5.4.2
Capture/Reconstitution of GPCRs

In realizing the potential of Biacore's approach to studying membrane-associated systems, the on-surface reconstitution was expanded to include a method referred to as "capture/reconstitution" (Fig. 5.8A) [36]. The scheme is started by engineering the receptor of interest to contain a specific peptide tag on the C-terminus of the protein. Fortunately, cloning and expression technologies are readily available today, which make this requirement less of a burden. When the tagged receptor had been expressed, the approach involved solubilizing the membranes and capturing the receptor onto the biosensor surface using a specific recognition molecule. Great success was achieved using a monoclonal antibody called 1D4, which recognized a nine-amino acid C-terminal tag. The advantage of this capturing approach was that the receptor was both purified and concentrated onto the sensor surface. The capturing step was then followed by an injection of lipids to reconstitute a bilayer (Fig. 5.8B), as performed previously by Karlsson and Löfås with rhodopsin. By using this approach, it was possible to demonstrate the binding of conformationally sensitive monoclonal antibodies and a chemokine ligand to the GPCRs, CXCR4 and CCR5 (Fig. 5.8C and D). Although control studies indicated that the receptor activity appeared less sensitive to the presence of the reconstituted lipid layer, the suggestion was clear that it might be possible to monitor the binding activity of some GPCRs in their solubilized state.

5.5
Interactions of Membrane-Solubilized Proteins

There are two main issues regarding the reconstitution of lipid layers around surface-tethered receptors. The first problem is that the reconstitution step takes time, which would lead to limited throughput when running screening assays. The second and more pressing problem is that analytes (particularly small molecules) may bind to the lipid layer as well as to the target receptor, such that the responses for the target/analyte interaction could be masked by the signal from analyte binding to the lipid layer. However, by tethering detergent-solubilized receptor to the surface, analyte binding to the background lipid layer could be minimized, and the assay speeded up by removing the reconstitution step.

Again, using rhodopsin as a model system, several research groups have demonstrated the viability of using the solubilization approach to prepare GPCR surfaces. For example, Komolov et al. captured solubilized dark-adapted rhodopsin through its N-terminal carbohydrate moiety onto a concanavalin A-coated sensor surface and tested the receptor's ability to bind transducin in the absence and presence of light (Fig. 5.9A) [37]. As shown in Fig. 5.9B, light initiated transducin binding to the rhodopsin, and this interaction could be disrupted by the addition of guanosine triphosphate (GTP), indicating that the solubilized receptor had retained activity. In complementing these studies, both Northup [38] and Rebois et al. [39] provided protocols for the covalent coupling of concanavalin A to

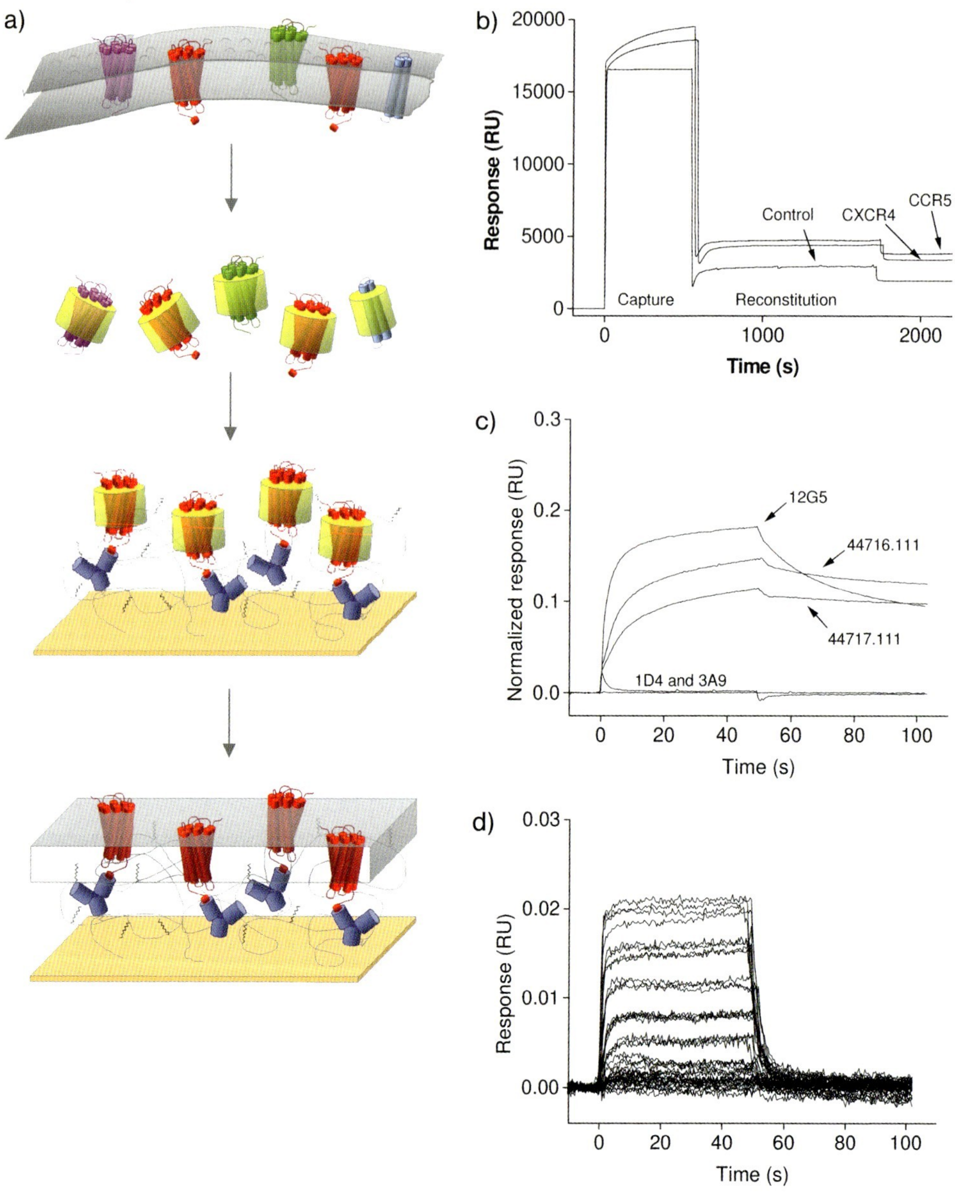

Fig. 5.8 Capture/reconstitution of G-protein-coupled receptors. (A) Cartoon of the steps involved in the capture/reconstitution method: tagged receptors are cloned/expressed, solubilized, captured on a 1D4 antibody-coated sensor chip, and then reconstituted in a lipid layer by the injection of mixed micelles. (B) Sensorgrams depicting the assembly of CXCR4, CCR5, and control surfaces. (C) Binding of three conformation-dependent antibodies and two negative-control antibodies to CXCR4 captured/reconstituted on a L1 chip surface. (D) SDF-1 binding to CXCR4 captured/reconstituted on a L1 chip surface. (Panels B–D reproduced from Ref. [36], with permission from Elsevier; © 2003.)

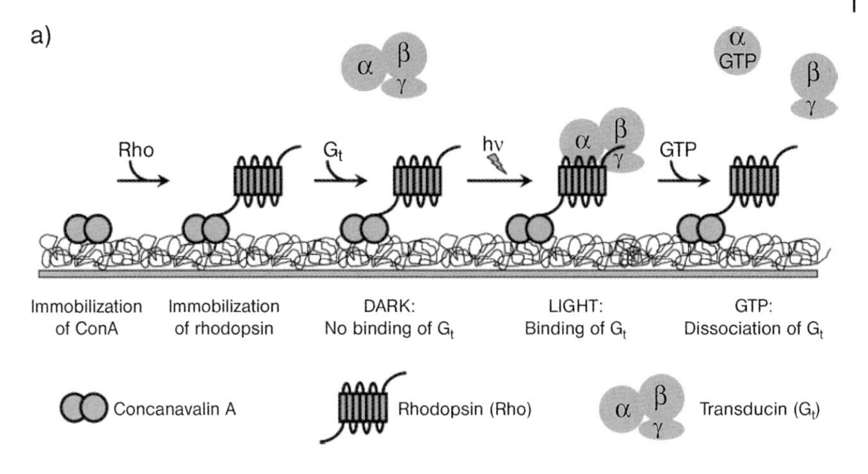

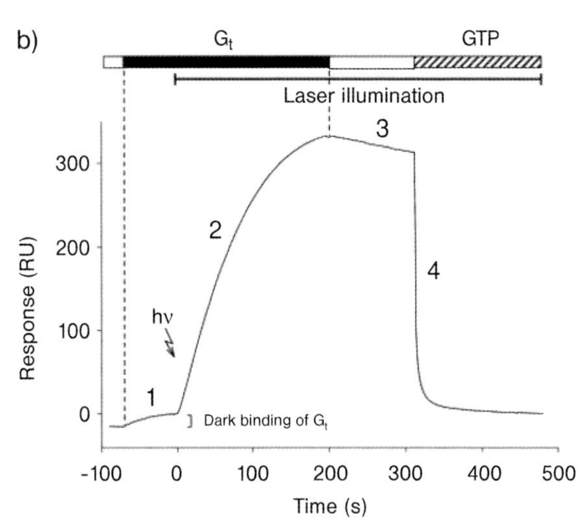

Fig. 5.9 (A) Cartoon of rhodopsin (Rho) capture on a concanavalin A-coated sensor chip and the receptor's subsequent light-induced interaction with transducin (G_t). (B) Sensorgram of G_t binding to rhodopsin: (1) minimal G_t binding to Rho in the absence of light (t=–100–0 s); (2) G_t binding to Rho exposed to light (t=0–180 s); (3) slow dissociation of G_t from light-activated Rho; (4) disruption of the G_t/Rho complex by GTP. (Reproduced from Ref. [37], with permission from the American Chemical Society; © 2006.)

surfaces, for solubilizing and capturing high densities [~5000–10 000 response units (RU)] of rhodopsin, and characterizing the receptor's interactions with G-protein subunits.

While biosensors have traditionally been used to investigate interactions in high resolution to obtain kinetic information, the technology can also be applied in a screening mode to optimize a GPCR preparation prior to its detailed characterization. For example, biosensors are uniquely suited to screen recombinant clones for GPCR expression levels and to identify buffer conditions that maintain the isolated receptor's stability and activity. Zeder-Lutz et al. developed a biosensor-based method to improve recombinant GPCR expression [40]. By using a *Pichia pastoris* expression vector, this group produced 11 GPCRs that were C-terminally biotinylated and N-terminally FLAG and His_{12}-tagged. Biosensors assays were incorporated into a methodology that quickly quantitated GPCR expression level by screening crude extracts from six to twelve clones of each receptor. In order to compare expression levels in the clone panel, the extracts were flowed across streptavidin-coated sensor chips, after which anti-FLAG and anti-His antibodies were flowed across the captured GPCRs. Antibody binding responses correlated with the GPCR expression level in each clone, and the best-expressing clones were selected for further investigation. Further downstream in the preparation of active GPCR isolates, it was realized that the biosensor could be used as an exceptionally powerful tool to screen receptor solubilization conditions, once it was found that the binding activity of solubilized receptors could be monitored without the reconstitution of lipid bilayers.

It is not difficult to imagine that some conditions may solubilize a large proportion of the receptor but render most of it inactive, whereas other conditions may solubilize only a small proportion but maintain high activity. Unfortunately, technologies such as radioligand assays, which are used routinely to study membrane-associated systems, are unable to quantitate the fraction of inactive material in a preparation. In contrast, one of the most powerful features of the biosensor approach is that the amount of receptor captured on the biosensor surface can be quantitated, as can the amount of analyte that it is capable of binding. In this way, it is possible to measure directly how much receptor is present, and how active it is. Because the biosensor quantitates the total amount of receptor present and then assesses the degree of activity, the effects of various solubilization conditions can be compared both rapidly and quantitatively.

A further advantage of the Biacore biosensors is that the majority are fully automated. In fact, the autosampler itself may be used to set up solubilization trials, thus standardizing the amount of time that cells are exposed to each solubilization condition [41]. In a typical screening experiment, approximately 4×10^6 canine thymocyte cells overexpressing a receptor of interest were placed into a 4-mL vial with solubilization buffer, but no detergent. The autosampler was then programmed to suspend the cells and transfer 100 µL of cell suspension into 100 µL of the solubilizing detergent. The autosampler was commanded to mix the sample and, after a selected time period, to inject the crude mixture directly over the sensor surface coated with capturing mAb. When the receptor had been captured, the autosampler then injected the analyte (a conformationally sensitive monoclonal antibody or native ligand) used to test for receptor activity.

By using this approach it was possible to screen a large panel of solubilization conditions efficiently and to assess how each condition affected receptor activity. How the choice of two detergents affected receptor activity in relation to conformational antibody binding is illustrated in Fig. 5.10A, while the effect of different detergent/lipid/cholesterol combinations on CCR5 activity is shown in Fig. 5.10B. The optimal buffer condition from these screens was used to establish how the two receptors bound native protein partners and small-molecule inhibitors (Fig. 5.10C and D). The ability of the biosensor to rapidly identify conditions that effectively solubilized the receptor, while maintaining its activity, represents a quantum leap forward in membrane receptor research.

Structural studies of membrane-associated receptors, however, require receptor fractions that are highly stable and active. Hence, the next common step in receptor optimization is the extraction of active material using affinity purification. The biosensor can also be used as a complement to affinity chromatography to improve receptor activity. As illustrated in Fig. 5.11, the biosensor technology was used to design a chromatography assay to isolate active receptor fraction from the crude solubilized milieu [42]. When crude solubilized CCR5 was passed over a gp120 surface, active CCR5 bound, inactive material washed through, and enriched CCR5 could be eluted with TAK-779 (Fig. 5.11A). The biosensor played an important role in identifying the suitable regeneration condition as well as fractions containing enriched CCR5. As shown in Fig. 5.11B, fractions 4 to 6 contained the most active receptor. Figure 5.11C shows the elution profiles determined by the biosensor and absorbance (A_{280nm}), and also illustrates the importance of using the biosensor to monitor receptor activity. If only the absorbance profile of TAK-779 had been used to identify fractions of interest, most of the receptor fractions would have been missed as the majority of TAK-779 eluted from the column prior to TAK-779/CCR5 complexes. This biosensor assay was also rapid, with the entire panel of column fractions being screened in little over 1 h; moreover, those fractions which contained active CCR5 were also identified. In contrast, other confirmation assays such as Western blot would not only have taken days to obtain but would also have failed to reveal the receptor activity in each fraction. The efficacy of this biosensor-based affinity purification strategy is demonstrated graphically in Fig. 5.11D, with gp120:CD4 binding to the purified CCR5 more than threefold that with unpurified CCR5.

Typically, crystallographic efforts have used a "shotgun" approach to determining the buffer conditions best suited for crystal growth, without considering how buffer components affect receptor stability and activity. However, when using the biosensor it is possible to identify those conditions which maintain receptor activity and – as importantly – those which destroy it. For example, by using the activity-enriched CCR5 preparation developed in Fig. 5.11, gp120:CD4 binding to CCR5 diluted into a panel of standard crystallization buffers was tested [43]. In this screen, the receptor retained activity most often when it was stored in buffers containing high-molecular-weight polyethylene glycols (PEGs). This screening provides valuable information for crystallization trials, in that whilst those conditions which best maintain activity may not necessarily promote crystallization, those that disrupt receptor activity are not worthy of further investigation.

Taken together, the findings of these recent investigations demonstrate the important role that biosensors can play in optimizing receptor solubilization

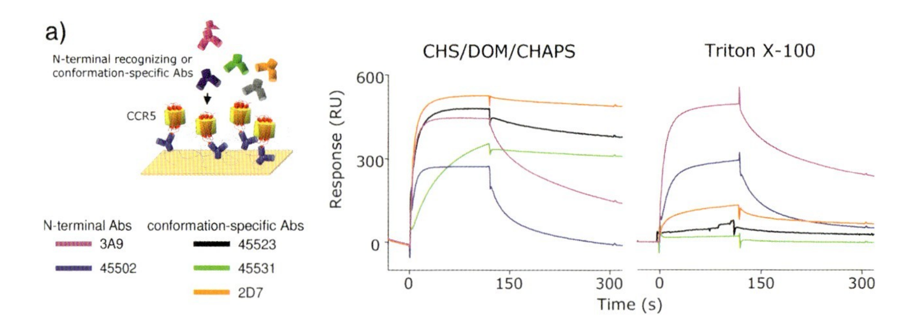

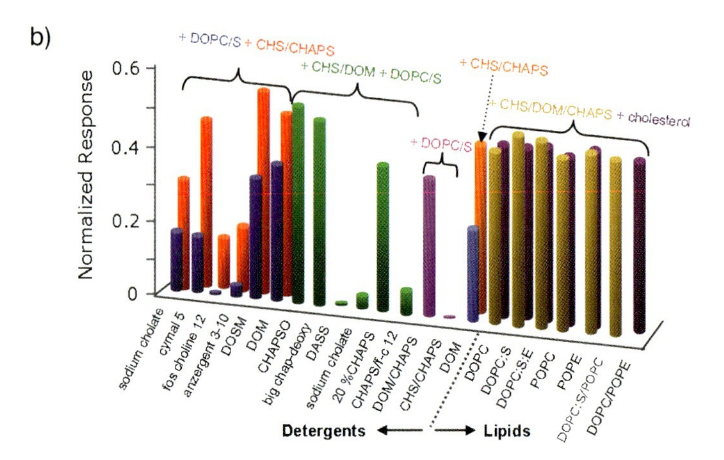

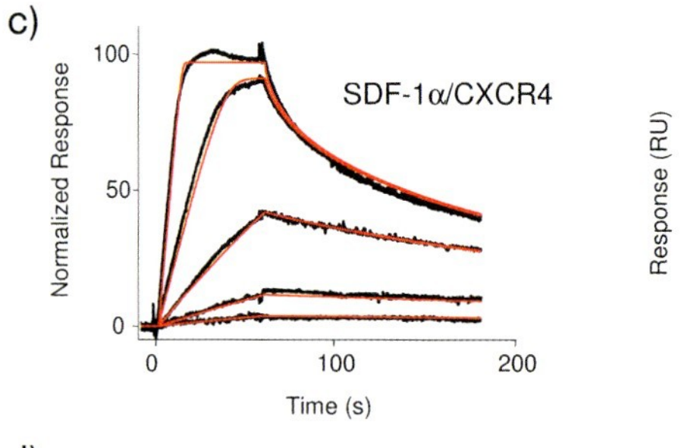

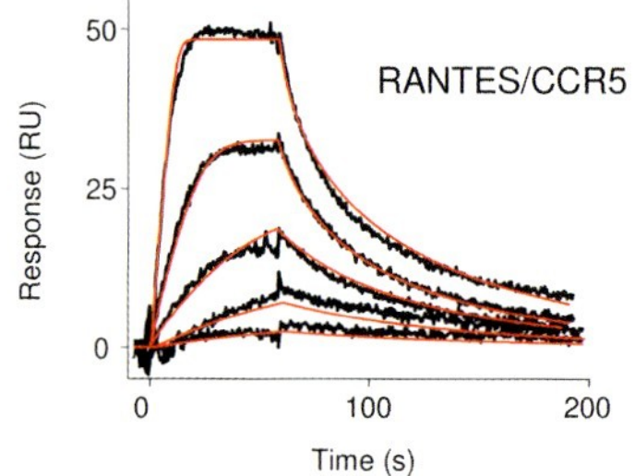

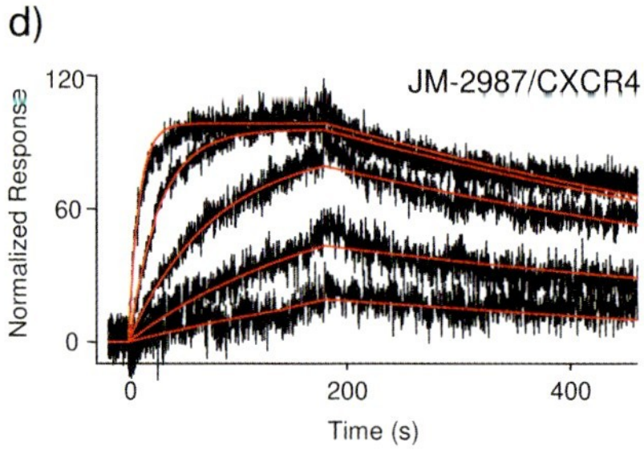

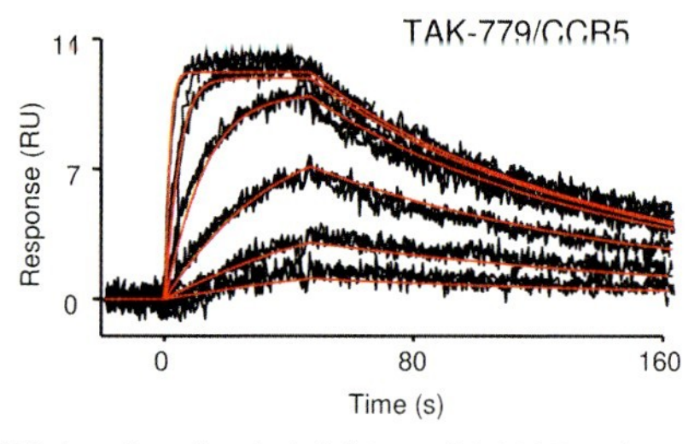

Fig. 5.10 Biosensor-based screen of CCR5 and CXCR4 solubilization conditions. (A) Effect of solubilization conditions on CCR5 binding of N-terminal-recognizing and conformation-specific antibodies. (B) CCR5 activity for a conformationally dependent mAb 2D7 under different solubilization conditions. (C) Native ligands SDF1-α and RANTES binding to CXCR4 and CCR5, respectively. (D) Small-molecule inhibitors JM-2987 and TAK-779 binding to CXCR4 and CCR5, respectively. In panels (C) and (D), the red lines represent the fit of a 1:1 interaction model to the experimental data (black lines). (Panels A and B reproduced from Ref. [41], with permission from Elsevier; © 2005. Panels C and D reproduced from Ref. [43], with permission from Elsevier; © 2006.)

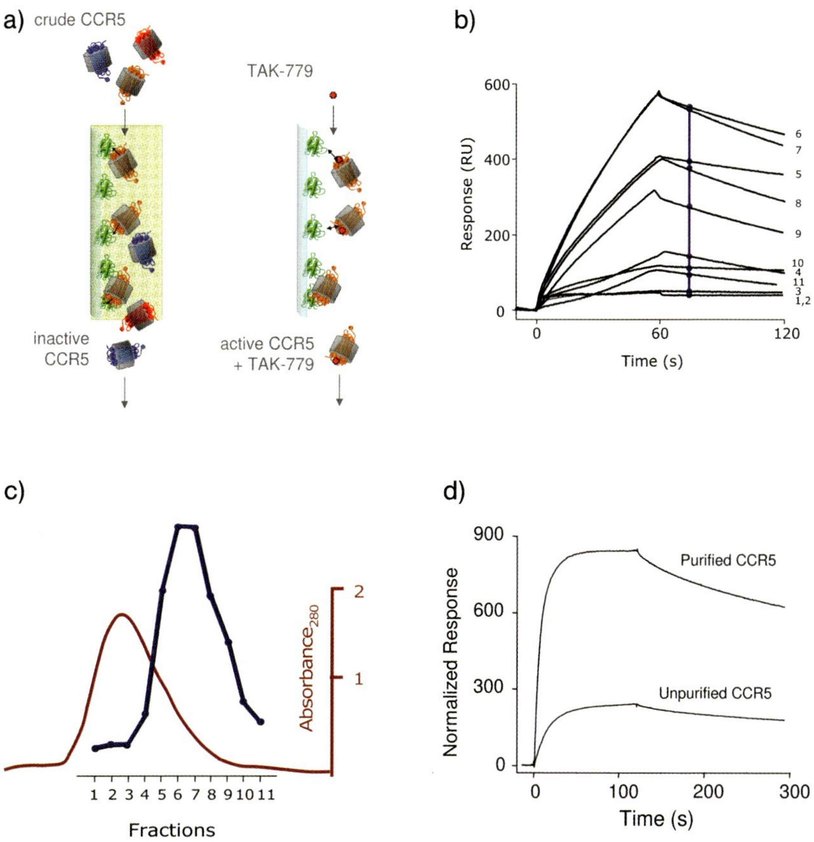

Fig. 5.11 Isolation of active CCR5. (A) Cartoon of chromatography steps. Left: Crude solubilized CCR5 was passed over a gp120 column. The active receptor bound while the inactive receptor flowed through. Right: Active receptor was eluted from the column with the small-molecule inhibitor TAK-779. (B) Responses for CCR5/TAK-779 column fractions 1 to 11 binding to an anti-CCR5 antibody surface. (C) Elution profiles determined by absorbance (A_{280nm}, red) and activity (measured at t = 75 s in panel B; blue) for CCR5/TAK-779 fractions eluted from the gp120 column. (D) 1:1 gp120:CD4 binding to purified and unpurified solubilized CCR5. (Panel D reproduced from Ref. [42], with permission from Elsevier; © 2006.)

conditions. In addition, such studies highlight the qualitative applications of biosensors that are often overlooked when the technology is most often used as a biophysical tool for kinetic analyses. Clearly, biosensors provide an opportunity to screen for receptor expression levels, to identify solubilization and crystallization buffer components that maintain receptor activity, and to isolate activity-enriched receptor fractions via affinity chromatography. Moreover, when suitable membrane targets are available, the biosensor can also excel in its more traditional role as a kinetic analysis tool.

5.6 Summary

In the past, membrane protein characterization has been hampered by difficulties associated with obtaining sufficient amounts of active material required for structural and functional analyses. Except for rhodopsin, GPCRs are produced naturally at low levels, and quite often represent the major challenge to expression in recombinant systems. An added challenge is to maintain the integrity of a membrane protein once removed from its native lipid environment. With regards to future progress, biosensor technology clearly can – and should – play a significant role in the examination of membrane-associated systems. Indeed, the ability to create – both readily and reproducibly – membrane environments on a chip surface offers unique opportunities to monitor the interaction of analytes with lipid surfaces. Also, the possibility of capturing lipo/protein vesicles offers exciting opportunities to design customized surfaces. As technology advances, biosensors will undoubtedly be applied at all stages of membrane protein research, from initial production through isolation to detailed biophysical characterization. It is, in fact, difficult to imagine biosensor technology – which is inherently surface-based – having a more appropriate area of application than in the study of membrane-associated systems.

References

1 Rich R. L., Myszka D. G. (2005) Survey of the year 2004 commercial optical biosensor literature, *J. Mol. Recog.* 18: 431–478.

2 Rich R. L., Myszka D. G. (2006) Survey of the year 2005 commercial optical biosensor literature, *J. Mol. Recog.* 19: 478–534.

3 Day Y. S. N., Baird C. L., Rich R. L., Myszka D. G. (2002) Direct comparison of equilibrium, thermodynamic, and kinetic rate constants determined by surface- and solution-based biophysical methods, *Protein Sci.* 11: 1017–1025.

4 Myszka D. G., Abdiche Y. N., Arisaka F., Byron O., Eisenstein E., Hensley P., Thomson J. A., Lombardo C. R., Schwarz F., Stafford W., Doyle M. L. (2003) The ABRF-MIRG'02 Study: Assembly state, thermodynamic, and kinetic analysis of an enzyme/inhibitor interaction, *J. Biomol. Techn.* 14: 247–269.

5 Drake A. W., Myszka D. G., Klakamp S. L. (2004) Characterizing high-affinity antigen/antibody complexes by kinetic- and equilibrium-based methods, *Anal. Biochem.* 328: 35–43.

6 Papalia G. A., Leavitt S., Bynum M. A., Katsamba P. S., Wilton R., Qiu H., Steukers M., Wang S., Bindu L., Phogat S., Giannetti A. M., Ryan T. E., Pudlak V. A., Matusiewicz K., Michelson K. M., Nowakowski A., Pham-Baginski A., Brooks J., Tieman B., Bruce B. D., Vaughn M., Baksh M., Cho Y. H., De Wit M., Smets A., Vandersmissen J., Michiels L., Myszka D. G. (2006) Comparative analysis of ten small molecules binding to carbonic anhydrase II by different investigators using Biacore technology, *Anal. Biochem.* 359: 94–105.

7 Navratilova I., Papalia G. A., Rich R. L., Bedinger D., Brophy S., Condon B., Deng T., Emerick A. W., Guan H.-W., Hayden T., Hcutmckcrs T., Hoorelbeke B., McCroskey M. C., Murphy M. M., Nakagawa T., Parmeggiani F., Qin X., Rebe S., Tomasevic N., Tsang T., Waddell M. B., Zhang F. F., Leavitt S., Myszka D. G. (2007) Thermodynamic benchmark

study using Biacore technology, *Anal. Biochem.* 364: 67–77.

8 Rich R. L., Myszka D. G. (2007) Higher-throughput, label-free, real-time molecular interaction analysis, *Anal. Biochem.* 361: 1–6.

9 Cannon M. J., Myszka D. G. (2005) Surface plasmon resonance, in: *Methods for Structural Analysis of Protein Pharmaceuticals*, Jiskoot W., Crommelin D. (Eds.), AAPS Press: Arlington, VA, pp. 527–544.

10 Cooper M. A. (2006) Optical biosensors: where next and how soon? *Drug Discov. Today* 11: 1061–1070.

11 Mukhopadhyay R. (2005) Surface plasmon resonance instruments diversify, *Anal. Chem.* 77: 313A–317A.

12 Cooper M. A. (2004) Advances in membrane receptor screening and analysis, *J. Mol. Recog.* 17: 286–315.

13 Beseničar M., Maček P., Lakey J. H., Anderluh G. (2006) Surface plasmon resonance in protein-membrane interactions, *Chem. Phys. Lipids* 141: 169–178.

14 Mozsolits H., Aguilar M.-I. (2002) Surface plasmon resonance spectroscopy: an emerging tool for the study of peptide-membrane interactions, *Biopolymers (Peptide Science)* 66: 3–18.

15 Pattnaik P. (2005) Surface plasmon resonance: applications in understanding receptor-ligand interaction, *Appl. Biochem. Biotechnol.* 126: 79–92.

16 Cunningham B. C., Wells J. A. (1993) Comparison of a structural and functional epitope, *J. Mol. Biol.* 234: 554–563.

17 Clackson T., Wells J. A. (1995) A hot spot of binding energy in a hormone-receptor interface, *Science* 267: 383–386.

18 Hensley P., Doyle M. L., Myszka D. G., Woody R. W., Brigham-Burke M. R., Erickson-Miller C. L., Griffin C. A., Jones C. S., McNulty D. E., O'Brien S. P., Amegadzie B. Y., MacKenzie L., Ryan M. D., Young P. R. (2000) Evaluating energetics of erythropoietin ligand binding to homodimerized receptor extracellular domains, *Methods Enzymol.* 323: 177–207.

19 Myszka D. G., Arulanantham P. R., Sana T., Wu Z., Morton T. A., Ciardelli T. L. (1996) Kinetic analysis of ligand binding to interleukin-2 receptor complexes created on an optical biosensor surface, *Protein Sci.* 5: 2468–2478.

20 Liparoto S. F., Myszka D. G., Wu Z., Goldstein B., Laue T. M., Ciardelli T. L. (2002) Analysis of the role of the interleukin-2 receptor γ chain in ligand binding, *Biochemistry* 41: 2543–2551.

21 Nagahama M., Hara H., Fernandez-Miyakawa M., Itohayashi Y., Sakurai J. (2006) Oligomerization of *Clostridium perfringens* ε-toxin is dependent upon membrane fluidity in liposomes, *Biochemistry* 45: 296–302.

22 Rebolj K., Ulrih N. P., Macek P., Sepčić K. (2006) Steroid structural requirements for interaction of ostreolysin, a lipid-raft binding cytolysin, with lipid monolayers and bilayers, *Biochim. Biophys. Acta* 1758: 1662–1670.

23 Mitra A. K., Célia H., Ren G., Luz J. G., Wilson I. A., Teyton L. (2004) Supine orientation of a murine MHC class I molecule on the membrane bilayer, *Curr. Biol.* 14: 718–724.

24 Geukens N., Frederix F., Reekmans G., Lammertyn E., Van Mellaert L., Dehaen W., Maes G., Anné J. (2004) Analysis of type I signal peptidase affinity and specificity for preprotein substrates, *Biochem. Biophys. Res. Commun.* 314: 459–467.

25 Cooper M. A., Hansson A., Löfås S., Williams D. H. (2000) A vesicle capture sensor chip for kinetic analysis of binding to membrane-bound receptors, *Anal. Biochem.* 277: 196–205.

26 Anderluh G., Beseničar M., Kladnik A., Lakey J. H., Maček P. (2005) Properties of nonfused liposomes immobilized on an L1 Biacore chip and their permabilization by a eukaryotic pore-forming toxin, *Anal. Biochem.* 344: 43–52.

27 Baird C. L., Courtenay E. S., Myszka D. G. (2002) Surface plasmon resonance characterization of drug-liposome interactions, *Anal. Biochem.* 310: 93–99.

28 Abdiche Y. N., Myszka D. G. (2004) Probing the mechanism of drug/lipid membrane interactions using Biacore, *Anal. Biochem.* 328: 233–243.

29 Santagata S., Boggon T. J., Baird C. L., Shan W.-S., Gomez C., Myszka D. G.,

Shapiro L. (2001) G-protein signaling through tubby proteins, *Science* 292: 2041–2050.

30 Amon M. A., Ali M., Bender V., Chan Y.-N., Toth I., Manolios N. (2006) Lipidation and glycosylation of a T cell antigen receptor (TCR) transmembrane hydrophobic peptide dramatically enhances in vitro and in vivo function, *Biochim. Biophys. Acta* 1763: 879–888.

31 Cocklin S., Jost M., Robertson N. M., Weeks S. D., Weber H.-W., Young E., Seal S., Zhang C., Mosser E., Lool P. J., Saunders A. J., Rest R. F., Chaiken I. M. (2006) Real-time monitoring of the membrane-binding and insertion properties of the cholesterol-dependent cytolysin anthrolysin O from Bacillus anthracis, *J. Mol. Recog.* 19: 354–363.

32 Stahelin R. V., Karathanassis D., Bruzik K. S., Waterfield M. D., Bravo J., Williams R. L., Cho W. (2006) Structural and membrane binding analysis of the phox homology domain of phosphoinositide 3-kinase-C2α, *J. Biol. Chem.* 281: 39396–39406.

33 He Y.-Y., Huang J.-L., Chignell C. F. (2006) Cleavage of epidermal growth factor by caspase during apoptosis is independent of its internalization, *Oncogene* 25: 1521–1531.

34 Hoffman T. L., Canziani G., Jia L., Rucker J., Doms R. W. (2000) A biosensor assay for studying ligand-membrane receptor interactions: Binding of antibodies and HIV-1 Env to chemokine receptors, *Proc. Natl. Acad. Sci. USA* 97: 11215–11220.

35 Karlsson O. P., Löfås S. (2002) Flow-mediated on-surface reconstitution of G-protein coupled receptors for applications in surface plasmon resonance biosensors, *Anal. Biochem.* 300: 132–138.

36 Stenlund P., Babcock G. J., Sodroski J., Myszka D. G. (2003) Capture and reconstitution of G protein-coupled receptors on biosensor surfaces, *Anal. Biochem.* 316: 243–250.

37 Komolov K. E., Senin I. I., Philippov P. P., Koch K.-W. (2006) Surface plasmon resonance study of G protein/receptor coupling in a lipid bilayer-free system, *Anal. Chem.* 78: 1228–1234.

38 Northup J. (2004) Measuring rhodopsin-G-protein interactions by surface plasmon resonance, *Methods Mol. Biol.* 261: 93–111.

39 Rebois V., Schuck P., Northup J. K. (2002) Elucidating kinetic and thermodynamic constants for interaction of G protein subunits and receptors by surface plasmon resonance spectroscopy, *Methods Enzymol.* 344: 15–42.

40 Zeder-Lutz G., Cherouati N., Reinhart C., Pattus F., Wagner R. (2006) Dot-blot immunodetection as a versatile and high-throughput assay to evaluate recombinant GPCRs produced in the yeast Pichia pastoris, *Prot. Exp. Purif.* 50: 118–127.

41 Navratilova I., Sodroski J., Myszka D. G. (2005) Solubilization, stabilization, and purification of chemokine receptors using Biacore technology, *Anal. Biochem.* 339: 271–281.

42 Navratilova I., Pancera M., Wyatt R. T., Myszka D. G. (2006) A biosensor-based approach toward purification and crystallization of G protein-coupled receptors, *Anal. Biochem.* 353: 278–273.

43 Navratilova I., Dioszegi M., Myszka D. G. (2006) Analyzing ligand and small molecule binding activity of solubilized GPCRs using biosensor technology, *Anal. Biochem.* 355: 132–139.

6
Atomic Force Microscopy: High-Resolution Imaging of Structure and Assembly of Membrane Proteins

Simon Scheuring, Nikolay Buzhynskyy, Rui Pedro Gonçalves and Szymon Jaroslawski

6.1
Atomic Force Microscopy

6.1.1
Sample Preparation

For membrane protein atomic force microscopy (AFM) analysis, it is possible to use as samples two-dimensional (2-D) crystals, densely packed membrane protein reconstitutions, and/or native membranes. Protocols to form highly ordered 2-D crystals and/or reconstitutions at high membrane protein density are readily available (Kühlbrandt, 1992; Rigaud et al., 1997, 2000; Stahlberg et al., 2001). Native membranes are obtained from cells disrupted by passage through a French pressure cell, while lysates are loaded onto sucrose gradients and centrifuged at high speed for several hours. Membranes sediment to their specific density, due to their specific protein content within the sucrose gradient, and are well separated from the cytoplasmic fractions. Membrane samples can often be derived from intermediate purification steps of membrane protein purification procedures (Scheuring and Sturgis, 2005; Buzhynskyy et al., 2007). Sucrose is removed from the membrane samples by dialysis against a sucrose-free buffer. In general, for AFM analysis, membranes are maintained at 4 °C.

6.1.2
Equipment and Experimental Procedure

For contact mode imaging, it is recommended to use mica as support, freshly cleaved before each experiment (Schabert and Engel, 1994). Immediately after cleavage, 40 μL of an adsorption buffer is pipetted onto the mica surface, after which a few microliters of membrane solution are injected into the adsorption buffer drop. After membrane adsorption (which may take between a few minutes and a few hours, depending on membrane concentration and exposed charges on the protein surfaces), the sample is rinsed with 10 volumes of a recording buffer.

Biophysical Analysis of Membrane Proteins. Investigating Structure and Function. Edited by Eva Pebay-Peyroula

ISBN: 978-3-527-31677-9

For imaging, buffers that contain 10 mM Tris–HCl, pH 7.5 and variable KCl-content (0 mM to 500 mM) should be prepared to allow the optimization of imaging conditions during experiments. The acquisition of the data presented in this chapter was performed with a commercial Nanoscope-E contact-mode AFM (from Digital Instruments, Santa Barbara, CA, USA) equipped with a low-noise laser, and a 160 µm scanner (J-scanner) using oxide-sharpened Si_3N_4 cantilevers with a length of 100 µm (k = 0.1 N m^{-1}; Olympus Ltd., Tokyo, Japan). However, today a variety of commercial atomic force microscopes and cantilevers allow the acquisition of high-resolution topographs of biological samples in buffer solution. For imaging, minimal loading forces of ~100 pN were applied at scan frequencies of 4 to 6 Hz using optimized feedback parameters. The piezo precision can be determined by analyzing bacteriorhodopsin membrane protein 2-D crystals (lattice dimensions: a = b = 62.5, γ = 120°).

6.1.3 Experimental Rationales

The key elements of an atomic force microscope (Binnig et al., 1986) are the cantilever with a pyramidal-shaped tip that touches the sample, an optical system consisting of a laser and a multi-faced photo-diode that allows detection of cantilever deflection, a piezo-electric scanner that translates the sample relative to the tip in x, y, z directions, and a computer that drives the microscope and stores the surface contours (Fig. 6.1A).

A topograph is recorded by raster scanning the sample below the tip, which is attached to the flexible cantilever, while the feedback loop drives the piezo in z-dimension compensating for the difference between the deflection value set by the user and the measured deflection, in order to keep the cantilever deflection (i.e., the force) constant. The mica support, the tip and the sample are permanently immersed in buffer solution throughout the experiment (Fig. 6.1B). The optical system resolves cantilever deflections of 0.1 nm, which corresponds to a force difference of typically 10 to 50 pN. With modern instruments, stable contact mode operation is possible at forces of ~100 pN, provided that the sample is in aqueous solution with optimized ionic strength (Müller et al., 1999b).

Tight sample adsorption is an important parameter for contact-mode AFM imaging. Operating the microscope in contact mode induces friction forces, and therefore samples need to adhere well to freshly cleaved mica supports. As demonstrated using a variety of biological samples with different surface charge densities (Butt, 1992), sample adsorption is controlled by the nature and concentration of electrolytes in the buffer solution. Immobilization therefore requires buffer conditions in which the Debye layer thickness is minimized to allow adsorption by attractive van der Waals forces (Müller et al., 1997). An adsorption buffer containing 10 mM Tris–HCl, pH 7.6, 150 mM KCl, 25 mM $MgCl_2$ was often successful. Non-flatly adsorbed membranes, or not completely collapsed vesicles, can present multiple technical problems, such that forces applied by the atomic force microscope tip might be wrongly estimated. Not-tightly-attached membranes can be

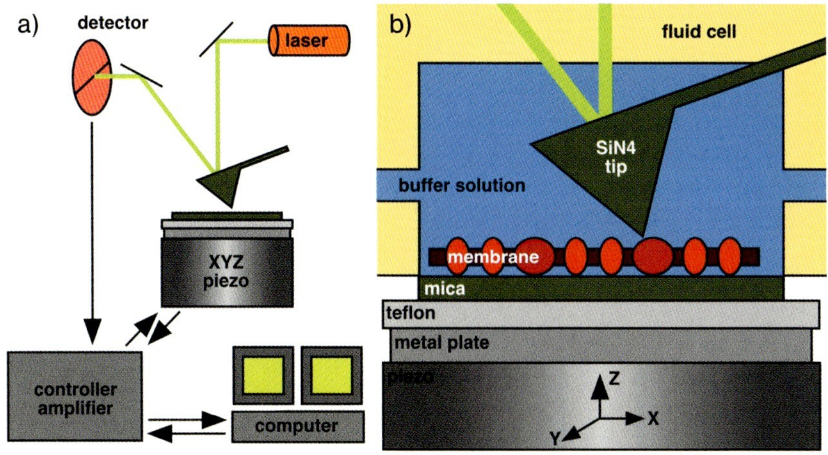

Fig. 6.1 A schematic representation of the atomic force microscope set-up. (A) A laser beam is focused and deflected on the rear-side of the microscope cantilever, monitoring the relative position of the cantilever on a multifield detector. This signal is processed by the controller that drives the piezo in x- and y-dimensions and compensates for the cantilever deflection through movements in z-dimension. This compensation of the tip deflection by z-movement of the piezo, which in turn will influence the tip deflection at each position (xy), is termed the feedback loop. The user interacts over a computer with the controller, feeds in the scanning parameters, and reads out the piezo movements corresponding to the image, to and from the controller. The faster the feedback loop reacts, the more precise is the surface contouring of the sample. B) A close-up view of the fluid cell. The sample is deposited on mica, which is glued to a Teflon plate (to protect the piezo), which in turn is glued to a magnetic plate. The tip and sample are permanently immerged in buffer solution during the measurements.

vertically "squeezed" by the tip; as a consequence, this "squeezing" of the object contributes to the apparent spring constant, and the feedback loop monitoring and compensating for the tip deflection will not drive the piezo appropriately to gain a highly contrasted image. Furthermore, membranes that are not tightly attached can provoke "smeared" images, most likely due to a faint displacement of the sample during scanning.

High-resolution contact-mode imaging involves electrostatic and van der Waals interactions between the sample and the tip (Israelachvili, 1991). For imaging, the electrolyte should be selected to balance electrostatic repulsions and van der Waals attraction. These forces can be quantitatively described and their integration evaluated in force–distance curves, allowing optimization of the recording conditions and to adjust the forces applied to the cantilever to ~100 pN (Müller et al., 1999b). For high-resolution contact-mode AFM imaging a buffer containing 10 mM Tris–HCl, pH 7.6, 150 mM KCl is a good starting point for subsequent buffer optimization during the experiment (Scheuring et al., 2005b).

Alternatively, the atomic force microscope can be operated in oscillating mode – that is, when the cantilever oscillates in resonance frequency, and the tip interacts with the sample solely at the end of its downward movement, which reduces the contact time and the friction forces compared to contact-mode AFM (Zhong et al., 1993; Hansma et al., 1994). Compared to contact-mode AFM, oscillating-mode AFM has the advantage of allowing imaging surfaces of macromolecules, even when they are only weakly attached to the support. Oscillating-mode AFM can provide well-resolved images of individual molecules (Goldsbury et al., 1999), of 2-D crystals (Möller et al., 1999), and of native membranes (Bahatyrova et al., 2004). The resolution acquired by oscillating-mode AFM remained, however, slightly inferior compared to contact-mode imaging, and the majority of high-resolution images were acquired using contact mode (Müller et al., 2006). In particular, those membranes which represent large, 2-D objects that can easily be adsorbed to AFM supports, are ideal for contact-mode imaging. Hence, the focus of this chapter is on contact-mode AFM imaging.

Problems related to AFM imaging may concern the imaging of strongly protruding protein structures. The feedback loop is more challenged when contouring strongly corrugated surfaces, but the mobility of those is also increased (Scheuring et al., 2003a). For both reasons, it is generally expected that the contouring precision will decrease with increasing height of surface-protruding structures. When strongly protruding globular domains are imaged, the tip geometry, in a first approximation a hemisphere convolutes the topography significantly: although the height measurement of a strongly protruding domain can be precise, the diameter may appear enlarged due to tip convolution, and hence it seems unreasonable to interpret structural details of large globular topographies (Scheuring et al., 2003b). As a consequence of the tip convolution and the feedback loop system, the imaging resolution within one AFM topograph is not identical on all image areas. A weakly protruding particle may be highly resolved, while a globular structure in proximity is strongly subjected to tip convolution and to non-precision of contouring. The problem becomes more pronounced when the tip geometry deviates from the shape of a perfect hemisphere, a phenomenon referred to as "asymmetric tip" or "double tip". This effect makes difficult the analysis of non-oligomeric molecules (in particular in native membranes, where no regular lattice or crystallographic symmetry can be used to estimate tip symmetry) or of molecules of which no alternative structural information is available. The situation becomes more complex if there is a need to consider particular physico-chemical surface properties interfering at each tip-sample contact position (Israelachvili, 1991; Israelachvili and Wennerström, 1996).

Another crucial limitation for high-resolution AFM is related to the tip radius. It must be stressed that only very sharp tips with almost nanometer-sized asperity can acquire topographs at resolution of ~10 Å. Based on the image analysis of barrel-shaped molecules with small diameters, such as LH2 (Scheuring et al., 2001, 2003a, 2004a; Gonçalves et al., 2005a,b; Scheuring and Sturgis, 2005), a high-resolution tip radius of (2.5 nm is estimated (Scheuring et al., 2005b). The finding of a tip with such a sharp apex can be considered as another limitation of high-resolution AFM imaging.

In principle, the atomic force microscope can provide atomic resolution on solid crystalline surfaces such as mica or graphite (Binnig et al., 1986, 1987). For membranes in buffer solution, imaging resolution of ~10 Å is most probably limited due to protein fluctuation at room temperature. In contact-mode, high-resolution AFM, the images are acquired at scan speeds of about $1\,nm\,ms^{-1}$. It seems reasonable to assume, that some protein fluctuation is averaged during this time scale, thus limiting the acquisition of more detailed structural information.

6.2 Combined Imaging and Force Measurements by AFM

Force–distance measurements using the atomic force microscope measure inter- and intra-molecular forces with high sensitivity. Attaching a molecule between the atomic force microscope tip and the surface allows the application of well-defined forces and measuring of the rupture of molecular bonds (Florin et al., 1994; Moy et al., 1994), or conformational changes in polymeric molecules (Rief et al., 1997). Fitting the worm-like-chain (WLC) model to force–distance curves measured on single molecules allows the determination of the polymer-specific parameters that describe their pure entropic elasticity at low forces and their enthalpic contribution at higher stretching forces (Bustamante et al., 1994; Marko and Siggia, 1994). Hence, the atomic force microscope may be used as a tool to study the transition from a folded protein into its stretched amino acid chain – that is, of protein unfolding (Rief et al., 1997; Oberhauser et al., 1998, 1999; Marszalek et al., 1999; Mehta et al., 1999; Fisher et al., 2000). The first protein for which unfolding was investigated by AFM was the muscle protein titin (Rief et al., 1997). From the elasticity measured for the unfolded amino acid strand, a persistence length of 0.4 nm was determined, and this served as a reference for other protein unfolding experiments. Force measurements on proteins can be performed in a controlled manner when the protein under investigation is specifically coupled to the atomic force microscope tip. When a specific interaction is employed, the force acts on a defined point on the protein. However, such coupling interfered with high-resolution imaging, as the release of the extracted proteins from the tip is a major problem. In contrast, the use of unspecific anchoring of the protein to the tip of the atomic force microscope allows repeated imaging and force measurements to be made over a long time period. In order to non-specifically anchor a membrane protein to the atomic force microscope tip, the tip is usually kept in contact with the protein surface for 0.1 s to 1 s at a loading force of between 100 pN and 1 nN in order to allow the proteins to adhere. Upon retraction of the tip, forces are applied to the adhered protein and measured as a function of the separation distance (Oesterhelt et al., 2000).

6.2.1 Imaging and Force Measurement of a Bacterial Surface Layer (S-Layer)

The surface layers (S-layers) represent the outermost cell wall layer of many bacteria and Archaea (Baumeister et al., 1989; Sleytr et al., 1993), and are regular 2-D protein networks (Baumeister et al., 1988). These layers withstand non-physiological pH, radiation, temperature, proteolysis, pressure and detergent treatment (Engelhardt and Peters, 1998), thus protecting the cell from such hostile factors. Moreover, they serve as molecular sieves as well as in phage recognition (Sleytr et al., 1997).

PS2 is the protein that forms the S-layer of *Corynebacterium glutamicum* (Peyret et al., 1993; Chami et al., 1995, 1997). The native S-layer of *C. glutamicum* was tightly adsorbed to the mica AFM support and imaged under native conditions. The S-layer revealed a unit cell dimensions of $a = b = 16.0 \pm 0.2$ nm and $\gamma = 60 \pm 1°$ (Fig. 6.2A). High-resolution imaging and force spectroscopy measurements provided information about the molecular architecture and the mechanical properties of the S-layer assembly (Scheuring et al., 2002).

On the S-layer, a large number of force–distance curves were acquired, and reproducibly force–distance curves were found with strong rupture peaks of ~270 pN interspersed by faint rupture peaks of ~70 pN (Fig. 6.2B). These force–distance curves represented the unzipping of a S-layer hexamer, as reported previously by high-resolution imaging (Fig. 6.2A) and after (Fig. 6.2C) force measurements (Scheuring et al., 2002). The fact that three times two subunits were pulled concomitantly out of the S-layer, rather than six individual subunits (Müller et al., 1999a), inferred a cooperativity between subunits in the *C. glutamicum* S-layer assembly that explained the extraordinary stability of the S-layer protein network.

When force spectroscopy measurements were applied to polytopic membrane proteins (Oesterhelt et al., 2000), the technique proved to be sufficiently sensitive to detect forces between transmembrane helices (Janovjak et al., 2003), and trace occupation or vacancy of active sites within transporters (Kedrov et al., 2005).

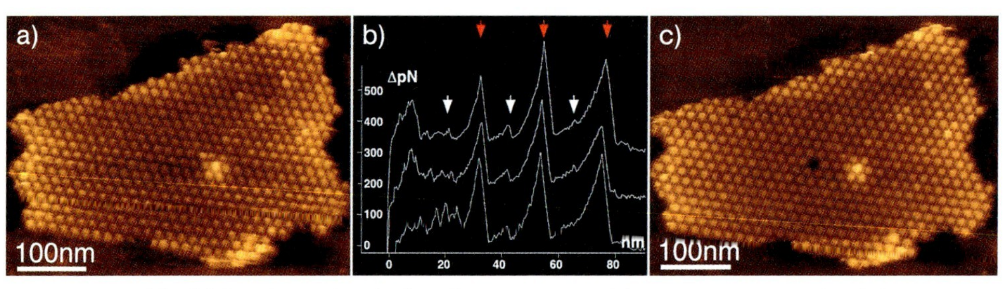

Fig. 6.2 Imaging and unzipping of *Corynebacterium glutamicum* S-layer proteins (adapted from Scheuring et al., 2002). (A) Medium-resolution AFM topograph of an intact S-layer patch. (B) Three examples of highly reproducible unzipping force–separation curves of S-layer hexamers. The weak unzipping peaks of the odd-numbered S-layer subunits are indicated by white arrows, and the strong rupture peaks of the even numbered subunits by red arrows. The *C. glutamicum* S-layer revealed a cooperative stabilization of dimeric units within the layer assembly. (C) The same membrane as shown in (A) but imaged after recording an unzipping event. An individual S-layer hexamer was removed by the atomic force microscope tip.

6.3
High-Resolution Imaging by AFM

6.3.1
High-Resolution AFM of Aquaporin-Z (AQPZ)

Aquaporins are ubiquitous membrane channels which occur in bacteria, fungi, plants, and animals. They are highly specific for water or small and uncharged hydrophilic solutes, and are involved in osmoregulation. Hydropathy analysis of the first sequenced members of this family indicated six membrane spans and two unusually long loops (Gorin et al., 1984; Agre et al., 1993; Jung et al., 1994). Meanwhile, almost 200 genes have been sequenced, including 10 in human, almost all of which have the highly conserved NPA motifs within these loops (Heymann and Engel, 2000). Approximately half of these channel proteins are exclusively water-selective. Other channels facilitate the passage of small hydrophilic molecules such as glycerol or urea (Agre, 2004). The first aquaporin structure to be solved was AQP1, by electron crystallography from highly ordered 2-D crystals (Murata et al., 2000), and the structure provided many answers concerning water selectivity within the channel. Later, the structure of a member of the glycerol facilitator family was solved, namely GlpF from *Escherichia coli* (Fu et al., 2000). Since then, additional AQP1 (Sui et al., 2001), AQP0 (Gonen et al., 2004b, 2005; Harries et al., 2004), AQP4 (Hiraki et al., 1981), AqpM (Lee et al., 2005), and SoPIP2;1 (Tornroth-Horsefield et al., 2006) structures have been presented from a variety of electron and X-ray crystallography studies.

In *E. coli*, a water channel – AqpZ – has been identified by homology cloning (Calamita et al., 1998), with 2-D crystals of sizes ranging up to 5 μm having been assembled from AqpZ tetramers (Ringler et al., 1999). The sidedness of AqpZ was determined by imaging 2-D crystals reconstituted from AqpZ bearing an N-terminal poly histidine tag, and imaging the identical 2-D crystals after proteolysis of the N-terminus (Scheuring et al., 1999). The lateral and vertical resolutions of the images were determined to be 8 Å and 1 Å, respectively (Fig. 6.3A). At this resolution and the high signal-to-noise ratio, individual surface-protruding loops on each monomer of the AqpZ tetramer are contoured. In lacking any aquaporin high-resolution structure at that date (Scheuring et al., 1999), the average topography (Fig. 6.3B) was compared with the sequence-based structure prediction of AqpZ. Three protrusions were found on the periplasmic surface of the AqpZ monomer, one close to the fourfold symmetry axis, and one small and one elongated protrusion at the periphery, that were assigned to surface-protruding loops (Scheuring et al., 1999). The structure of AqpZ was solved at 2.5 Å resolution using X-ray crystallography (Savage et al., 2003). The AFM topography was seen to compare well to the molecular surface representation of the AqpZ structure (Fig. 6.3C). On the periplasmic surface, topographical features corresponding to the loop A peaking at Pro30 (ranging from Ala26 to Gly35), the end of helix 3 and beginning of the loop C peaking at Thr107 (ranging from Gly105 to Asp110), and

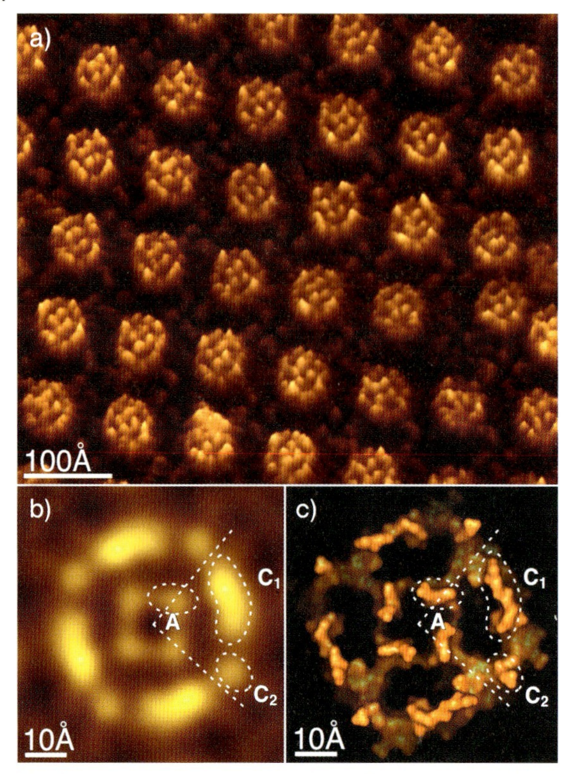

Fig. 6.3 High-resolution AFM analysis of *Escherichia coli* waterchannel aquaporin-Z (AqpZ) (adapted from Scheuring et al., 1999). (A) High-resolution AFM topograph of an AqpZ 2-D crystal. (B) Average topography of the extracellular surface of the AqpZ tetramer, showing three protrusions per AqpZ monomer, two peripheral C_1 and C_2, and one close to the fourfold axis, denoted A. (C) Surface representation of the atomic structure of AqpZ (Savage et al., 2003). The surface topography found in the high-resolution AFM analysis (B) compares favorably with all major surface-protruding structures of the atomic structure, loop A and the beginning, C_1, and the end, C_2, of loop C.

the long domain on the periphery of the tetramer (ranging from Ser114 to Ser130), the entire loop C ranges from Gly105 to Ser130, were reliably contoured.

6.3.2
High-Resolution AFM of Aquaporin-0 (AQP0)

High-resolution topographs (Fig. 6.4A) of the extracellular surface of junctional AQP0 revealed three topographical features per monomer, two on the periphery and one central. Comparison of the topography average (Fig. 6.4B) with the surface representation (Fig. 6.4C) of the atomic structure of the junctional AQP0 (Gonen et al., 2004b, 2005) allowed an assignment to be made that the predominant protrusion on the periphery C_1 comprised the first amino acids of loop C (ranging from

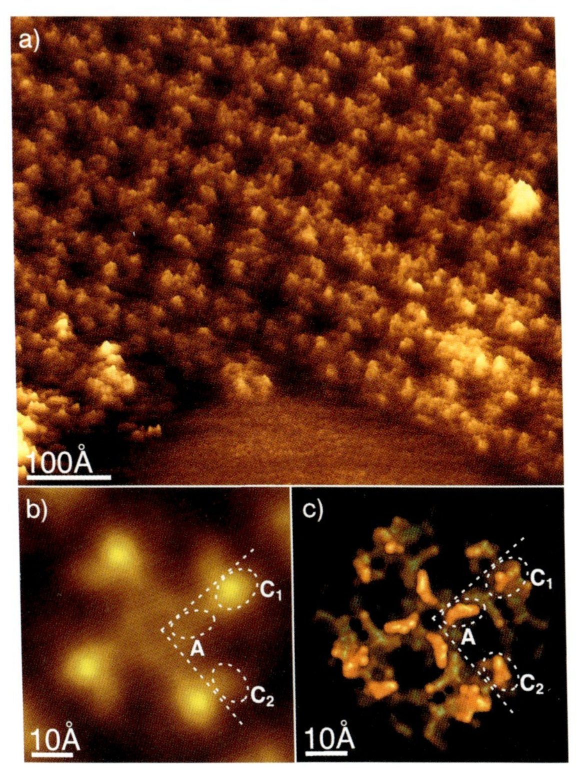

Fig. 6.4 High-resolution AFM analysis of junctional aquaporin-0 (AQP0) in native sheep lens fiber cell membranes (adapted from Buzhynskyy et al., 2007). (A) High-resolution AFM topograph of junctional AQP0 in a tetragonal array. Molecules at the array edge are slightly disordered with respect to the AQP0 array lattice. (B) Average topography of the AQP0 tetramer, showing three protrusions per AQP0 monomer, two peripheral C_1 and C_2, and one close to the fourfold axis, denoted A. (C) Surface representation of the atomic structure of AQP0 (Gonen et al., 2004b, 2005). The surface topography found in the high-resolution topography (B) compares favorably with all major surface-protruding structures of the atomic structure, loop A and the beginning, C_1, and the end, C_2, of loop C.

Pro109 to Val112). Loop C then descends towards the channel and forms a second protrusion C_2 (ranging from His122 to Val125), with the entire loop C ranging from Thr108 to Ser126. These two features are well resolved in the topography average and compared favorably with the AQP0 structure (Gonen et al., 2004b, 2005; Harries et al., 2004) surface representation (Buzhynskyy et al., 2007).

The extracellular loop A connecting helices 1 and 2 peaks at Pro36 (ranging from Ser31 to His40) was identified as the conformational switch to change the protein from a water channel to a junctional adhesion protein (Gonen et al., 2005). The central protrusion in the AFM topography compares favorably with the junctional AQP0 conformation (Buzhynskyy et al., 2007), and no topographical feature was detected at the position of the a-loop in the non-junctional AQP0 structure (Harries et al., 2004).

6.3.3
Comparison Between AQPZ and AQP0 Topographies

The structures of AqpZ (1RC2; Savage et al., 2003), junctional AQP0 (2B6O; Gonen et al., 2005) and non-junctional AQP0 (1YMG; Harries et al., 2004) are very similar. Aquaporin amino acid sequences are highly conserved, assuring faithful water channeling, while the least-preserved regions are the membrane-exposed loops. Loop A comprises 10 amino acids in both AqpZ and AQP0, while loop C comprises 26 amino acids in AqpZ and 19 amino acids in AQP0. However, the structures of these loops are different in AqpZ and AQP0. These differences are accurately contoured in the high-resolution AFM topographs in both proteins (cf. Fig. 6.3B and C with Fig. 6.4B and C). This means on the one hand that the atomic force microscope is capable of contour folding differences between homologous proteins with high accuracy, but on the other hand that it is very difficult to interpret structural features without knowledge derived from other techniques, for example loop C in AQP0 which folds back towards the membrane in the middle of its sequence (Fig. 6.4B and C).

6.3.4
The Supramolecular Assembly of Photosynthetic Complexes in Native Membranes of *Rhodospirillum photometricum* by AFM

The supramolecular organization of the photosynthetic apparatus of *Rhodospirillum* (*Rsp.*) *photometricum*, was studied in detail and yielded striking novel findings concerning antenna heterogeneity, antenna domain formation, and complex assembly (Scheuring et al., 2004a,b; Scheuring and Sturgis, 2005). The majority of LH2 assemble in nonameric rings with ~50 Å diameter (Fig. 6.5A). However, on closer examination LH2 of various sizes were found within native membranes. Diameter distribution and image processing analysis showed heterogeneity of the LH2 complex stoichiometry around the general nonameric assembly (~70%), with smaller octamers (~15%) and larger decamers (~15%). This finding was qualitatively corroborated by examination of individual complexes in raw data images (Scheuring et al., 2004a). It seems probable that LH2 stoichiometry heterogeneity is an inherent feature of LH2, as it has also been observed in *Phaeospirillum* (*Phsp.*) *molischianum* (Gonçalves et al., 2005a) and *Rhodospirillum palustris* (Scheuring et al., 2006). In contrast to the heterogeneity found for LH2 complexes, the reaction center (RC)–LH1 core complexes appeared uniform in size: monomeric RC surrounded by a closed elliptical $LH1_{16}$ assembly, with long and short axes of 95 Å and 85 Å, following the long RC axis (Fig. 6.5A). Analysis of the distribution of the photosynthetic complexes of *Rsp. photometricum* showed significant clustering of both antenna complexes and core complexes. Membranes, with domains densely packed with photosynthetic proteins and protein-free lipid bilayers were found. Complex clustering is a functional demand, as each light-harvesting component that segregates away from the system is non-functional as it cannot pass its harvested energy to a neighboring complex and eventually to the RC (Scheuring et al., 2004a).

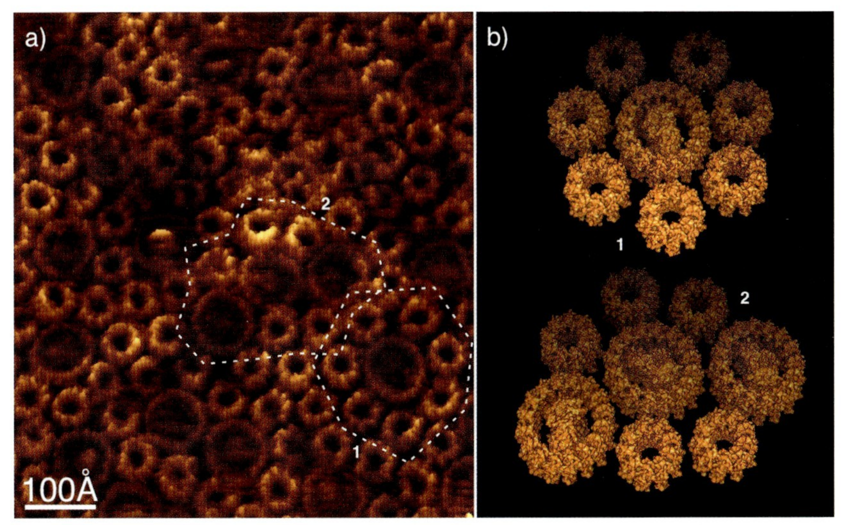

Fig. 6.5 High-resolution AFM analysis of photosynthetic membranes of *Rhodospirillum (Rsp.) photometricum* (adapted from Scheuring and Sturgis, 2005; Scheuring et al., 2007). (A) High-resolution AFM topograph of a high-light-adapted photosynthetic membrane of *Rsp. photometricum*. The ratio of LH2 complexes to core complexes (containing LH1 surrounding the RC) is ~3.5. No fixed assembly pattern between LH2 and core complexes was found. (B) Structural models of supramolecular assemblies of photosynthetic complexes corresponding to the assemblies 1 and 2 outlined in the topograph shown in (A).

There is no fixed structural assembly of LH2 and core complexes: core complexes completely surrounded by LH2 (Fig. 6.5A and B; 1) and core complexes making multiple core–core contacts (Fig. 6.5A and B; 2) were found (Scheuring et al., 2004a; Scheuring and Sturgis, 2005). However, detailed pair correlation function analysis showed that the most frequent assembly was two core complexes separated by an intercalated LH2 (Scheuring and Sturgis, 2005). The supramolecular assembly of the photosynthetic complexes in *Rsp. photometricum* membranes from cells grown under different light intensities were studied and compared. In membranes from low-light-adapted cells, increased quantities of peripheral LH2 were found. Additional LH2 were not randomly inserted into the membrane but rather formed para-crystalline hexagonally packed antenna domains. Core complexes remained in domains, in which they were locally much higher concentrated (LH2 rings/core complex=~3.5) than the average density under low-light cell growth (LH2 rings/core complex=~7). Indeed, these domains in the low-light-adapted membranes resembled the high-light-adapted membranes in terms of protein composition and complex distribution (Scheuring and Sturgis, 2005). This indicated that complex assembly followed an eutectic phase behavior with an ideal LH2 rings/core complex ratio ~3.5 independent of the growth conditions, and additional LH2 being synthesized under low-light conditions, were integrated in specialized antenna domains (Scheuring and Sturgis, 2005). The LH2 packing in antenna domains was evidenced to be rigid and possibly to exclude quinone/quinol diffusion (Scheuring and Sturgis, 2006). Reaction centers that are grouped together, independent of the growth conditions, and formation of antenna domains under low-light conditions, prevent photo-damage under high-light conditions and ensure efficient photon capture under low-light conditions (Scheuring and Sturgis, 2005). These high-resolution AFM topographs have shown the supramolecular assembly of the bacterial photosynthetic complexes in native membranes in detail (Fig. 6.5A). In these images, the translational and rotational degrees of freedom of the complexes in multiprotein assemblies was determined (Scheuring et al., 2007), in order to build realistic atomic models of supramolecular assemblies by docking high-resolution structures into the topographs (Fig. 6.5B).

6.3.5
AQP0–Connexon Junction Platforms in Native Sheep Lens Membranes

Eye lens membranes contain mainly two membrane proteins, the lens-specific aquaporin-0 (AQP0) and gap junction connexins. AQP0, a water channel in lens cortex cell membranes, is a junctional adhesion molecule in mature lens core membranes without water channel function (Gonen et al., 2004a,b, 2005). Although connexons of adjacent cells equally form junctions, these feature a pore sufficiently large to allow the passage of metabolites and ions. The native assembly of AQP0 and connexons in core lens membranes were analyzed using high-resolution AFM (Buzhynskyy et al., 2007). AQP0 and connexons together formed junctional microdomains in planar lipid bilayers in native core membranes where AQP0 formed 2-D arrays surrounded by connexons.

In medium-resolution topographs (Fig. 6.6A), strongly protruding protein structures at the edges of the AQP0 junction patches were found. These protruding structures eventually entered the patches, if the patches consisted of differently oriented AQP0 lattices. High-resolution topographs revealed submolecular structure on the AQP0 surface (see Section 6.3.2) and the flower-shaped structure with the central channel of the connexons (Fig. 6.6B). In agreement with the structural analysis of connexons (Unger et al., 1997) the molecules were ~80 Å in size, with a top ring diameter of ~40 Å and a central cavity of ~20 Å. These connexons lined up at the peripheries of AQP0 junction patches. A particularly stunning structure was represented by five individual connexons that lined up and terminated an AQP0 patch (Fig. 6.6B; arrows). The center-to-center distances between these five connexon rings were (from left to the right) ~105 Å, ~95 Å, ~103 Å, and ~85 Å. Since, connexon rings in contact have a center-to-center distance of 77 Å (Unger et al., 1997), it was concluded that theses connexons must partially register with the AQP0 lattice and/or be interspaced by AQP0 subunits and lipids. Even when connexons were grouped, they did not form hexagonal lattices (Fig. 6.6B), in contrast to connexons from other tissues (Kistler et al., 1994). These AQP0–connexon junction microdomains in the membranes assure adhesion and nutrition exchange between neighboring stacked lens fiber cells (Buzhynskyy et al., 2007).

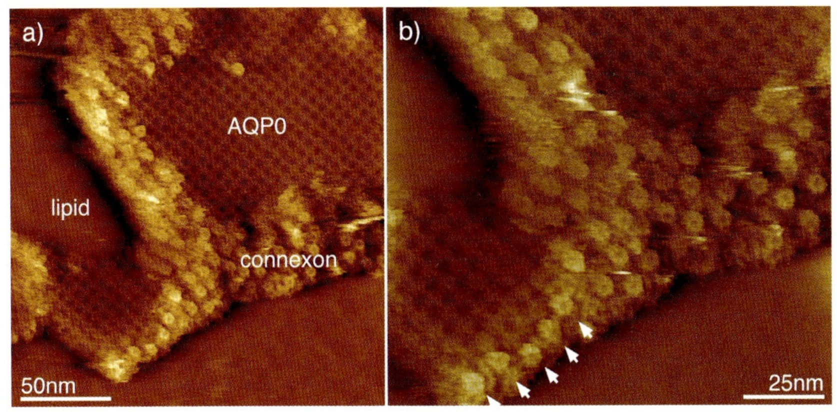

Fig. 6.6 High-resolution analysis of an AQP0–connexon junction microdomain (adapted from Buzhynskyy et al., 2007). (A) Medium-resolution AFM topograph of an AQP0–connexon junction microdomain within a large lipid membrane. Connexons line up between the AQP0 patches and at the edges of the junction microdomains towards the lipid bilayer. Connexons are non-ordered, interspersed by AQP0 and lipid. (B) High-resolution AFM topograph of a connexon seal at a patch edge (the arrows indicate five individual connexons).

6.4 Conclusions

High-resolution AFM imaging can be combined with force measurements to provide structural and mechanical information about a protein assembly (Section 6.2.1). High-resolution topographs of membranes proteins reveal sufficient structural details to characterize the fold of surface-protruding domains of approximately five amino acids in size (Sections 6.3.1–6.3.3). The supramolecular assembly of membrane proteins can be studied in specialized native membranes from prokaryotic (Section 6.3.4) and eukaryotic cells (Section 6.3.5). The imaged structures of individual proteins compare favorably with atomic structures (Sections 6.3.1 and 6.3.2) and the imaged supramolecular assemblies of membrane proteins can successfully be docked by atomic structures (Section 6.3.4).

6.5 Feasibilities, Limitations, and Outlook

Overall, the atomic force microscope represents a fairly poor screening technique. The first problem is that scanners which are precise enough to be used for high-resolution image acquisition are limited in their full scan range ability to somewhat more than 100 μm^2, thereby providing an analysis of only a limited number of membranes per experiment. This problem may be overcome by the precise adjustment of sample quantity used per experiment, in order to ensure that as many membranes as possible are adsorbed per unit area, without them stacking or aggregating. A second problem is that AFM imaging requires tightly adsorbed membranes. Adsorption is related to surface-exposed charges, and can be triggered by ionic strength in the adsorption buffer solution (Müller et al., 1997). This is not always trivial, as sample nativeness must also be considered as a determining factor for adsorption buffer conditions during screening. Although vesicular samples can be imaged, the procedure is technically more demanding than when imaging sheet-like membranes. In case artificial reconstitutions are imaged, the size and shape of the membranes may be influenced through the sample preparation procedure. Normally, when imaging native membranes the size and shape of the membrane sample can scarcely be influenced. However, the smaller the membranes, the more vesicular is the preparation, and the trickier the AFM imaging becomes (Gonçalves et al., 2005a), most likely because the surface tension of small vesicles hampers the flattening and adsorption of the spherical object to the support. This problem may be resolved by extensive adsorption buffer screening, in order to identify conditions where small vesicles tightly adsorb and open upon support adsorption. Another possibility is to use the tip as a nano-dissector to open vesicles mechanically (Scheuring et al., 2004b), although both approaches are invasive. Attempts may also be made to create larger membranes from small vesicles by vesicle fusion, either through detergent addition (Bahatyrova et al., 2004) or freeze–thawing cycles (Scheuring et al., 2005a). However, these procedures are generally undesirable as they may significantly alter the native complex assembly.

Clearly, the "trump card" of AFM is the outstanding signal-to-noise ratio, which manifests as a capability of imaging single molecules. This feature renders AFM, to date, the unique technique for visualizing non-ordered supramolecular membrane protein assemblies in native membranes, where techniques that imply averaging or ensemble measurements fail (Buzhynskyy et al., 2007; Scheuring, 2006).

Currently, a number of technical developments suggest that AFM will strengthen its abilities for the analysis of membrane proteins in the near future. The initial improvement is that atomic force microscope tips with nano-tubes may provide improved reproducibility of high-resolution image acquisition and higher resolution (Cheung et al., 2000). The second improvement is that fast-scanning AFMs capable of acquiring images at rates sufficient to track the motion of individual molecules, may allow the visualization of flexible and diffusing molecules, and of enzymatic activity (Viani et al., 2000; Ando et al., 2001).

Acknowledgments

The authors thank former and present collaborators. This study was supported by the Institut Curie and the INSERM (Institut National de la Santé et la Recherche Médicale).

References

1 Agre, P. (2004) Nobel lecture. Aquaporin water channels. *Biosci. Rep.*, **24**, 127–163.

2 Agre, P., Preston, G. M., Smith, B. L., Jung, J. S., Raina, S., Moon, C., Guggino, W. B. and Nielsen, S. (1993) Aquaporin chip: The archetypal molecular water channel. *Am. J. Physiol.*, **265**, F463–F476.

3 Ando, T., Kodera, N., Takai, E., Maruyama, D., Saito, K. and Toda, A. (2001) A high-speed atomic force microscope for studying biological macromolecules. *Proc. Natl. Acad. Sci. USA*, **98**, 12468–12472.

4 Bahatyrova, S., Frese, R. N., Siebert, C. A., Olsen, J. D., van Der Werf, K. O., Van Grondelle, R., Niederman, R. A., Bullough, P. A. and Hunter, C. N. (2004) The native architecture of a photosynthetic membrane. *Nature*, **430**, 1058–1062.

5 Baumeister, W., Wildhaber, I. and Engelhardt, H. (1988) Bacterial surface proteins: Some structural, functional and evolutionary aspects. *Biophys. Chem.*, **29**, 39–49.

6 Baumeister, W., Wildhaber, I. and Phipps, B. M. (1989) Principles of organization in eubacterial and archebacterial surface proteins. *Can. J. Microbiol.*, **35**, 215–227.

7 Binnig, G., Quate, C. F. and Gerber, C. (1986) Atomic force microscope. *Phys. Rev. Lett.*, **56**, 930–933.

8 Binnig, G., Gerber, C., Stoll, E., Albrecht, T. R. and Quate, C. F. (1987) Atomic resolution with atomic force microscopy. *Europhys. Lett.*, **3**, 1281–1286.

9 Bustamante, C., Marko, J. F., Siggia, E. D. and Smith, S. (1994) Entropic elasticity of lambda-phage DNA. *Science*, **265**, 1599–1600.

10 Butt, H.-J. (1992) Measuring local surface charge densities in electrolyte solutions with a scanning force microscope. *Biophys. J.*, **63**, 578–582.

11 Buzhynskyy, N., Hite, R. K., Walz, T. and Scheuring, S. (2007) The supramolecular architecture of junctional microdomains in native lens membranes. *EMBO Rep.*, **8**, 51–55.

12 Calamita, G., Kempf, B., Bonhivers, M., Bishai, W. R., Bremer, E. and Agre, P. (1998) Regulation of the *Escherichia coli* water channel gene aqpz. *Proc. Natl. Acad. Sci. USA*, **95**, 3627–3631.

13 Chami, M., Bayan, N., Dedieu, J.-C., Leblon, G., Shechter, E. and Gulik-Krzywicki, T. (1995) Organization of the outer layers of the cell envelope of *Corynebacterium glutamicum*: A combined freeze-etch electron microscopy and biochemical study. *Biol. Cell*, **83**, 219–229.

14 Chami, M., Bayan, N., Peyret, J. L., Gulik-Krzywicki, T., Leblon, G. and Shechter, E. (1997) The s-layer protein of *Corynebacterium glutamicum* is anchored to the cell wall by its c-terminal hydrophobic domain. *Mol. Microbiol.*, **23**, 483–492.

15 Cheung, C. L., Hafner, J. H. and Lieber, C. M. (2000) Carbon nanotube atomic force microscopy tips: Direct growth by chemical vapor deposition and application to high-resolution imaging. *Proc. Natl. Acad. Sci. USA*, **97**, 3809–3813.

16 Engelhardt, H. and Peters, J. (1998) Structural research on surface layers: A focus on stability, surface layer homology domains, and surface layer-cell wall interactions. *J. Struct. Biol.*, **124**, 276–302.

17 Fisher, T. E., Marszalek, P. E. and Fernandez, J. M. (2000) Stretching single molecules into novel conformations using the atomic force microscope. *Nature Struct. Biol.*, **7**, 719–724.

18 Florin, E.-L., Moy, V. T. and Gaub, H. E. (1994) Adhesion forces between individual ligand-receptor pairs. *Science*, **264**, 415–417.

19 Fu, D., Libson, A., Miercke, L. J., Weitzman, C., Nollert, P., Krucinski, J. and Stroud, R. M. (2000) Structure of a glycerol-conducting channel and the basis for its selectivity. *Science* **290**, 481–486.

20 Goldsbury, C., Kistler, J., Aebi, U., Arvinte, T. and Cooper, G. J. S. (1999) Watching amyloid fibrils grow by time-lapse atomic force microscopy. *J. Mol. Biol.*, **285**, 33–39.

21 Gonçalves, R. P., Bernadac, A., Sturgis, J. N. and Scheuring, S. (2005a) Architecture of the native photosynthetic

apparatus of *Phaeospirillum molischianum*. *J. Struct. Biol.*, **152**, 221–228.

22 Gonçalves, R. P., Busselez, J., Lévy, D., Seguin, J. and Scheuring, S. (2005b) Membrane insertion of *Rhodopseudomonas acidophila* light harvesting complex 2 (lh2) investigated by high resolution AFM. *J. Struct. Biol.*, **149**, 79–86.

23 Gonçalves, R. P., Agnus, G., Sens, P., Houssin, C., Bartenlian, B. and Scheuring, S. (2006) 2-chamber-AFM: Probing membrane proteins separating two aqueous compartments. *Nature Methods*, **3**, 1007–1012.

24 Gonen, T., Cheng, Y., Kistler, J. and Walz, T. (2004a) Aquaporin-0 membrane junctions form upon proteolytic cleavage. *J. Mol. Biol.*, **342**, 1337–1345.

25 Gonen, T., Sliz, P., Kistler, J., Cheng, Y. and Walz, T. (2004b) Aquaporin-0 membrane junctions reveal the structure of a closed water pore. *Nature*, **13**, 193–197.

26 Gonen, T., Cheng, Y., Sliz, P., Hiroaki, Y., Fujiyoshi, Y., Harrison, S. C. and Walz, T. (2005) Lipid-protein interactions in double-layered two-dimensional aqp0 crystals. *Nature*, **438**, 633–638.

27 Gorin, M. B., Yancey, S. B., Cline, J., Revel, J. P. and Horwitz, J. (1984) The major intrinsic protein (mip) of the bovine lens fiber membrane: Characterization and structure based on cdna cloning. *Cell*, **39**, 49–59.

28 Hansma, P. K., Cleveland, J. P., Radmacher, M., Walters, D. A., Hillner, P. E., Bezanilla, M., Fritz, M., Vie, D., Hansma, H. G., Prater, C. B., Massie, J., Fukunaga, L., Gurley, J. and Elings, V. (1994) Tapping mode atomic force microscopy in liquids. *Appl. Phys. Lett.*, **64**, 1738–1740.

29 Harries, W. E., Akhavan, D., Miercke, L. J., Khademi, S. and Stroud, R. M. (2004) The channel architecture of aquaporin 0 at a 2.2-Å resolution. *Proc. Natl. Acad. Sci. USA*, **101**, 14045–14050.

30 Heymann, J. B. and Engel, A. (2000) Structural clues in the sequences of the aquaporins. *J. Mol. Biol.*, **295**, 1039–1053.

31 Hiraki, K., Hamanaka, T. and Kito, Y. (1981) Phase transitions of the purple membrane and the brown halo-membrane: X-ray diffraction, circular dichroism spectrum and absorption spectrum studies. *Biochim. Biophys. Acta*, **647**, 18–28.

32 Israelachvili, J. (1991) *Intermolecular and Surface Forces*. Academic Press Limited, London.

33 Israelachvili, J. and Wennerstrom, H. (1996) Role of hydration and water structure in biological and colloidal interactions. *Nature*, **379**, 219–225.

34 Janovjak, H., Kessler, M., Oesterheld, D., Gaub, H. and Müller, D. J. (2003) Unfolding pathways of native bacteriorhodopsin depend on temperature. *EMBO J.*, **22**, 5220–5229.

35 Jung, J., Preston, G., Smith, B., Guggino, W. and Agre, P. (1994) Molecular structure of the water channel through aquaporin chip. The hourglass model. *J. Biol. Chem.*, **269**, 14648–14654.

36 Kedrov, A., Krieg, M., Ziegler, C., Kuhlbrandt, W. and Muller, D. J. (2005) Locating ligand binding and activation of a single antiporter. *EMBO Rep.*, **6**, 668–674.

37 Kistler, J., Goldie, K., Donaldson, P. and Engel, A. (1994) Reconstitution of native-type noncrystalline lens fiber gap junctions from isolated hemichannels. *J. Cell Biol.*, **126**, 1047–1058.

38 Kühlbrandt, W. (1992) Two-dimensional crystallization of membrane proteins. *Q. Rev. Biophys.*, **25**, 1–25.

39 Lee, J. K., Kozono, D. E., Remis, J., Kitagawa, Y., Agre, P. and Stroud, R. M. (2005) Structural basis for conductance by the archaeal aquaporin aqpm at 1.68 Å. *Proc. Natl. Acad. Sci. USA*, **102**, 18932–18937.

40 Marko, J. F. and Siggia, E. D. (1994) Fluctuations and supercoiling of DNA. *Science*, **265**, 506–508.

41 Marszalek, P. E., Pang, Y. P., Li, H., El Yazal, J., Oberhauser, A. F. and Fernandez, J. M. (1999) Atomic levers control pyranose ring conformations. *Proc. Natl. Acad. Sci. USA*, **96**, 7894–7898.

42 Mehta, A. D., Rief, M., Spudich, J. A., Smith, D. A. and Simmons, R. M. (1999) Single-molecule biomechanics with optical methods. *Science*, **283**, 1689–1695.

43 Möller, C., Allen, M., Elings, V., Engel, A. and Müller, DJ (1999) Tapping mode atomic force microscopy produces faithful high-resolution images of protein surfaces. *Biophys. J.*, **77**, 1050–1058.

44 Moy, V. T., Florin, E.-L. and Gaub, H. E. (1994) Intermolecular forces and energies between ligands and receptors. *Science*, **266**, 257–259.

45 Müller, D. J., Amrein, M. and Engel, A. (1997) Adsorption of biological molecules to a solid support for scanning probe microscopy. *J. Struct. Biol.*, **119**, 172–188.

46 Müller, D. J., Baumeister, W. and Engel, A. (1999a) Controlled unzipping of a bacterial surface layer with atomic force microscopy. *Proc. Natl. Acad. Sci. USA*, **96**, 13170–13174.

47 Müller, D. J., Fotiadis, D., Scheuring, S., Müller, S. A. and Engel, A. (1999b) Electrostatically balanced subnanometer imaging of biological specimens by atomic force microscopy. *Biophys. J.*, **76**, 1101–1111.

48 Müller, D. J., Sapra, T., Scheuring, S., Kedrov, A., Frederix, P., Fotiadis, D. and Engel, A. (2006) Single molecule studies of membrane proteins. *Curr. Opin. Struct. Biol.*, **16**, 489–495.

49 Murata, K., Mitsuoka, K., Hirai, T., Walz, T., Agre, P., Heymann, J. B., Engel, A. and Fujiyoshi, Y (2000) Structural determinants of water permeation through aquaporin-1. *Nature*, **407**, 599–605.

50 Oberhauser, A. F., Marszalek, P. E., Erickson, H. P. and Fernandez, J. M. (1998) The molecular elasticity of the extracellular matrix protein tenascin. *Nature*, **393**, 181–185.

51 Oberhauser, A. F., Marszalek, P. E., Carrion-Vazquez, M. and Fernandez, J. M. (1999) Single protein misfolding events captured by atomic force microscopy. *Nat. Struct. Biol.*, **6**, 1025–1028.

52 Oesterhelt, F., Oesterhelt, D., Pfeiffer, M., Engel, A., Gaub, H. E. and Müller, D. J. (2000) Unfolding pathways of individual bacteriorhodopsins. *Science*, **288**, 143–146.

53 Peyret, J. L., Bayan, N., Joliff, G., Gulik-Krzywicki, T., Mathieu, L., Schechter, E. and Leblon, G. (1993) Characterization of the cspb gene encoding ps2, an ordered surface-layer protein in *Corynebacterium glutamicum*. *Mol. Microbiol.*, **9**, 97–109.

54 Rief, M., Gautel, M., Oesterhelt, F., Fernandez, J. M. and Gaub, H. E. (1997) Reversible unfolding of individual titin immunoglobulin domains by AFM. *Science*, **276**, 1109–1112.

55 Rigaud, J. L., Mosser, G., Lacapere, J. J., Oloffson, A., Lévy, D. and Ranck, J. L. (1997) Bio-beads: An efficient strategy for 2d crystallization of membrane proteins. *J. Struct. Biol.*, **118**, 226–235.

56 Rigaud, J., Chami, M., Lambert, O., Lévy, D. and Ranck, J. (2000) Use of detergents in two-dimensional crystallization of membrane proteins. *Biochim. Biophys. Acta*, **23**, 112–128.

57 Ringler, P., Borgnia, M. J., Stahlberg, H., Maloney, P. C., Agre, P. and Engel, A. (1999) Structure of the water channel aqpz from *Escherichia coli* revealed by electron crystallography. *J. Mol. Biol.*, **291**, 1181–1190.

58 Savage, D. F., Egea, P. F., Robles-Colmenares, Y., O'Connell, J. D. and Stroud, R. M. (2003) Architecture and selectivity in aquaporins: 2.5 Å X-ray structure of aquaporin z. *PLoS Biol.*, **1**, 334–340.

59 Schabert, F. A. and Engel, A. (1994) Reproducible acquisition of *Escherichia coli* porin surface topographs by atomic force microscopy. *Biophys. J.*, **67**, 2394–2403.

60 Scheuring, S. (2006) AFM studies of the supramolecular assembly of bacterial photosynthetic core-complexes. *Curr. Opin. Chem. Biol.*, **10**, 387–393.

61 Scheuring, S. and Sturgis, J. N. (2005) Chromatic adaptation of photosynthetic membranes. *Science*, **309**, 484–487.

62 Scheuring, S. and Sturgis, J. N. (2006) Dynamics and diffusion in photosynthetic membranes from *Rhodospirillum photometricum*. *Biophys. J.*, **91**, 3707–3717.

63 Scheuring, S., Ringler, P., Borgina, M., Stahlberg, H., Müller, D. J., Agre, P. and Engel, A. (1999) High resolution topographs of the *Escherichia coli* waterchannel aquaporin z. *EMBO J.*, **18**, 4981–4987.

64 Scheuring, S., Reiss-Husson, F., Engel, A., Rigaud, J.-L. and Ranck, J.-L. (2001) High resolution topographs of the *Rubrivivax* *gelatinosus* light-harvesting complex 2. *EMBO J.*, **20**, 3029–3035.

65 Scheuring, S., Stahlberg, H., Chami, M., Houssin, C., Rigaud, J. and Engel, A. (2002) Charting and unzipping the surface-layer of *Corynebacterium glutamicum* with the atomic force microscope. *Mol. Microbiol.*, **44**, 675–684.

66 Scheuring, S., Seguin, J., Marco, S., Lévy, D., Breyton, C., Robert, B. and Rigaud, J.-L. (2003a) Afm characterization of tilt and intrinsic flexibility of *Rhodobacter sphaeroides* light harvesting complex 2 (lh2). *J. Mol. Biol.*, **325**, 569–580.

67 Scheuring, S., Seguin, J., Marco, S., Lévy, D., Robert, B. and Rigaud, J. L. (2003b) Nanodissection and high-resolution imaging of the *Rhodopseudomonas viridis* photosynthetic core-complex in native membranes by AFM. *Proc. Natl. Acad. Sci. USA*, **100**, 1690–1693.

68 Scheuring, S., Rigaud, J.-L. and Sturgis, J. N. (2004a) Variable lh2 stoichiometry and core clustering in native membranes of *Rhodospirillum photometricum*. *EMBO J.*, **23**, 4127–4133.

69 Scheuring, S., Sturgis, J. N., Prima, V., Bernadac, A., Lévy, D. and Rigaud, J.-L. (2004b) Watching the photosynthetic apparatus in native membranes. *Proc. Natl. Acad. Sci. USA*, **101**, 11293–11297.

70 Scheuring, S., Busselez, J. and Levy, D. (2005a) Structure of the dimeric pufx-containing core complex of *Rhodobacter blasticus* by in situ AFM. *J. Biol. Chem.*, **180**, 1426–1431.

71 Scheuring, S., Levy, D. and Rigaud, J.-L. (2005b) Watching the components of photosynthetic bacterial membranes and their "in situ" organization by atomic force microscopy. *Biochim. Biophys. Acta*, **1712**, 109–127.

72 Scheuring, S., Gonçalves, R. P., Prima, V. and Sturgis, J. N. (2006) The photosynthetic apparatus of *Rhodopseudomonas palustris*: Structures and organization. *J. Mol. Biol.*, **358**, 83–96.

73 Scheuring, S., Boudier, T. and Sturgis, J. N. (2007) From high-resolution AFM topographs to atomic models of supramolecular membrane protein assemblies. *J. Struct. Biol.*, doi:10.1016/j.jsb.2007.01.021.

74 Sleytr, U. B., Messner, P., Pum, D. and Sára, M. (1993) Crystalline bacterial cell surface layers: General principles and application potential. *J. Appl. Bacteriol. Symp.* Suppl., 74, 21S–32S.

75 Sleytr, U. B., Bayley, H., Sara, M., Breitwieser, A., Kupcu, S., Mader, C., Weigert, S., Unger, F. M., Messner, P., Jahn-Schmid, B., Schuster, B., Pum, D., Douglas, K., Clark, N. A., Moore, J. T., Winningham, T. A., Levy, S., Frithsen, I., Pankovc, J., Beale, P., Gillis, H. P., Choutov, D. A. and Martin, K. P. (1997) Applications of s-layers. *FEMS Microbiol. Rev.*, **20**, 151–175.

76 Stahlberg, H., Fotiadis, D., Scheuring, S., Remigy, H., Braun, T., Mitsuoka, K., Fujiyoshi, Y. and Engel, A. (2001) Two-dimensional crystals: A powerful approach to assess structure, function and dynamics of membrane proteins. *FEBS Lett.*, **504**, 166–172.

77 Sui, H., Han, B. G., Lee, J. K., Walian, P. and Jap, B. K. (2001) Structural basis of water-specific transport through the aqp1 water channel. *Nature*, **414**, 872–878.

78 Tornroth-Horsefield, S., Wang, Y., Hedfalk, K., Johanson, U., Karlsson, U., Tajkhorshid, E., Neutze, R. and Kjellbom, P. (2006) Structural mechanism of plant aquaporin gating. *Nature*, **439**, 688–694.

79 Unger, V. M., Hargrave, P. A., Baldwin, J. M. and Schertler, G. F. X. (1997) Arrangement of rhodopsin transmembrane α-helices. *Nature*, **389**, 203–206.

80 Viani, M. B., Pietrasanta, L. I., Thompson, J. B., Chand, A., Gebeshuber, I. C., Kindt, J. H., Richter, M., Hansma, H. G. and Hansma, P. K. (2000) Probing protein-protein interactions in real time. *Nature Struct. Biol.*, **7**, 644–647.

81 Zhong, Q., Inniss, D., Kjoller, K. and Elings, V. B. (1993) Fractured polymer/silica fiber surface studied by tapping mode atomic force microscopy. *Surf. Sci. Lett.*, **290**, 688–692.

Part IV
Dynamics

7
Molecular Dynamics Studies of Membrane Proteins: Outer Membrane Proteins and Transporters

Syma Khalid, John Holyoake and Mark S. P. Sansom

7.1
Introduction

Structural biology [especially X-ray diffraction (XRD), and to a lesser extent cryo-electron microscopy] provides high-resolution – but essentially static – structures of membrane proteins. Molecular dynamics (MD) simulations enable us to extend such structural data by exploring the conformational dynamics of membrane proteins in a lipid bilayer environment. In this chapter, such simulations are illustrated via their application to two major classes of membrane (transport) proteins: (i) β-barrel proteins from the outer membranes of Gram-negative bacteria [1]; and (ii) α-helical membrane transport proteins (www.tcdb.org). To facilitate this discussion, focus is centered on examples from the authors' own laboratory and related studies. For more general reviews of membrane protein simulations, the reader is referred to Refs. [2, 3]. All discussion of simulations of channel proteins and aquaporins has been omitted, as these are treated elsewhere and have been the subject of a number of recent reviews (e.g., Refs. [4, 5]).

The focus here is on simulations from a functional perspective – that is, as an aid to the present understanding of structure–function relationships of membrane proteins. Attention is restricted to atomistic simulations, in which protein, lipid, and water atoms are all treated explicitly. The recent years have also seen advances in the use of continuum solvent models [6] and coarse-grained [7, 8] simulations to study membrane protein insertion into bilayers, and of atomistic simulations [9, 10] to explore lipid–protein interactions.

7.1.1
Molecular Dynamics Simulations

Molecular dynamics simulations describe the dynamics of a membrane protein on a 10- to 50-ns timescale. The components of the simulation system (protein, lipids, water, etc.) and their interactions are described by a suitable molecular mechanics forcefield. This enables the potential energy of the system to be

Biophysical Analysis of Membrane Proteins. Investigating Structure and Function. Edited by Eva Pebay-Peyroula

ISBN: 978-3-527-31677-9

calculated as a function of the atomic coordinates of the component atoms. Numerical integration of Newton's equations of motions for every atom in the system enables the positions of the atoms to be recorded every, for example 1 ps (10^{-12} s), yielding a trajectory (i.e., a movie) of the protein, lipid, and water molecules. Analysis of the trajectory in terms of, for example, protein conformational drift and fluctuations, protein–ligand and protein–lipid interactions, may then be performed offline.

Molecular dynamics simulations therefore provide a useful complement to experimental approaches (such as XRD and cryoelectron microscopy) which provide static, time-, and spatially-averaged structures for membrane proteins at low temperatures. Molecular dynamics simulations can yield a picture of the dynamics of a membrane protein in a lipid bilayer environment at physiologically relevant temperatures. Such simulations of membrane proteins are a relatively recent addition to an established history of MD simulations of proteins in general [11, 12]. A membrane protein system studied by MD would typically consist of a protein embedded in a lipid bilayer [13] with water molecules and ions on either side (Fig. 7.1). Typically, such a system would consist of between ~50 000 and 100 000 atoms. Earlier simulations of membrane proteins [14, 15] were somewhat restricted in their timescale by computational considerations. However, continued advances in computational power and the development of more scalable (i.e., parallelizable) simulation algorithms have now enabled longer atomistic simulations to be performed. By using modest computational resources (e.g., a small linux cluster) and current simulation codes, simulation times of the order of 10 ns can be achieved in within a few weeks. Thus, multiple simulations can be performed to, for example, compare the dynamical behavior of proteins with/without bound ligands or of wild-type and mutant proteins.

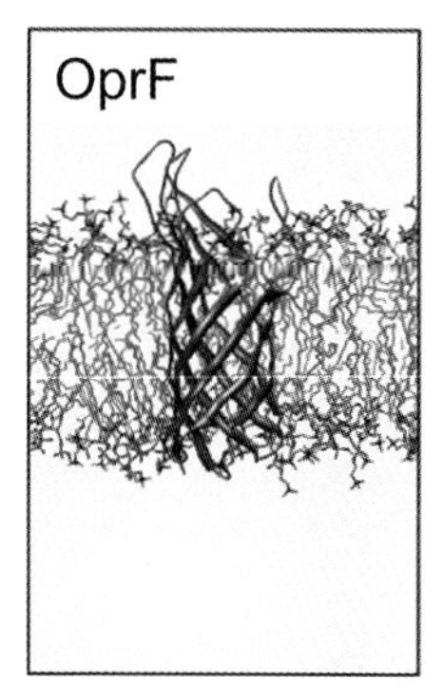

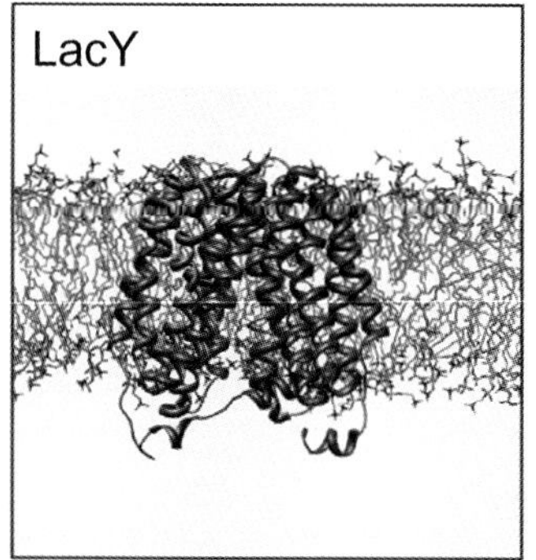

Fig. 7.1 Examples of membrane protein molecular dynamics simulations for OprF (a homology model of a simple bacterial outer membrane protein [41]) and LacY, a bacterial inner membrane transport protein [87]. In both cases, the protein is shown embedded in a phospholipid (DMPC) bilayer.

7.2 Outer Membrane Proteins

The outer membrane (OM) of Gram-negative bacteria serves as a protective barrier against the external environment, and controls the influx and efflux of solutes. Outer membrane proteins (OMPs) confer a variety of functions on the membrane, including passive and active transport, host/pathogen recognition, signal transduction, and enzymatic catalysis. It has been predicted that 2 to 3% of the genes in Gram-negative bacteria encode OMPs [16, 17]. In contrast to the α-helical nature of inner membrane proteins [18], most OMPs have a β-barrel architecture [1, 19, 20]. The barrels are composed of anti-parallel β-strands that are connected by short turns on the periplasmic side of the membrane, and by more extended loops on the extracellular side. The OM is asymmetric in nature. The inner leaflet (i.e., that facing the periplasmic space) is similar in phospholipid composition to the inner (cytosolic) membrane. In contrast, the outer leaflet is composed of complex lipopolysaccharides (LPS) [21]. These are anionic oligosaccharides crosslinked by divalent cations with multiple saturated fatty acid tails. The structure of LPS varies substantially from species to species, and can be modified within a single cell in response to changes in the local environment. The low baseline permeability of the OM is due to the combination of highly charged sugars and tightly ordered hydrocarbon chains, enabling it to protect the cell against toxic agents. Although some modeling and simulation studies of LPS have been conducted [22, 23], these have yet to be combined with simulations of OMPs.

In order to enable the influx and efflux of solutes across the OM, it is rendered selectively permeable to molecules smaller than ~600 Da by the presence of *porins* [24, 25]. In addition to non-specific porins, solute-specific porins (e.g., for sugars [26, 27] and for phosphate [28]), and a number of other OMPs with varying transport functions are also found in bacterial OMs. The non-porin OMPs range in function from the transport of specific solutes (e.g., transport of siderophores by TonB-coupled OMPs [29]), through to peptide autotransport [30]. There also are a number of OMP enzymes which do not perform a transport role. Other OMPs play a role in target cell recognition by pathogenic bacteria. A summary of earlier simulation studies is provided in Ref. [31]. In this chapter, attention will be focused on simulations of OMPs involved in transport, and on simulations of related OMPs in different environments.

7.2.1 OmpA

The small and relatively simple protein OmpA has provided something of a testbed for simulations of more complex OMPs. The structure of the N-terminal transmembrane (TM) domain of OmpA has been solved by XRD [32, 33] and by NMR (the latter in detergent micelles) [34, 35]. OmpA is a small, monomeric protein composed of an N-terminal, eight-stranded β-barrel (the TM domain) and a globular C-terminal located in the periplasm. The β-barrel is connected by large loops on the extracellular side that point away from the barrel, and by short turns on the periplasmic side. In the crystal structure, the β-barrel does not form a single continuous channel extending from one mouth of to the other, but rather several cavities are formed, separated by polar and charged side chains pointing into the β-barrel interior.

Despite the absence of a continuous pore in the X-ray structure, several functional studies have shown that OmpA forms low-conductance pores when reconstituted into planar lipid bilayers [36–38]. A combination of modeling and simulation studies of the OmpA N-terminal domain in a dimyristoylphosphatidylcholine (DMPC) bilayer [39] suggested that small conformational changes in charged side chains within the center of the β-barrel could provide a gate which controlled the transition between the closed OmpA pore seen in the crystal structure and an open-pore state that corresponds to the 60 pS conductance pores observed experimentally. These simulations revealed a degree of flexibility in the side chains lining the aqueous cavities within the β-barrel. The proposed gate was an R138-E52 salt bridge, which was stable throughout the 5 ns simulation, and prevented the passage of, for example, water molecules along the pore. An alternative gate conformation, in which the R138 side chain was oriented towards the side chain of E128 rather than E52, yielded an open conformation of the pore. Simulations of this putative open state revealed that water molecules were able to diffuse along the full length of the β-barrel on a nanosecond timescale. The radius profile of the resultant pore was shown to be in agreement with experimental conductance data. Thus, it was suggested that that changes in rotameric state of the R138 side chain could provide gating mechanism, switching OmpA between a closed and open pore state (Fig. 7.2). This model has received support from recent mutagenesis and functional studies of OmpA [40].

The results of recent studies have suggested that the gating mechanism proposed for OmpA may be extended to OmpA homologues from other bacterial species. For example, aqueous cavities formed by pore-occluding residues in the barrel were seen in a homology model of OprF, the main OMP of *Pseudomonas aeruginosa*. Molecular dynamics simulations of the OprF model (see Fig. 7.1) displayed remarkably similar dynamics to OmpA [41]. Water molecules failed to pass from one mouth of the barrel to the other in any of the six simulations reported. In particular, a persistent internal salt bridge formed by E8-K121, prevented water permeation in all of the simulations, and was identified as a likely gate, analogous to the R138-E52 salt bridge in OmpA.

Modeling and simulation studies of an OmpA homologue have also been used to demonstrate that computational approaches may be applied to more complex, multidomain OMPs, rather than just to TM β-barrels [42]. PmOmpA is a two-domain OMP from *Pasteurella multocida*; the N-terminal domain of PmOmpA is a homologue of the transmembrane β-barrel domain of OmpA from *Escherichia coli*, whilst the C-terminal domain of PmOmpA is a homologue of the extra-membrane *Neisseria meningitidis* RmpM C-terminal domain. This enabled a model of a complete two-domain PmOmpA to be constructed and its conformational dynamics to be explored via MD simulations. A degree of water penetration into

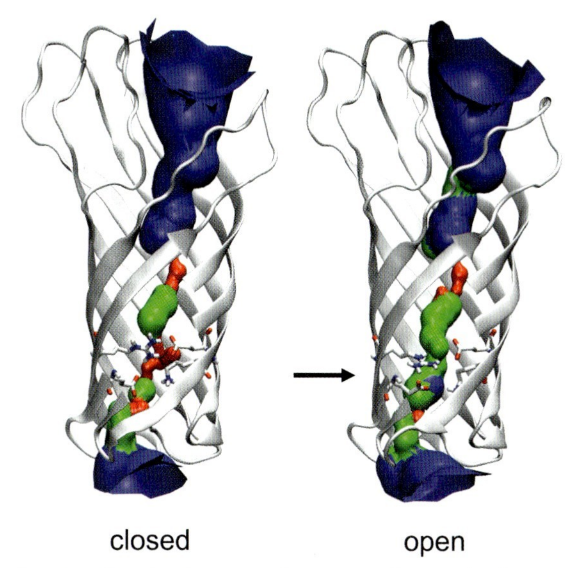

Fig. 7.2 Closed and open conformations of OmpA, as revealed by molecular dynamics simulations. The pore-lining surface is drawn using HOLE [111], color-coded such that: red = pore radius less than that of a water molecule; green = pore radius approximately that of a water molecule; and blue = pore radius greater than that of a water molecule. The arrow indicates the charged residue cluster, which forms the gate.

the interior of the β-barrel suggested the formation of a TM pore. The PmOmpA model was conformationally stable over a 20-ns simulation, but substantial flexibility was observed in the short linker region between the N- and the C-terminal domains. The C-terminal domain was observed to interact with the lipid bilayer headgroups.

7.2.2
Simulations of OMPs in Diverse Environments

Most OMP simulations have been performed in phospholipid bilayers, as a first approximation to the in-vivo environment. However, simulations of OMPs have proved useful in exploring the relationship between conformational dynamics and diverse experimental environments. For example, OmpA has been studied experimentally in lipid bilayers (e.g., functional pore studies [38] and electron paramagnetic resonance (EPR) spectroscopy [43]), in detergent micelles (NMR studies [34]), and in crystals containing the protein plus a small number of bound detergent molecules [33]. Comparisons of MD simulations in these three environments (Fig. 7.3) have revealed differences in the conformational dynamics of the protein.

The dynamics of OmpA in a DMPC lipid bilayer and in a detergent (dodecylphosphocholine; DPC) micelle were compared for 10-ns MD simulations [44]. A greater flexibility of the protein (~1.5× from Cα atom fluctuations) was seen in the micelle

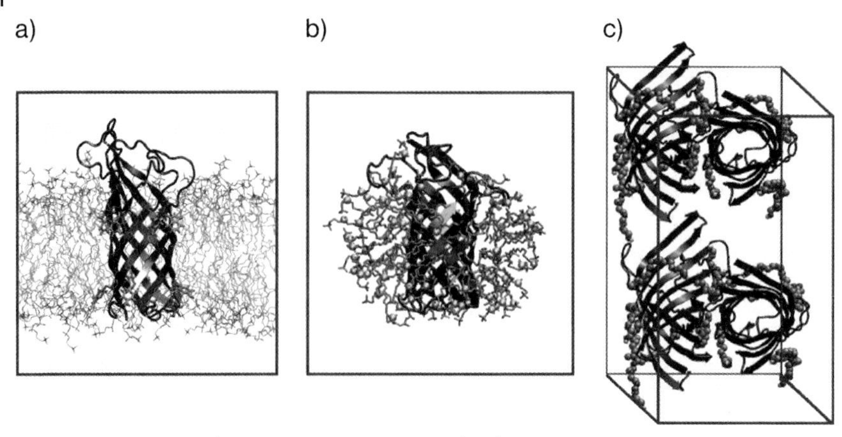

Fig. 7.3 An outer membrane protein (OmpA) simulated in three different environments. (A) A phospholipid bilayer [39]; (B) a detergent micelle [44]; (C) a crystal unit cell [45].

environment compared to the lipid bilayer. The increased mobility probably arises from the reduced packing constraints in the micellar environment, although the slight differences in the hydrophobic chain properties of the detergents and lipids may also contribute. A functional consequence of the OmpA/micelle simulations was that in this environment side chain conformational rearrangements led to the formation of a continuous pore through the center of the OmpA barrel. This reinforced the proposal (see above) of a gating mechanism for OmpA pores involving breaking and reformation of salt bridges within the β-barrel.

OmpA has also been used in simulations of a membrane protein in a crystal [45]. The simulation unit cell contained four protein molecules, plus detergent molecules and water. A good correlation was observed between simulated protein conformational fluctuations and experimental estimates (as derived from crystallographic *B* factors). The behavior of the detergent molecules in a unit cell revealed the interactions important in the formation and stabilization of the OmpA crystal. In particular, the detergent molecules formed a dynamic, extended micellar structure spreading over adjacent OmpA monomers within the crystal. The OmpA β-barrel in the crystalline environment was found to be more flexible than in a phospholipid bilayer, and similar in flexibility to in a micellar environment.

A number of OMP structures (e.g., OmpA [34], OmpX [46], and PagP [47]) have been determined by NMR as well as by X-ray crystallography. Often, these structures exhibit conformational differences, but MD simulations may play a key role in exploring the dynamics of these structures to identify any significant differences. A comparison of the OmpA, OmpX and PagP X-ray and NMR structure of the TM domain by a set of 15-ns simulations of the protein embedded in DMPC bilayers showed a dependence of simulation behavior on the "quality" of the initial structure [48]. The overall mobilities of the residues were qualitatively similar for the corresponding X-ray and NMR structures. However, all three proteins were generally more mobile in the simulated based on NMR structures.

Changes in OMP conformational dynamics have also been used in predictions of function. For example, OpcA is an adhesion protein from *N. meningitidis* which forms a 10-stranded β-barrel [49]. The X-ray structure reveals the β-barrel to be rather wide (mean radius ~0.2 nm) and water-filled; however, extracellular loop L2 traverses the barrel axis and this prevents formation of a continuous pore. OpcA was crystallized in the presence of Zn^{2+} ions, three of which are identified in the X-ray structure – one in the central cavity and two in the loops on the extracellular surface. Molecular dynamics simulations of OpcA in a DMPC bilayer in the *absence* of bound Zn^{2+} [50] revealed a striking increase in the flexibility of the loop regions. The largest conformational changes were seen in loop L2, which resulted in opening of the putative pore. Thus, the Zn^{2+} ions binding loops L2 and L4 together in the crystal structure may hold the protein into a non-physiological conformation. The conformational changes observed in the MD studies also suggest that the ligand (proteoglycan) binding site formed by the extracellular loops may be quite dynamic, which in turn suggests an induced-fit mechanism of binding. Thus, these simulations have not only provided evidence suggesting a non-physiological conformation of the OpcA protein in the crystal structure, but have also shed light on the mechanism of ligand binding.

7.2.3
Porins

One of the first MD simulations of an OMP [51] was of the *E. coli* porin OmpF embedded in a palmitoyl-oleoyl-phosphatidylethanolamine (POPE) bilayer. Although, by current standards this simulation was short (1 ns), it provided valuable insights into the functional dynamics of OmpF. In particular, reduced mobility of the water molecules within the protein pore relative to bulk solvent was observed. This slowing of water diffusion within pores agreed with earlier empirical approaches [52], and has since been observed in a number of simulations of OM channels. Further MD simulations of OmpF have provided insights into the origin of its cation selectivity [53, 54]. Whilst single potassium ions were free to permeate the pore, chloride ions could only pass the constriction zone when paired with potassium ions. Steered MD simulations [55] have been used to simulate the transport of dipolar molecules through OmpF. More recently, MD studies of OmpF have revealed the influence of side chain protonation states on the conformational dynamics of the protein [56]. In particular, varying the protonation state of a key side chain (D312) substantially influenced the cross-sectional area profile of the pore. This further emphasizes the importance of electrostatic interactions within the central core of the β-barrel domains of OMPs.

7.2.4
More Complex Outer Membrane Transporters

While most OMPs are β-barrels, the size and shape of these β-barrels can vary substantially. A feature of some of the more complex OMPs is the presence of additional domains located within the β-barrel. For example, the TonB-dependent transporter family contain a globular "plug" or "cork" domain within the β-barrel [57]. Furthermore, the two structures of bacterial autotransporters [30] reveal α-helices located within their β-barrels [58, 59]. Simulation studies enable the conformational dynamics of interactions between the β-barrel and the "inserted" domain to be probed. This is important given that several proposed transport mechanisms invoke a degree of repacking of these domains relative to one another [30].

7.2.4.1 TonB-Dependent Transporters

Passive diffusion is not the only mechanism of transport across the OM. Rather, a number of ions and their chelates are transported actively via a complex mechanism involving outer membrane transporters coupled to a periplasmic protein (TonB), which in turn is coupled to an energy-transducing protein complex in the inner membrane [29]. Thus, TonB-dependent transporters in the outer membrane mediate high-affinity binding and active transport of iron-chelating siderophores or vitamin B_{12} (in the case of BtuB) into the periplasm. The crystal structures of a number of TonB-dependent transporters, with and without bound ligand, are known (see http://blanco.biomol.uci.edu/Membrane_Proteins_xtal.html for a summary listing). All of these structures are 22-stranded β-barrels with a globular "cork" domain within the barrel that occludes the pore. Their mechanism of solute transport is thought to be a multistep process involving several conformational changes. Some insights into aspects of this mechanism were achieved through MD simulations of the ligand-free and iron/siderophores-bound states of the transporter FhuA embedded in a DMPC bilayer [60].

Two functionally important findings were identified from these simulations. First, substantial conformational changes were observed in loop L8 which, as the most mobile of the extracellular loops, exhibited a binding-state dependency in its flexibility. When the ligand is bound, L8 blocks access to the binding site, effectively "closing" it. A similar loop rearrangement when ligand is bound was seen in the crystal structure of the ferric citrate receptor, FecA [61, 62]. The second key finding concerns the dynamics of the water molecules within FhuA. The water permeability of FhuA was reduced by the presence of the plug domain within the pore. These results support the view that translocation of the siderophore, either actively or passively requires a substantial conformational change in the plug domain.

Simulations of a number of TonB transporters (FepA [63], FecA [61, 62], BtuB [64], and FpvA [65]) have been performed and compared in terms of, for example, lipid–protein interactions (S. Khalid and M. S. P. Sansom, unpublished results). Although a simple phospholipid (DMPC) was used during the simulations, it forms relatively long-lasting interactions at sites homologous to that which binds Lipid A in the crystal structure of FhuA [66]. This supports the findings of other studies, which suggest that simulations are able to reveal lipid binding sites on OMPs [67] and other membrane proteins [10].

7.2.4.2 Autotransporters

OM autotransporters may be divided into two subfamilies: conventional autotransporters, and trimeric autotransporters. Typically, autotransporters are expressed as precursor proteins with three basic functional domains: an N-terminal signal peptide, an internal passenger domain, and a C-terminal translocator domain (a β-barrel TM domain). The structures of two outer membrane autotransporters have been determined [58, 59]. A key feature of both proteins is the presence of α-helices within the β-barrel. NalP from *N. meningitidis* is a conventional autotransporter; it is a 10-stranded β-barrel, with a single helix attached located centrally within the hydrophilic β-barrel interior. In contrast, HiA from *Haemophilus influenzae* is a trimeric autotransporter in which three monomers each contribute four β-strands and one α-helix to form a 12-stranded β-barrel surrounding three α-helices.

Molecular dynamics simulations in a DMPC bilayer were used to compare the intact TM domain of NalP with the same domain from which the central the central helix was removed [68] (Fig. 7.4). Both, the intact and the helix-removed TM domains were conformationally stable on a 10-ns timescale, although removal of the central α-helix resulted in a degree of narrowing of the pore at both mouths of the β-barrel. This narrowing of the pore was largely a consequence of the increased flexibility of the loops and turns in the absence of the α-helix. At the extracellular mouth, loops L2 and L5 folded in part into the pore region. In particular, a persistent H-bond was detected between residues Y1017 and E873. In the presence of the α-helix, the reduced mobility of the loops prevented formation of this H-bond, resulting in a slightly wider pore. The reduced dimensions of the pore at the mouths of the β-barrel did not prevent entry of water into the pore region. Indeed, water molecules were seen to enter the pore when the helix was removed, an observation that suggests a plug-like role for the helix. Estimations of the conductance based on the pore radius profile were in good agreement with experimental values. Overall, these simulations support a (auto)transport role for the *monomeric* form of NalP.

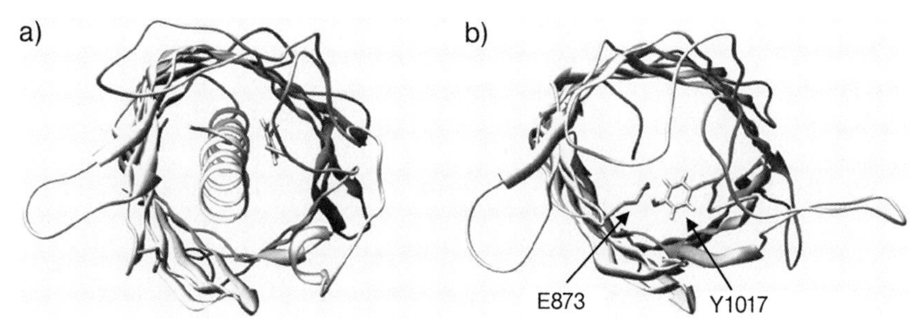

Fig. 7.4 Snapshots from simulations of the outer membrane autotransporter protein NalP [68], viewed along the bilayer normal. (A) The intact protein; (B) taken from a simulation of the β-barrel with the central α-helix removed.

As noted above, HiA is a trimeric autotransporter where, in contrast to NalP, the loops connecting the helices to the β-strands are large, extending over one-third of the way into the interior of the β-barrel. Molecular dynamics simulations of HiA (S. Khalid and M. S. P. Sansom, in preparation) have revealed a degree of β-barrel distortion when the helices were removed from the barrel interior. Further distortion was observed when the loops connecting the helices to the barrel were also removed. Thus, in agreement with experimental studies [59], simulations of HiA suggest that these loops play a key role in maintaining the structure of the trimeric β-barrel. A glycine residue from each loop, located near the center of the barrel, forms two H-bonds to the corresponding glycine residue from the two other loops. Thus, the triangular arrangement of glycines is stabilized by six H-bonds which, in turn, help to stabilize a rather flexible trimeric β-barrel.

7.2.4.3 **TolC**

A different type of OM transport systems is provided by the TolC family of proteins, which play a key role in the type I secretion of toxins, small peptides and drug molecules by Gram-negative bacteria [69]. Type I secretion involves a tripartite arrangement of proteins in an efflux complex. The complex comprises TolC (located in the outer membrane and extending into the periplasmic space), a perisplasmic "adaptor" protein, and an inner membrane efflux pump protein. This arrangement of the three proteins enables the direct passage of the solute from the cytoplasm to the external medium. This is supported by the X-ray structures of TolC [70] and of its homologues OprM [71] and VceC [72], of the inner membrane protein, AcrB [73, 74], and of the adaptor protein, AcrA [75] and its homologue MexA [76, 77]. Taken together, TolC, AcrB and AcrA are capable of forming a extended (length ~260 Å) molecular tunnel.

Focusing on the outer membrane component, TolC is a cylindrical trimer with a TM β-barrel domain and a periplasmic α-helical domain. The TM region of TolC is a 12-stranded β-barrel, whilst the periplasmic domain is composed of 12 α-helices. The main barrier to the export solutes appears to be at the periplasmic mouth. Thus, the X-ray structure of the TM region shows the β-barrel to be in an open conformation, whereas the pore is closed at its periplasmic mouth. Gating of TolC is therefore a key aspect of the function of the efflux complex, although it remains incompletely understood. A mechanism of opening has been proposed that involves a twisting motion of the helices accompanied by the disruption of H-bonds and electrostatic interactions [70].

Extended MD simulations of TolC embedded within a DMPC lipid bilayer have been performed to explore the intrinsic flexibility of the protein [78] (Fig. 7.5). Multiple 20-ns simulations of TolC were performed; these included simulations of the isolated TM (i.e., β-barrel) domain, of the intact TolC protein, and of OprM (a TolC homologue from *P. aeruginosa*). These simulations revealed that the extracellular loop regions of TolC exhibited substantial flexibility. During the simulations, inward collapse of the loops resulted in closure of the pore at its extracellular mouth. A network of H-bonds between residues located on the loops leads to a constriction of the pore, reducing the pore radius to <2 Å. A comparable role of

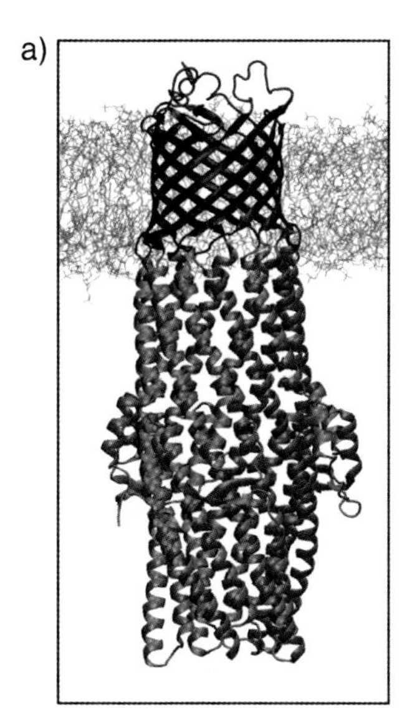

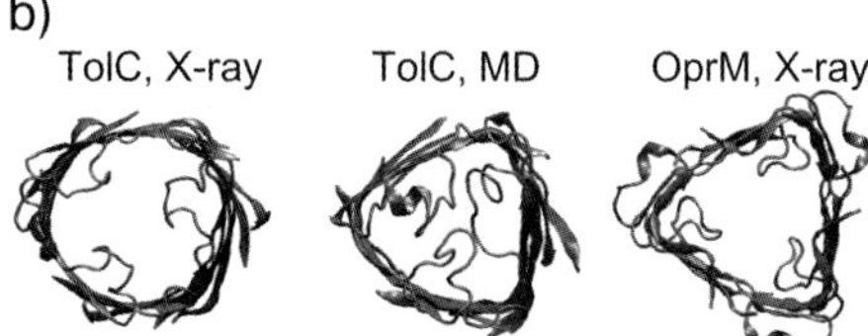

Fig. 7.5 Molecular dynamics (MD) simulations of the outer membrane drug efflux system protein TolC [78]. (A) TolC in a DMPC bilayer, showing the TM domain (dark gray) and the periplasmic domain (pale gray). (B) Conformational changes in the TM β-barrel of TolC, viewed from the extracellular face. The cylindrical conformation is seen in the X-ray structure of TolC, whereas a triangular prism conformation is seen in the 20-ns snapshot from the MD simulations, and also in the X-ray structure of the homologous protein OprM.

loops in gating has been also been suggested based on recent X-ray structures of OmpG at different pH values [79].

In addition to closure of the TM pore by the extracellular loops, changes in the cross-sectional conformation of the β-barrel domain of TolC are seen in the simulations. Comparing X-ray structures shows that the β-barrel domain of TolC is cylindrical in cross-section, whereas that of its homologue, OprM, is closer to a triangular prism. During simulations of TolC in a lipid bilayer, the β-barrel switches from a cylindrical to a prism conformation, whilst retaining the overall pore. This conformational change is more marked in the simulation of the intact protein in comparison with the simulation of the isolated TM domain. Thus, the intact TolC switches to a conformation similar to that of its homologue OprM. The more pronounced change in the intact TolC suggests that the transition between the cylindrical and prism conformations may be coupled to the presence of the periplasmic domain.

The α-helical domains of both TolC and OprM protrude ~100 Å into the periplasmic space, where they interact with the inner membrane transport protein (e.g., AcrB). The X-ray structure of TolC shows an internal pore in the tunnel wide enough to allow unimpeded passage of the solute. However, the MD simulations have shown that the dynamics of this region may be rather complex, involving both a global contraction ("breathing") of the pore and an iris-like motion of the whole protein.

Overall, these simulations have shown that the conformational dynamics of TolC are rather more complex than was anticipated. In particular, models in which the gating of TolC is restricted to an iris-like motion at the periplasmic mouth appears to be a little too simple. The intrinsic flexibility of TolC during simulations suggests that such as motion is combined with a twisting motion of the upper half of the periplasmic tunnel. The extracellular mouth appears to act as a gating region in which the mobility of the loops plays a key role. Thus, MD simulations have revealed a complex range of conformational dynamics in TolC, suggesting that it may be gated at both ends of the molecule, with complex "breathing" motions in the intermediate domains.

7.3
Cytoplasmic Membrane Transport Proteins

The other group of membrane transport proteins for which a significant number of simulation studies have been carried out are the α-helical membrane transport proteins (see www.tcdb.org for a database and classification of transporters). These include several major classes of membrane proteins, including the MFS proteins [80–82] and the ATP-binding cassette (ABC) transporters [83–85], for which a number of X-ray structures have been determined.

At this point, rather than attempt an exhaustive survey of simulations of α-helical transport proteins, they will be discussed in the context of strategies which may be employed in simulation studies to explore mechanistically relevant dynamic behavior of transport proteins. These can be summarized as: simulated state transitions; extrapolated intrinsic flexibilities; and non-equilibrium simulation methods. The use of simulations to evaluate models of mammalian transport proteins based on bacterial transporter structures is also considered.

7.3.1
Simulated State Transitions

Transport mechanisms for MFS and ABC transporter proteins involve cyclic interconversion of a transporter between a set of intermediate states. One complete cycle results in one round of solute transport. The X-ray structure of a transporter in general provides a single intermediate state of the complete cycle. Simulations of state transitions involve changing a structural feature of a given intermediate state to that found in a proposed adjacent state. Information on the transport mechanism is thus provided by the way in which the transporter structure responds to the perturbation. Simulated state transitions have been used to investigate the conformational dynamics of members of the two largest transporter superfamilies: an ABC transporter member, BtuCD (86), and an MFS member, LacY (87).

7.3.1.1 BtuCD

The ABC transporters comprise a large family of ATP-driven transporters, with diverse functions ranging from nutrient uptake in bacteria to antigen presentation in the immune system of mammals [83]. Each member consists of two nucleotide binding domains (NBDs) and two transmembrane domains (TMDs), either as separate subunits or combined within the same polypeptide chain. The NBDs are associated with the cytoplasmic face of the TMDs. The NBDs exhibit a high degree of sequence and structural conservation, and are responsible for the binding and hydrolysis of ATP that drives the conformational changes involved in transport. The TMDs form the pathway through the membrane for the solute and, unlike the NBDs, they show considerable sequence and structural diversity. This probably reflects the wide range of solutes transported by different members of the ABC family.

Three bacterial ABC transporter structures have been determined using XRD: the vitamin B_{12} transporter, BtuCD [86], the multidrug transporter SAV1866 [85], and the putative metal-chelate transporter HI1470/1 [88]. Of these, only BtuCD has so far been the subject of simulation studies. (There have also been simulations of the lipid A exporter MsbA, but as the initial X-ray structures of MsbA have been retracted [89] these studies will not be discussed further). Interestingly, comparison of the BtuCD structure and the HI1470/1 structure supports a switch between an outward-facing (in BtuCD) and an inward-facing (in HI1470/1) permeation pathway conformation as the basis of the solute transport mechanism.

BtuCD transports vitamin B_{12} from the periplasm into the cytosol. Its two TMDs consist of 10 transmembrane α-helices, with a central vitamin B_{12} transport pathway formed by the interface of the two TMDs. The structure has been determined for an ATP-free semi-open state, with a central cavity sufficient to hold a B_{12} molecule and open to the periplasm, but closed to the cytoplasm by a constriction (gate) formed by the TM4-5 loops of each subunit.

A simulation study of the coupling of nucleotide binding to a putative conformational changes of BtuCD was carried out by Tieleman et al. [90]. In this study, a simulation following docking of MgATP into each NBD of the ATP-free X-ray structure was used to represent a state transition from the semi-open ATP-free state to the/an MgATP-bound closed state. The presence of MgATP bound to the NBDs caused a clear difference in behavior compared to an equivalent MgATP-free simulation. MgATP was seen progressively to draw the two NBDs together. Significantly, tight binding across the dimer interface was seen for only one MgATP molecule, while the binding site of the other MgATP opened (however, see below). Movement of the NBDs was coupled to conformational changes in the TMDs, with consequent closure of the periplasmic end of the channel. Changes in the translocation pathway occurred in a manner likely to move vitamin B_{12} toward the cytoplasmic exit.

More recently, simulations of BtuCD have been compared with simulations of BtuD (i.e., just the NBD dimer) and of BtuCDF (i.e., of the ABC transporter BtuCD docked with the periplasmic binding protein BtuF [91]). These simulations [92] suggest a somewhat more complex picture than envisaged by the previous study. Thus, MD simulations of the BtuCD and BtuCDF complexes (both in a lipid bilayer), and of the isolated BtuD and BtuF proteins (in water) have been used to explore the conformational dynamics of this complex transport system. As in the previous simulation study, the presence of bound ATP induces closure of the NBDs. This occurred in a symmetrical fashion, but only in the BtuD dimer, and not in BtuCD. It seems that ATP constrains the flexibility of the NBDs in BtuCD, such that their closure may only occur upon binding of BtuF to the complex. Upon the introduction of BtuF, and concomitant with NBD association, one ATP-binding site displays a closure, while the opposite site remains relatively unchanged. This asymmetry may reflect an initial step in the "alternating hydrolysis" mechanism and is consistent with measurements of nucleotide-binding stoichiometries.

Taken together, these studies demonstrate the ability of MD simulations to explore the dynamic properties of the BtuCD structure (Fig. 7.6A) in relation to mechanisms proposed on the basis of biochemical studies. Indeed, both studies support an asymmetrical ATP-binding mechanism, though the details differ. This

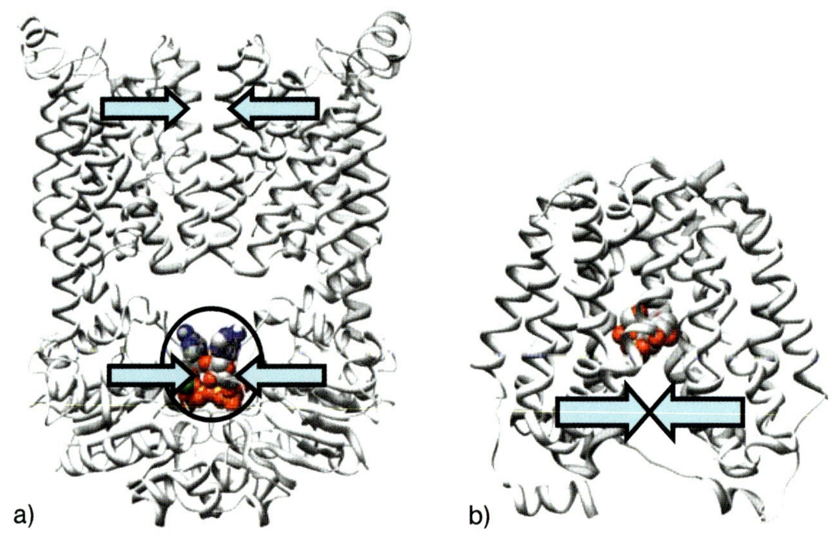

Fig. 7.6 Domain motions observed in simulations of membrane transport proteins. (A) The ABC transporter BtuCD (based on [90]). (B) The MFS protein LacY (based on [93]). In each case the arrows indicate the approximate direction of the domain motions observed in the simulations.

provides a basis on which to interpret further experimental data, with iterative contributions from computational and biochemical investigations.

7.3.1.2 LacY

State transition simulations [93] have also been used to study the coupling of proton translocation to lactose translocation in the proton/lactose symporter lactose permease (LacY). LacY is the archetypal member of the major facilitator superfamily (MFS), a group of transporters second in number of members to only the ABC transporters [80]. Members of this superfamily share a common fold, as indicated by sequence-based predictions and the three currently available crystal structures of MFS members, LacY [87], GlpT [94] and EmrD [95]. Despite low sequence identities (<15%), these structures have the same arrangement of helices, with each being composed of two domains of 6TM helices in each. X-ray crystallography has provided a hint of two states of MFS transporters: LacY and GlpT were crystallized in an inward-facing conformation, allowing intracellular access to their central solute binding site; while EmrD is in a closed conformation (similar to that seen in cryoelectron microscopy of OxlT [96]). However, as the X-ray structures alone cannot provide information on the conformational dynamics of interconversion between these states, it is here that simulation studies of LacY have provided valuable insights.

In bacteria, LacY uses the energy associated with movement of a proton down its electrochemical gradient to drive the accumulation of lactose. Extensive biochemical studies have identified six residues essential to function. E325 and R302 are directly involved in proton translocation, H322 and E269 couple proton translocation and sugar binding, whilst R144 and E126 are required for substrate binding. Detailed transport mechanisms have been proposed and refined based on biochemical data and more recently the X-ray structures [82, 97].

The X-ray structure of LacY represents the inwardly open, ligand-bound, E325-protonated state of a conformationally "locked" C154G mutant of LacY. Schulten et al. [93] performed simulations of this state ($E269^-/E325^H$), and of two adjacent states that mimic proton transfers, namely $E269^H/E325^-$ and $Apo/E269^-/E325^-$. Except for the $Apo/E269^-/E325^-$ simulation, bound galactose was included at the binding site in these simulations, docked onto the thiodigalactoside (TDG) molecule resolved in the X-ray structure. These simulations revealed mechanistically significant differences. The $E269^-/E325^H$ simulation remained in a conformation close to that of the X-ray structure. A salt bridge was observed between R144 and E269. Furthermore, E269 interacted strongly with the galactose molecule. Protonation of E269 as in simulation $E269^H/E325^-$ resulted in a dramatic difference in behavior. The R144–E269 salt bridge was broken, while the galactose remained in the binding site, but shifted towards the periplasmic (closed) gate. Most significantly, protonation of E269 triggered a substantial closure of the intracellular entrance. In contrast, simulation $Apo/E269^-/E325^-$ exhibited similar behavior to the $E269^-/E325^H$ simulation. Together, these simulations provide important information on the translocation mechanism of LacY. Closure of the intracellular entrance was achieved by motions of TM4 and TM5, which is noteworthy given the location of the inactivating C154G mutation in TM5. Furthermore, the greater flexibility of the N-terminal domain reported in this study is consistent with the findings of previous studies [98]. The significant closure of the intracellular entrance shows that, while the overall transport cycle occurs on a millisecond time scale, key conformational events may be captured by simulations of substantially shorted time scales. Simulations such as these represent a means of testing transport cycles based on biochemical data.

A second recent set of simulations [99] explored the conformational dynamics of LacY by starting with simulations of the state captured in the crystal and moving "outwards" to approximations of subsequent steps in the translocation mechanism. Integral to this study was a cautious methodical approach by which simulations were extended from more conservative initial configurations to more exploratory configurations. At each stage, multiple simulations were performed to validate the reproducibility of the observations. Furthermore, bias from initial system configuration was removed by the use of multiple systems, varying in the size of lipid bilayer.

In total, 10 different simulations were performed with all of the "locked" C154G mutant of LacY. Simulations varied in the presence/absence of bound ligand (TDG) and in the protonation state of the E325 side chain. They also differed in the initial configuration of the lipid bilayer, and repeat (random velocity seed) simulations were performed. The initial TDG-bound, E325-protonated simulations confirmed the overall stability of the conformation seen in the LacY crystal structure, with little drift in structure from the crystal structure. Such a result lends confidence to the quality of the structure and the simulation protocol.

Observations of the dynamic behavior of the LacY structure in response to removal of the bound ligand can provide information about the conformational events involved in the translocation cycle, albeit limited by the relatively short time scales available to atomistic MD simulations. Such apo (i.e., no bound solute) simulations therefore represent a crude approximation to the events after ligand dissociation into the cell. All of the apo simulations exhibited similar behavior in which there was a significant and reproducible closure of the intracellular entrance to the ligand-binding cavity (Fig. 7.6B). This closure might be considered surprising given the short time of the simulations compared to the turnover rate of the protein. However, it is likely that individual conformational transitions may occur on such time scales. Of greater interest is the way in which this closure occurs. Rather than the conformational changes involved in translocation being rigid-body domain movements, in which the intra-domain structure remains relatively static, substantial *intra*-domain movements of α-helices TM5 and TM11 were seen to be involved in the closure. This echoes the involvement of TM5 in the earlier simulations by Schulten and colleagues (see above).

7.3.2
Intrinsic Flexibilities

Although equilibrium MD simulations of membrane proteins are limited in terms of the extent to which conformational events observed in the accessible time scales are directly related to transport processes, the patterns of intrinsic flexibility seen in such simulations may be extrapolated to provide information on the larger (and presumably slower) conformational events involved in the function of a transport protein. An example of this is provided by a simulation study of the functional implications of the intrinsic flexibility of the protein translocation channel SecY [100].

The biosynthesis of secreted and membrane proteins involves the transport of the nascent polypeptide across the endoplasmic reticulum in eukaryotes, or across the cytoplasmic membrane in prokaryotes. Such proteins are targeted to the membrane by a signal peptide at the start of their sequence. After targeting, the protein is transported across the membrane through a protein-conducting channel. This channel is a conserved heterotrimeric, $\alpha\beta\gamma$ complex, called SecY in eubacteria and archaea, and Sec61 in eukaryotes. The α subunit (SecY in eubacteria and archaea; Sec61α in mammals) forms a central membrane-spanning channel. Biochemical studies suggest that lateral opening of this channel allows release of nascent hydrophobic transmembrane segments into the membrane.

The structure of SecYEβ from the archaea *Methanococcus jannaschii* has been determined using XRD [101]. The structure indicates that a single complex can serve as a functional channel; this complex consists of 12 TM helices in total, 10 provided by the α subunit, SecY, and one each from the two remaining subunits. SecY has two linked halves, each of five consecutive TM helices (TM1-5 and TM6-10), arranged with twofold pseudo-symmetry. A large, funnel-shaped cavity is apparent, open to the cytoplasm, and constricted at the middle of the membrane. This indicates that the structure is in a closed conformation. A plug structure formed from TM2a causes the constriction. "Outward" displacement of this helical plug is proposed to open the channel for polypeptide translocation. A ring of hydrophobic residues may form a flexible hydrophobic gate (or "gasket") to prevent small molecule and ion leakage during polypeptide translocation. Flexibility is required to accommodate the passage of different polypeptide sequences of varying bulk. On the basis of the X-ray structure, it was proposed that the lateral gate required for TM segment release was between TM helices M2b and M7, and involved a clamshell-like opening motion.

Motions of an isolated SecY subunit in equilibrium MD simulations provided some information on SecY gating [100]. Despite time scales that were too short to observe full conformational transitions of the plug, or of the lateral gate, the dominant intrinsic motions as revealed by principal component analysis (PCA) support the "plug and clamshell" model of SecY channel gating. The first eigenvector motion of the plug was out of the pore, away from the intracellular cytoplasmic side. The proposed lateral gate helices M2b and M7 exhibited motions away from each other, consistent with an opening of the lateral gate. Furthermore, the simulation demonstrated that water molecules were able to approach the hydrophobic pore ring, but were unable to cross it, confirming that this gate (Fig. 7.7) could prevent leakage of water and ions in the absence of a translocating polypeptide.

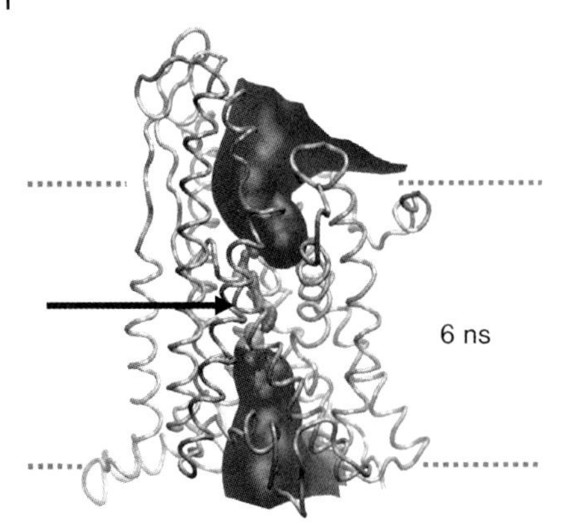

Fig. 7.7 Simulation of the SecY protein translocation pore [100]: a snapshot (at 6 ns) showing the pore-lining surface. The pore ring which acts as a hydrophobic gate is indicated by the horizontal arrow.

7.3.3 Non-Equilibrium Methods

Transporter/membrane protein function may also be investigated by non-equilibrium simulations in which conformational events are forced, using for example steered MD simulations [102]. A steered MD study of polypeptide translocation in SecY [103] explored the forced passage of a deca-alanine helix and of an alanine/leucine (AL19) helix through the protein-translocating channel. The α-helical peptides were positioned at the cytoplasmic mouth and "pulled" through the channel in <2 ns. Movement of the peptide led to the displacement of the M2a plug out of the channel. The plug moved as a structured body, exploiting the flexibility of the connecting coil region. After plug displacement, the channel was open, but the passage of water molecules was restricted (and ion permeation prevented) by the behavior of the hydrophobic pore ring (see above). Indeed, the pore ring was seen to be flexible, expanding to accommodate the translocating deca-alanine helix, and thus providing an elastic, yet tight, seal. Equilibrium simulations starting from the *perturbed* SecY structure, subsequent to the peptide translocation, demonstrated the ability of the plug region to return to a position near that seen in the X-ray structure.

7.3.4 Homology Models

The structures of bacterial membrane transport proteins may also be used as templates for homology models of their mammalian counterparts. A number of studies of, for example, mammalian K channels, have demonstrated that simulations based on homology models may be used to test the conformational stability of the

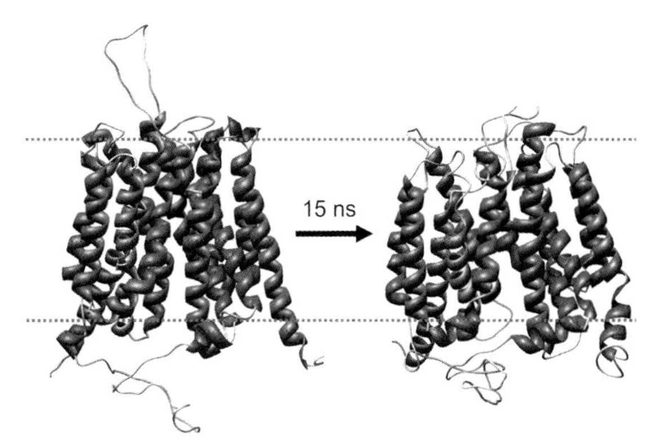

Fig. 7.8 Simulation of a homology model of a mammalian MFS protein (the glucose transporter GLUT1) based on the structure of the bacterial transporter GlpT [108]. Snapshots taken at the start and end of a 15-ns MD simulation. The approximate location of the lipid bilayer (omitted for clarity) is indicated by the horizontal lines.

model (albeit on a limited time scale) and to explore the relationship between conformational dynamics and function in the mammalian channel protein [104, 105].

Several studies have been conducted (e.g., [106, 107]) using bacterial MFS transporter structures (mainly LacY and GlpT) as templates for homology models of mammalian MFS proteins. Before employing such models, for example in mutational or docking studies, it is essential to develop a measure of their quality. A recent study [108] indicated that simulation studies may be used as a tool to aid in the evaluation of MFS homology models. Two MFS members (NupG, a bacterial nucleoside transporter; and GLUT1, a human glucose transporter) were modeled using GlpT and LacY, respectively, as templates. In addition, control models were created with shuffled sequences, to mimic "poor quality" homology models. These models, and the template crystal structures, were compared in 15-ns duration MD simulations in DMPC bilayers (Fig. 7.8). Comparison of the behavior of modeled structures with the crystal structures in these MD simulations provided a metric for model quality, based on conformational drift and loss of α-helicity in TM domains. Combined with studies of docking of the inhibitor forskolin to GLUT1 and to a control model, the simulations suggested that it might be possible to identify accurate homology models, despite low sequence identity between target sequences and templates.

7.4
Conclusions

These studies have provided a snapshot of the current status of biomolecular simulations of membrane proteins. It is evident that such simulations can now be performed with a reasonable degree of accuracy, and can provide insights into the relationship between membrane protein structure and function. Indeed, we are now at the stage where one might consider simulations as a standard tool for analysis of new membrane protein structures. Thus, they will continue to provide valuable information on mechanisms of membrane proteins, although limitations of sampling and time scale remain. As more structures and simulations emerge, sampling issues [109] may be addressed by *comparative* simulation studies, an approach which has been explored, for example, in membrane protein–lipid interactions [9]. In terms of time scales, both coarse-grained simulations [8] and elastic network models [110] offer promise for addressing longer time scale events, albeit more approximately. By integrating such approaches with atomistic simulations, a more integrative computational biology of membrane proteins will emerge.

Acknowledgments

Research in the laboratory of M.S.P.S. is funded by grants from the BBSRC, the EPSRC, the MRC, and the Wellcome Trust. The authors thank Peter Bond, Shozeb Haider, Anthony Ivetac, and Loredana Vaccaro for their help in preparing the figures.

References

1 Koebnik, R., Locher, K. P., and Van Gelder, P. (2000) Structure and function of bacterial outer membrane proteins: barrels in a nutshell. *Mol. Microbiol.*, **37**, 239–253.

2 Ash, W. L., Zlomislic, M. R., Oloo, E. O., and Tieleman, D. P. (2004) Computer simulations of membrane proteins. *Biochim. Biophys. Acta*, **1666**, 158–189.

3 Gumbart, J., Wang, Y., Aksimentiev, A., Tajkhorshid, E., and Schulten, K. (2005) Molecular dynamics simulations of proteins in lipid bilayers. *Curr. Opin. Struct. Biol.*, **15**, 423–431.

4 Roux, B., and Schulten, K. (2004) Computational studies of membrane channels. *Structure*, **12**, 1343–1351.

5 Noskov, S. Y., and Roux, B. (2006) Ion selectivity in potassium channels. *Biophys. Chem.*, **124**, 279–291.

6 Im, W., and Brooks, C. L. (2005) Interfacial folding and membrane insertion of designed peptides studied by molecular dynamics simulations. *Proc. Natl. Acad. Sci. USA*, **102**, 6771–6776.

7 Bond, P. J., and Sansom, M. S. P. (2006) Insertion and assembly of membrane proteins via simulation. *J. Am. Chem. Soc.*, **128**, 2697–2704.

8 Bond, P. J., Holyoake, J., Ivetac, A., Khalid, S., and Sansom, M. S. P. (2007) Coarse-grained molecular dynamics simulations of membrane proteins and peptides. *J. Struct. Biol.*, **157**, 593–605.

9 Deol, S. S., Bond, P. J., Domene, C., and Sansom, M. S. P. (2004) Lipid-protein interactions of integral membrane proteins: a comparative simulation study. *Biophys. J.*, **87**, 3737–3749.

10 Deol, S. S., Domene, C., Bond, P. J., and Sansom, M. S. P. (2006) Anionic phospholipids interactions with the potassium channel KcsA: simulation studies. *Biophys. J.*, **90**, 822–830.

11 Karplus, M. J., and McCammon, J. A. (2002) Molecular dynamics simulations

of biomolecules. *Nature Struct. Biol.*, **9**, 646–652.

12 Adcock, S. A., and McCammon, J. A. (2006) Molecular dynamics: survey of methods for simulating the activity of proteins. *Chem. Rev.*, **106**, 1589–1615.

13 Faraldo-Gómez, J. D., Smith, G. R., and Sansom, M. S. P. (2002) Setup and optimisation of membrane protein simulations. *Eur. Biophys. J.*, **31**, 217–227.

14 Forrest, L. R., and Sansom, M. S. P. (2000) Membrane simulations: bigger and better? *Curr. Opin. Struct. Biol.*, **10**, 174–181.

15 Domene, C., Bond, P., and Sansom, M. S. P. (2003) Membrane protein simulation: ion channels and bacterial outer membrane proteins. *Adv. Prot. Chem.*, **66**, 159–193.

16 Molloy, M. P., Herbert, B. R., Slade, M. B., Rabilloud, T., Nouwens, A. S., Williams, K. L., and Gooley, A. A. (2000) Proteomic analysis of the *Escherichia coli* outer membrane. *Eur. J. Biochem.*, **267**, 2871–2881.

17 Wimley, W. C. (2002) Toward genomic identification of β-barrel membrane proteins: composition and architecture of known structures. *Protein Sci.*, **11**, 301–312.

18 Popot, J. L., and Engelman, D. M. (2000) Helical membrane protein folding, stability, and evolution. *Annu. Rev. Biochem.*, **69**, 881–922.

19 Buchanan, S. K. (1999) β-Barrel proteins from bacterial outer membranes: structure, function and refolding. *Curr. Opin. Struct. Biol.*, **9**, 455–461.

20 Wimley, W. C. (2003) The versatile β-barrel membrane protein. *Curr. Opin. Struct. Biol.*, **13**, 404–411.

21 Doerrler, W. T. (2006) Lipid trafficking to the outer membrane of Gram-negative bacteria. *Mol. Microbiol.*, **60**, 542–552.

22 Lins, R. D., and Straatsma, T. P. (2001) Computer simulation of the rough lipopolysaccharide membrane of *Pseudomonas aeruginosa*. *Biophys. J.*, **81**, 1037–1046.

23 Shroll, R. M., and Straatsma, T. P. (2002) Molecular structure of the outer bacterial membrane of *Pseudomonas aeruginosa* via classical simulation. *Biopolymers*, **65**, 395–407.

24 Schirmer, T. (1998) General and specific porins from bacterial outer membranes. *J. Struct. Biol.*, **121**, 101–109.

25 Achouak, W., Heulin, T., and Pages, J. M. (2001) Multiple facets of bacterial porins. *FEMS Microbiol. Lett.*, **199**, 1–7.

26 Schirmer, T., Keller, T. A., Wang, Y. F., and Rosenbusch, J. P. (1995) Structural basis for sugar translocation through maltoporin channels at 3.1 Å resolution. *Science*, **267**, 512–514.

27 Forst, D., Welte, W., Wacker, T., and Diederichs, K. (1998) Structure of the sucrose-specific porin ScrY from *Salmonella typhimurium* and its complex with sucrose. *Nature Struct. Biol.*, **5**, 37–46.

28 Moraes, T. F., Bains, M., Hancock, P. E. W., and Strynadka, N. C. J. (2007) An arginine ladder in OprP mediates phosphate-specific transfer across the outer membrane. *Nature Struct. Mol. Biol.*, **14**, 85–87.

29 Faraldo-Gómez, J. D., and Sansom, M. S. P. (2003) Acquisition of siderophores in Gram-negative bacteria. *Nature Rev. Mol. Cell. Biol.*, **4**, 105–115.

30 Henderson, I. R., Navarro-Garcia, F., Desvaux, M., Fernandez, R. C., and Ala'Aldeen, D. (2004) Type V protein secretion pathway: the autotransporter story. *Microbiol. Mol. Biol. Rev.*, **68**, 692–744.

31 Bond, P. J., and Sansom, M. S. P. (2004) The simulation approach to bacterial outer membrane proteins. *Mol. Membr. Biol.*, **21**, 151–162.

32 Pautsch, A., and Schulz, G. E. (1998) Structure of the outer membrane protein A transmembrane domain. *Nature Struct. Biol.*, **5**, 1013–1017.

33 Pautsch, A., and Schulz, G. E. (2000) High-resolution structure of the OmpA membrane domain. *J. Mol. Biol.*, **298**, 273–282.

34 Arora, A., Abildgaard, F., Bushweller, J. H., and Tamm, L. K. (2001) Structure of outer membrane protein A transmembrane domain by NMR spectroscopy. *Nature Struct. Biol.*, **8**, 334–338.

35 Fernandez, C., Hilty, C., Bonjour, S., Adeishvili, K., Pervushin, K., and Wüthrich, K. (2001) Solution NMR studies of the integral membrane proteins OmpX and OmpA from *Escherichia coli*. *FEBS Lett.*, **504**, 173–178.

36 Saint, N., De, E., Julien, S., Orange, N., and Molle, G. (1993) Ionophore properties of OmpA of Escherichia coli. *Biochim. Biophys. Acta*, **1145**, 119–123.

37 Sugawara, E., and Nikaido, H. (1992) Pore-forming activity of OmpA protein of *Escherichia coli*. *J. Biol. Chem.*, **267**, 2507–2511.

38 Arora, A., Rinehart, D., Szabo, G., and Tamm, L. K. (2000) Refolded outer membrane protein A of *Escherichia coli* forms ion channels with two conductance states in planar lipid bilayers. *J. Biol. Chem.*, **275**, 1594–1600.

39 Bond, P. J., Faraldo-Gómez, J. D., and Sansom, M. S. P. (2002) OmpA – A pore or not a pore? Simulation and modelling studies. *Biophys. J.*, **83**, 763–775.

40 Hong, H., Szabo, G., and Tamm, L. K. (2006) Electrostatic couplings in OmpA ion-channel gating suggest a mechanism for pore opening. *Nature Chem. Biol.*, **2**, 627–635.

41 Khalid, S., Bond, P. J., Deol, S. S., and Sansom, M. S. P. (2005) Modelling and simulations of a bacterial outer membrane protein: OprF from *Pseudomonas aeruginosa*. *Proteins: Struct. Funct. Bioinf.*, **63**, 6–15.

42 Carpenter, T., Khalid, S., and Sansom, M. S. P. (2006) A multidomain outer membrane protein from *Pasteurella multocida*: modelling and simulations studies on PmOmpA. *Biochim. Biophys. Acta* (submitted).

43 Ramakrishnan, M., Pocanschi, C. L., Kleinschmidt, J. H., and Marsh, D. (2004) Association of spin-labeled lipids with β-barrel proteins from the outer membrane of *Escherichia coli*. *Biochemistry*, **43**, 11630–11636.

44 Bond, P. J., and Sansom, M. S. P. (2003) Membrane protein dynamics vs. environment: simulations of OmpA in a micelle and in a bilayer. *J. Mol. Biol.*, **329**, 1035–1053.

45 Bond, P. J., Faraldo-Gómez, J. D., Deol, S. S., and Sansom, M. S. P. (2006) Membrane protein dynamics and detergent interactions within a crystal: a simulation study of OmpA. *Proc. Natl. Acad. Sci. USA*, **103**, 9518–9523.

46 Fernandez, C., Hilty, C., Wider, G., Güntert, P., and Wüthrich, K. (2004) NMR structure of the integral membrane protein OmpX. *J. Mol. Biol.*, **336**, 1211–1221.

47 Hwang, P. M., Choy, W. Y., Lo, E. I., Chen, L., Forman-Kay, J. D., Raetz, C. R. H., Privé, G. G., Bishop, R. E., and Kay, L. E. (2002) Solution structure and dynamics of the outer membrane enzyme PagP by NMR. *Proc. Natl. Acad. Sci. USA*, **99**, 13560–13565.

48 Cox, K., Bond, P. J., Grottesi, A., Baaden, M., and Sansom, M. S. P. (2007) Outer membrane proteins: comparing X-ray and NMR structures by MD simulations in lipid bilayers. *Eur. Biophys. J.* (submitted).

49 Prince, S. M., Achtman, M., and Derrick, J. P. (2002) Crystal structure of the OpcA integral membrane adhesin from Neisseria meningitides. *Proc. Natl. Acad. Sci. USA*, **99**, 3417–3421.

50 Bond, P. J., Derrick, J. P., and Sansom, M. S. P. (2007) Membrane simulations of OpcA: gating in the loops? *Biophys. J.*, **92**, L23–L25.

51 Tieleman, D. P., and Berendsen, H. J. C. (1998) A molecular dynamics study of the pores formed by *Escherichia coli* OmpF porin in a fully hydrated palmitoyloleoylphosphatidylcholine bilayer. *Biophys. J.*, **74**, 2786–2801.

52 Smart, O. S., Breed, J., Smith, G. R., and Sansom, M. S. P. (1997) A novel method for structure-based prediction of ion channel conductance properties. *Biophys. J.*, **72**, 1109–1126.

53 Im, W., and Roux, B. (2002) Ions and counterions in a biological channel: A molecular dynamics simulation of OmpF porin from *Escherichia coli* in an explicit membrane with 1 M KCl aqueous salt solution. *J. Mol. Biol.*, **319**, 1177–1197.

54 Im, W., and Roux, B. (2002) Ion permeation and selectivity of OmpF porin: A theoretical study based on molecular dynamics, Brownian dynamics, and continuum electrodiffusion theory. *J. Mol. Biol.*, **322**, 851–869.

55 Robertson, K. M., and Tieleman, D. P. (2002) Orientation and interactions of dipolar molecules during transport through OmpF porin. *FEBS Lett.*, **528**, 53–57.

56 Varma, S., Chiu, S. W., and Jakobsson, E. (2006) The influence of amino acid protonation states on molecular dynamics simulations of the bacterial porin OmpF. *Biophys. J.*, **90**, 112–123.

57 Wiener, M. C. (2005) TonB-dependent outer membrane transport: going for Baroque? *Curr. Opin. Struct. Biol.*, **15**, 394–400.

58 Oomen, C. J., van Ulsen, P., van Gelder, P., Feijen, M., Tommassen, J., and Gros, P. (2004) Structure of the translocator domain of a bacterial autotransporter. *EMBO J.*, **23**, 1257–1266.

59 Meng, G., Surana, N. K., St. Geme, J. W., and Waksman, G. (2006) Structure of the outer membrane translocator domain of the *Haemophilus influenzae* Hia trimeric autotransporter. *EMBO J.*, **25**, 2297–2304.

60 Faraldo-Gómez, J. D., Smith, G. R., and Sansom, M. S. P. (2003) Molecular dynamics simulations of the bacterial outer membrane protein FhuA: a comparative study of the ferrichrome-free and bound states. *Biophys. J.*, **85**, 1–15.

61 Ferguson, A. D., Chakraborty, R., Smith, B. S., Esser, L., van der Helm, D., and Deisenhofer, J. (2002) Structural basis of gating by the outer membrane transporter FecA. *Science*, **295**, 1715–1719.

62 Yue, W. W., Grizot, S., and Buchanan, S. K. (2003) Structural evidence for iron-free citrate and ferric citrate binding to the TonB-dependent outer membrane transporter FecA. *J. Mol. Biol.*, **332**, 353–368.

63 Buchanan, S. K., Smith, B. S., Venkatramani, L., Xia, D., Essar, L., Palnitkar, M., Chakraborty, R., van der Helm, D., and Deisenhofer, J. (1999) Crystal structure of the outer membrane active transporter FepA from *Escherichia coli*. *Nature Struct. Biol.*, **6**, 56–63.

64 Chimento, D. P., Mohanty, A. K., Kadner, R. J., and Wiener, M. C. (2003) Substrate-induced transmembrane signaling in the cobalamin transporter BtuB. *Nat. Struct. Biol.*, **10**, 394–401.

65 Cobessi, D., Celia, H., Folschweiller, N., Schalk, I. J., Abdallah, M. A., and Pattus, F. (2005) The crystal structure of the pyoverdine outer membrane receptor FpvA from *Pseudomonas aeruginosa* at 3.6 angstroms resolution. *J. Mol. Biol.*, **347**, 121–134.

66 Ferguson, A. D., Hofmann, E., Coulton, J. W., Diederichs, K., and Welte, W. (1998) Siderophore-mediated iron transport: crystal structure of FhuA with bound lipopolysaccharide. *Science*, **282**, 2215–2220.

67 Baaden, M., and Sansom, M. S. P. (2004) OmpT: molecular dynamics simulations of an outer membrane enzyme. *Biophys. J.*, **87**, 2942–2953.

68 Khalid, S., and Sansom, M. S. P. (2006) Molecular dynamics simulations of a bacterial autotransporter: NalP from *Neisseria meningitidis*. *Mol. Membr. Biol.*, **23**, 499–508.

69 Buchanan, S. K. (2001) Type I secretion and multidrug efflux: transport through the TolC channel-tunnel. *Trends Biochem. Sci.*, **26**, 3–6.

70 Koronakis, V., Sharff, A., Koronakis, E., Luisi, B., and Hughes, C. (2000) Crystal structure of the bacterial membrane protein TolC central to multidrug efflux and protein export. *Nature*, **405**, 914–919.

71 Akama, H., Kanemaki, M., Tsukihara, T., Nakagawa, A., and Nakae, T. (2004) Crystal structure of the drug discharge outer membrane protein, OprM, of *Pseudomonas aeruginosa*: dual modes of membrane anchoring and occluded cavity end. *J. Biol. Chem.*, **279**, 52816–52819.

72 Federici, L., Du, D., Walas, F., Matsumura, H., Fernandez-Recio, J., McKeegan, K. S., Borges-Walmsley, M. I., Luisi, B. F., and Walmsley, A. R. (2005) The crystal structure of the outer membrane protein VceC from the bacterial pathogen *Vibrio cholerae* at 1.8 Å resolution. *J. Biol. Chem.*, **280**, 15307–15314.

73 Murakami, S., Nakashima, R., Yamashita, E., and Yamaguchi, A. (2002) Crystal structure of bacterial multidrug efflux transporter AcrB. *Nature*, **419**, 587–593.

74 Murakami, S., Nakashima, R., Yamashita, E., Matsumoto, T., and Yamaguchi, A. (2006) Crystal structures of a multidrug transporter reveal a functionally rotating mechanism. *Nature*, **443**, 173–179.

75 Mikolosko, J., Bobyk, K., Zgurskaya, H. I., and Ghosh, P. (2006) Conformational flexibility in the multidrug efflux system protein AcrA. *Structure*, **14**, 577–587.

76 Higgins, M. K., Bokma, E., Koronakis, E., Hughes, C., and Koronakis, V. (2004) The structure of the periplasmic component of a drug efflux pump. *Proc. Natl. Acad. Sci. USA*, **101**, 9994–9999.

77 Akama, H., Matsuura, T., Kashiwagi, S., Yoneyama, H., Tsukihara, T., Nakagawa, A., and Nakae, T. (2004) Crystal structure of the membrane fusion protein, MexA, of the multidrug transporter in *Pseudomonas aeruginosa*. *J. Biol. Chem.*, **279**, 25939–25942.

78 Vaccaro, L., and Sansom, M. S. P. (2007) Gating at both ends and breathing in the middle: conformational dynamics of TolC. *Biophys. J.* (to be submitted).

79 Yildiz, O., Vinothkumar, K. R., Goswami, P., and Kühlbrandt, W. (2006) Structure of the monomeric outer-membrane porin OmpG in the open and closed conformation. *EMBO J.*, **25**, 3702–3713.

80 Saier, M. H., Jr., Beatty, J. T., Goffeau, A., Harley, K. T., Heijne, W. H., Huang, S. C., Jack, D. L., Jahn, P. S., Lew, K., Liu, J., Pao, S. S., Paulsen, I. T., Tseng, T. T., and Virk, P. S. (1999) The major facilitator superfamily. *J. Mol. Microbiol. Biotechnol.*, **1**, 257–279.

81 Abramson, J., Kaback, H. R., and Iwata, S. (2004) Structural comparison of lactose permease and the glycerol-3-phosphate antiporter: members of the major facilitator superfamily. *Curr. Opin. Struct. Biol.*, **14**, 413–419.

82 Guan, L., and Kaback, H. R. (2006) Lessons from lactose permease. *Annu. Rev. Biophys. Biomol. Struct.*, **35**, 67–91.

83 Holland, I. B., Cole, S. P. C., Kuchler, K., and Higgins, C. F. (2003), *ABC proteins: From Bacteria to Man*, Academic Press, London, UK.

84 Locher, K. P. (2004) Structure and mechanism of ABC transporters. *Curr. Opin. Struct. Biol.*, **14**, 426–431.

85 Dawson, R. J., and Locher, K. P. (2006) Structure of a bacterial multidrug ABC transporter. *Nature*, **443**, 180–185.

86 Locher, K. P., Lee, A. T., and Rees, D. C. (2002) The *E. coli* BtuCD structure: a framework for ABC transporter architecture and mechanism. *Science*, **296**, 1091–1098.

87 Abramson, J., Smirnova, I., Kasho, V., Verner, G., Kaback, H. R., and Iwata, S. (2003) Structure and mechanism of the lactose permease of *Escherichia coli*. *Science*, **301**, 610–615.

88 Pinkett, H. W., Lee, A. T., Lum, P., Locher, K. P., and Rees, D. C. (2007) An inward-facing conformation of a putative metal-chelate-type ABC transporter. *Science*, **315**, 373–377.

89 Chang, G., Roth, C. B., Reyes, C. L., Pornillos, O., Chen, Y. J., and Chen, A. P. (2006) Retraction. *Science*, **314**, 1875.

90 Oloo, E. O., and Tieleman, D. P. (2004) Conformational transitions induced by the binding of MgATP to the vitamin B12 ATP-binding cassette (ABC) transporter BtuCD. *J. Biol. Chem.*, **279**, 45013–45019.

91 Borths, E. L., Locher, K. P., Lee, A. T., and Rees, D. C. (2002) The structure of *Escherichia coli* BtuF and binding to its cognate ATP binding cassette transporter. *Proc. Natl. Acad. Sci. USA*, **99**, 16642–16647.

92 Ivetac, A., Campbell, J. D., and Sansom, M. S. P. (2007) Dynamics and function in a bacterial ABC transporter: simulation

studies of the BtuCDF system and its components. *Biochemistry*, **46**, 2767–2778.
93 Yin, Y., Jensen, M., Tajkhorshid, E., and Schulten, K. (2006) Sugar binding and protein conformational changes in lactose permease. *Biophys. J.*, **91**, 3972–3985.
94 Huang, Y., Lemieux, M. J., Song, J., Auer, M., and Wang, D. N. (2003) Structure and mechanism of the glycerol-3-phosphate transporter from *Escherichia coli*. *Science*, **301**, 616–620.
95 Yin, Y., He, X., Szewczyk, P., Nguyen, T., and Chang, G. (2006) Structure of the multidrug transporter EmrD from *Escherichia coli*. *Science*, **312**, 741–744.
96 Heymann, J. A., Sarker, R., Hirai, T., Shi, D., Milne, J. L., Maloney, P. C., and Subramaniam, S. (2001) Projection structure and molecular architecture of OxlT, a bacterial membrane transporter. *EMBO J.*, **20**, 4408–4413.
97 Guan, L., Smirnova, I. N., Verner, G., Nagamori, S., and Kaback, H. R. (2006) Manipulating phospholipids for crystallization of a membrane transport protein. *Proc. Natl. Acad. Sci. USA*, **103**, 1723–1726.
98 Bennett, M., D'Rozario, R., Sansom, M. S. P., and Yeagle, P. L. (2006) Asymmetric stability among transmembrane helices of lactose permease. *Biochemistry*, **45**, 8088–8095.
99 Holyoake, J., and Sansom, M. S. P. (2007) MD simulations of an MFS protein: LacY. *Structure* (submitted).
100 Haider, S., Hall, B. A., and Sansom, M. S. P. (2006) Simulations of a protein translocation pore: SecY. *Biochemistry*, **45**, 13018–13024.
101 van den Berg, B., Clemons, W. M., Collinson, I., Modis, Y., Hartmann, E., Harrison, S. C., and Rapoport, T. A. (2004) X-ray structure of a protein-conducting channel. *Nature*, **427**, 36–44.
102 Isralewitz, B., Baudry, J., Gullingsrud, J., Kosztin, D., and Schulten, K. (2001) Steered molecular dynamics investigations of protein function. *J. Mol. Graphics Model.*, **19**, 13–25.
103 Gumbart, J., and Schulten, K. (2006) Molecular dynamics studies of the archael translocon. *Biophys. J.*, **90**, 2356–2367.
104 Capener, C. E., Shrivastava, I. H., Ranatunga, K. M., Forrest, L. R., Smith, G. R., and Sansom, M. S. P. (2000) Homology modelling and molecular dynamics simulation studies of an inward rectifier potassium channel. *Biophys. J.*, **78**, 2929–2942.
105 Haider, S., Khalid, S., Tucker, S., Ashcroft, F. M., and Sansom, M. S. P. (2007) Molecular dynamics simulations of inwardly rectifying (Kir) potassium channels: a comparative study. *Biochemistry*, **46**, 3643–3652.
106 Salas-Burgos, A., Iserovich, P., Zuniga, F., Vera, J. C., and Fischbarg, J. (2004) Predicting the three-dimensional structure of the human facilitative glucose transporter Glut1 by a novel evolutionary homology strategy. *Biophys. J.*, **87**, 2990–2999.
107 Vardy, E., Arkin, I. T., Gottschalk, K. E., Kaback, H. R., and Schuldiner, S. (2004) Structural conservation in the major facilitator superfamily as revealed by comparative modeling. *Protein Sci.*, **13**, 1832–1840.
108 Holyoake, J., Caulfeild, V., Baldwin, S. A., and Sansom, M. S. P. (2006) Modelling, docking and simulation of the major facilitator superfamily. *Biophys. J.*, **91**, L84–L86.
109 Faraldo-Gómez, J. D., Forrest, L. R., Baaden, M., Bond, P. J., Domene, C., Patargias, G., Cuthbertson, J., and Sansom, M. S. P. (2004) Conformational sampling and dynamics of membrane proteins from 10-nanosecond computer simulations. *Proteins: Struct. Funct. Bioinf.*, **57**, 783–791.
110 Shrivastava, I. H., and Bahar, I. (2006) Common mechanism of pore opening shared by five different potassium channels. *Biophys. J.*, **90**, 3929–3940.
111 Smart, O. S., Neduvelil, J. G., Wang, X., Wallace, B. A., and Sansom, M. S. P. (1996) Hole: A program for the analysis of the pore dimensions of ion channel structural models. *J. Mol. Graph.*, **14**, 354–360.

8
Understanding Structure and Function of Membrane Proteins Using Free Energy Calculations

Christophe Chipot and Klaus Schulten

8.1
Introduction

Computational investigations of membrane proteins have greatly benefited from the spectacular increase in computational power witnessed in recent years. In particular, distributing the work load on arrays of processors of massively parallel architectures, molecular dynamics (MD) simulations [1, 2] have played an important role in this research area by handling large assemblies of atoms formed by the protein and its lipid environment – *viz.* on the order of 10^5 to 10^6 particles [3–5]. They have contributed to the understanding of the molecular mechanisms whereby these proteins function, whenever their three-dimensional structure was available. In the absence of a well-resolved structure, MD simulations have also helped interpreting inferences accrued from experimental sources, such as structure–activity relationships, or site-directed mutagenesis.

Although MD simulations can offer a detailed, atomic picture of membrane proteins, they span time scales over which relevant biophysical phenomena cannot be easily captured. This can be readily understood by considering the infinitesimal time step utilized to integrate numerically the equations of motion – *viz.* on the order of 10^{-15} s, whereas most significant biological processes in membrane proteins occur over the 10^{-6} to 10^{-3} s time scale. Moreover, equilibrium MD simulations are plagued by Boltzmann sampling, which favors low-energy configurations, thereby precluding the exploration of regions of phase space separated by appreciable free energy barriers. Noteworthily, whereas the parallelization of MD codes has opened new horizons for exploring complex, sizeable biological systems, it has only moderately increased the time scales over which these systems can be investigated. This explains why coarse-grained approaches [6–8] have recently become fashionable in the field of theoretical and computational biophysics, removing the fine atomic detail to retain only the quintessential structural features of the molecular assemblies.

In spite of the fact that MD simulations *per se* are not capable of capturing rare events characterized by significant free energy barriers, they can be used,

Biophysical Analysis of Membrane Proteins. Investigating Structure and Function. Edited by Eva Pebay-Peyroula

ISBN: 978-3-527-31677-9

nonetheless, to generate ensembles of configurations from which free energy differences can be determined. In a nutshell, free energy calculations can be classified into three categories: (i) the free energy is estimated from probability distributions; (ii) free energy is computed directly; and (iii) the free energy derivative is evaluated over some reaction coordinate and further integrated (strictly speaking, reference should be made to the more general vocabulary *order parameter*, for which a *reaction coordinate* constitutes a special case). It is remarkable that most of these approaches are rooted in a few basic ideas, which have been known for several years and were contributed by such pioneers in the field as John Kirkwood [9, 10], Robert Zwanzig [11], Benjamin Widom [12], John Valleau [13], and Charles Bennett [14]. For the purpose of this chapter, the above distinction will be narrowed down to two categories, free energy methods aimed at: (i) modeling point mutations; and (ii) determination of the free energy change along a well-delineated reaction coordinate.

In the first section of this chapter, the methodology and the theoretical foundations of free energy calculations are described from a practical perspective. Next, selected applications of these simulations to membrane proteins are presented, distinguishing between *in-silico* site-directed mutagenesis experiments, assisted transport processes across the biological membrane, and recognition and association of transmembrane segments in lipid environments. The chapter closes with a retrospective view and an outlook on the evolution of free energy calculations, and how the latter can help dissecting molecular mechanisms in membrane proteins through a link between function and energetics.

8.2 Theoretical Underpinnings of Free Energy Calculations

In this section, the physical principles underlying free energy calculations are outlined. The important results are provided without detailed demonstration. The reader is referred to the specialized literature for further information [15].

8.2.1 Alchemical Transformations

Just like the proverbial alchemist, hoping to transmute lead into gold, the modeler may wish to modify biological systems *in silico* by means of computational site-directed mutagenesis experiments. In order to perform such so-called "alchemical transformations", definition of a reference state, **0**, and a target state, **1**, is crucial. These alternate states are described respectively by Hamiltonian $\mathcal{H}_0(\mathbf{x}, \mathbf{p}_x)$ and Hamiltonian $\mathcal{H}_1(\mathbf{x}, \mathbf{p}_x)$, such that $\mathcal{H}_1(\mathbf{x}, \mathbf{p}_x) = \mathcal{H}_0(\mathbf{x}, \mathbf{p}_x) + \Delta\mathcal{H}(\mathbf{x}, \mathbf{p}_x)$, where $\{x\}$ denotes the atomic coordinates of the system, and $\{p_x\}$ its conjugated momenta. $\Delta\mathcal{H}(\mathbf{x}, \mathbf{p}_x)$ can be viewed as a perturbation between the initial and the final states of the transformation. The free energy difference between states **0** and **1** can be expressed in terms of a ratio of the corresponding partition functions, which after appropriate substitution, reads:

$$\Delta A = -\frac{1}{\beta} \ln \langle \exp[-\beta \Delta \mathcal{H}(\mathrm{x}, \mathbf{p}_x)] \rangle_0 \qquad (1)$$

Here, $\langle \cdots \rangle_0$ stands for an ensemble average over configurations representative of the reference system. $\beta = 1/k_B T$, where k_B is the Boltzmann constant and T is the temperature of the system. Equation (1) is the fundamental free energy perturbation (FEP) formula [11], which embodies a remarkable result: Free energy differences may be determined by sampling *only* equilibrium configurations of the initial state, **0**. In principle, this equation is "exact", in the sense that it is expected to converge regardless of $\Delta\mathcal{H}(\mathbf{x}, \mathbf{p}_x)$, which would obviously be true in the limit of infinite sampling. In practice, however, validity of the perturbation formula [Eq. (1)] only holds for sufficiently small changes between **0** and **1**.

8.2.1.1 What is Usually Implied by Small Changes?

The answer is often misconstrued, as the free energies characteristic of the reference and the target states need not necessarily be close. It is, however, pivotal that the corresponding configurational ensembles overlap appropriately to guarantee the desired accuracy [16, 17] (see Fig. 8.1). By and large, single-step transformations between utterly different states only rarely fulfill this requirement. To cir-

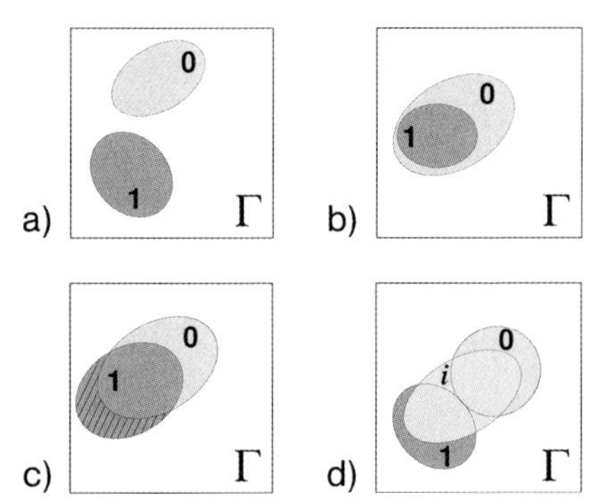

Fig. 8.1 Condition of overlapping ensembles in the perturbation theory. The thermodynamic ensembles representative of the reference, **0**, and the target, **1**, states do not overlap. Equation (1) will not converge towards the expected free energy difference **(a)**. In idealistic scenarios, low-energy configurations of the target state are also configurations of the reference state. The ensemble of configurations characteristic of state **1** is a subset of the ensemble representative of state **0**. The perturbation formula will converge towards the correct free energy difference **(b)**. Qualitative, possibly quantitative information about the accuracy of the calculation may be inferred from an incomplete overlap of the thermodynamic ensembles, highlighted here in the hatched region **(c)**. Disparate reference and target systems may be connected by means of an intermediate state that satisfies the criterion of mutual overlap between contiguous ensembles **(d)**.

cumvent this difficulty, the reaction pathway connecting state **0** and state **1** may be broken down into a number of intermediate, non-physical states, satisfying the requirement of mutual overlap of the ensembles between any two contiguous states [18]. To achieve this goal, the Hamiltonian is expressed as a function of a reaction coordinate, or "coupling parameter", λ, that controls the transformation [9]. Conventionally, λ varies between 0 and 1 when the system goes from the reference state to the target state.

The interval separating the intermediate states, k, of the transformation between the reference and the target systems – which corresponds to selected fixed values of the coupling parameter, λ – is often referred to as the "window". The total free energy change over the complete reaction pathway now reads:

$$\Delta A = -\frac{1}{\beta}\sum_k \ln\langle \exp[-\beta \Delta \mathcal{H}_{k,k+1}(\mathrm{x}, \mathbf{p}_x; \lambda)]\rangle_k \tag{2}$$

where $\mathcal{H}_{k,k+1}(\mathbf{x}, \mathbf{p}_x; \lambda)$ stands for the difference in the Hamiltonian between adjacent intermediates k and $k + 1$; and $\langle\cdots\rangle$ denotes the equilibrium average in intermediate k. Assessing an ideal number of intermediates, $N - 2$, between state **0** and state **1** evidently depends upon the nature of the system that undergoes the transformation. The condition of overlapping ensembles should be kept in mind when setting N, remembering that the choice of $\delta\lambda = \lambda_{k+1} - \lambda_k$ ought to correspond to a finite perturbation of the system. A natural choice may consist in defining a number of windows that guarantees reasonably similar free energy changes between contiguous substates. It follows from this criterion that the width of the consecutive windows connecting state **0** to state **1** may be uneven.

8.2.1.2 How is the Coupling Parameter Defined?

In practice, λ may correspond to a variety of order parameters – possibly true reaction coordinates. In the special example of "alchemical transformations", it can be used to scale the interaction of the perturbed system with its environment. Performing *in silico* point mutations requires the definition of molecular topologies that describe the reference and the target states of the transformation. Alternative schemes can be employed towards this end. In the single-topology paradigm, a topology common to states **0** and **1** is sought (see Fig. 8.2), and the non-bonded terms of the perturbed moieties – that is, the charges and the van der Waals parameters – are expressed, in general, as a linear function of λ. For instance, the charge, $q(\lambda)$, is written as a linear combination of the reference and the target states, that is $\lambda q_1 + (1 - \lambda)q_0$. In the dual-topology paradigm, however, the topologies representative of the reference and the target states are defined concomitantly, but never interact mutually throughout the simulation. The interaction energy of these topologies with their surroundings, $\mathcal{H}(\mathbf{x}, \mathbf{p}x; \lambda)$, is usually scaled linearly with λ, so that $\mathcal{H}(\mathbf{x}, \mathbf{p}x; \lambda) = \lambda\mathcal{H}_1(\mathbf{x}, \mathbf{p}x) + (1 - \lambda)\mathcal{H}_0(\mathbf{x}, \mathbf{p}x)$. This paradigm is prone to singularities when λ approaches 0 or 1, which can be partially circumvented by introducing soft-core potentials [19].

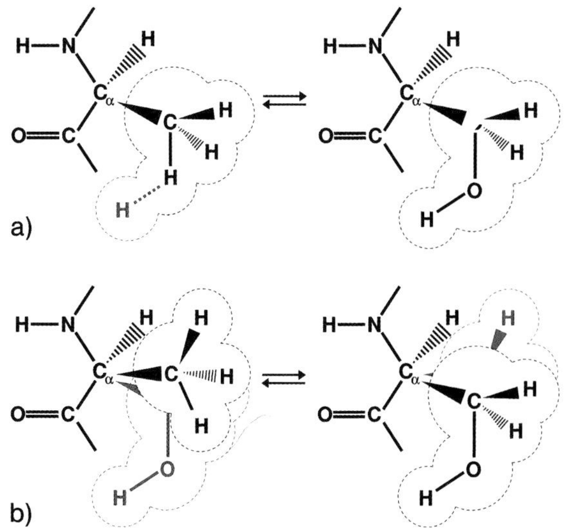

Fig. 8.2 Comparison of single-and dual-topology paradigms in the case of an alanine to serine "alchemical transformation". In the single-topology approach, a topology common to alanine and serine is sought. The hydrogen of the hydroxyl moiety is defined as a ghost atom, which will be grown subsequently as λ varies from 0 to 1. At the same time, the non-bonded parameters of the aliphatic carbon of the alanine side chain are transformed into those of an oxygen atom, while the C–H bond length is extended to that characteristic of a C–O chemical bond **(a)**. In the dual-topology approach, the topologies of both alanine and serine coexist, albeit never interact with each other. The interaction energy of these topologies with their environment is scaled as λ goes from 0 to 1 **(b)**.

Although the perturbation formula [Eq. (1)] had been established over 50 years ago [11], pioneering "alchemical transformations" in chemically relevant molecular systems had to wait for the availability of sufficient computational power. In 1984, employing a rudimentary, simplistic model, Tembe and McCammon demonstrated that FEP calculations could be applied successfully to model ligand–receptor assemblies [20]. To a large extent, this seminal work paved the road towards the determination of protein–ligand binding affinities based on thermodynamic perturbation theory. The first genuine *in-silico* point mutation on a concrete chemical system was the transformation of methanol into ethane by Jorgensen and Ravimohan, in 1985 [21]. Though impressive at the time, the agreement between the computationally and the experimentally determined relative solvation free energies aroused only moderate interest. Kollman, however, rapidly realized the remarkable potential of these calculations for predicting, essentially from first principles, the propensity of chemical species to interact and associate. In 1987, Kollman and coworkers filed a series of reports that opened new vistas for the modeling of site-directed mutagenesis experiments [22, 23]. Using the FEP machinery, these authors computed the free energy changes associated with the point mutations of the side chains of naturally occurring amino acids, as well as relative binding affinities in protein–inhibitor complexes.

8.2.1.3 Thermodynamic Integration

Closely related to the perturbation formula, thermodynamic integration (TI) restates the free energy difference between the reference and the target states as an integral, in which the integrand is an ensemble average [9, 24]:

$$\Delta A = \int_0^1 \left\langle \frac{\partial \mathcal{H}(\mathrm{x}, \mathbf{p}_x; \lambda)}{\partial \lambda} \right\rangle_\lambda \mathrm{d}\lambda \tag{3}$$

Here, $\langle \cdots \rangle_\lambda$ denotes an ensemble average over configurations representative of the transformation at λ. The criterion of an appropriate overlap of the thermodynamic ensembles, which is stringent in the FEP approach, is of somewhat lesser importance in the TI method. Convergence of the simulation is governed principally by the smoothness of free energy as a function of the coupling parameter. "Alchemical transformations" can be performed equivalently using either FEP or TI. Conceptual differences in Eqs. (1) or (2), and (3) suggest that these calculations will possess distinct convergence properties.

Interestingly enough, assuming that the variation of the kinetic energy between the reference, **0**, and the target, **1**, states can be neglected, the derivative of ΔA with respect to some reaction coordinate, ξ, is equal to $\langle \nabla_\xi \mathcal{V}(\mathrm{x}) \rangle_\xi$, that is, in other words, $-\langle F_\xi \rangle_\xi$ – *i.e.*, the average of the force exerted along ξ. This rationalizes the vocabulary of potential of mean force (PMF) used to characterize the free energy measured along a given reaction coordinate. Strictly speaking, the PMF refers to the reversible work involved in the process of moving two tagged particles from an infinite to a relative separation [25]. Shortly, it will be seen that this concept has been extended to a variety of reaction coordinates other than a simplistic interatomic distance.

8.2.2
Free Energy Changes Along a Reaction Coordinate

In sharp contrast to "alchemical transformations", in which one chemical species is transmuted into an alternate one by altering the interaction energy, methods targeted at the determination of free energy profiles $\Delta A(\xi)$ rely on the knowledge of the probability $\mathcal{P}(\xi)$ to find the system along a reaction coordinate ξ:

$$\Delta A(\xi) = -\frac{1}{\beta} \ln \mathcal{P}(\xi) + \Delta A_0 \tag{4}$$

where ΔA_0 is a constant. As has been commented on in the Introduction of this chapter, MD is plagued by Boltzmann sampling, which favors sampling of the low-energy regions of phase space. It follows that over realistic simulations, exploration of rugged free energy surfaces will be largely incomplete, emphasizing free energy minima at the expense of barriers. As a result, the free energy change of Eq. (4) is expected to be inaccurate, because $\mathcal{P}(\xi)$ is truncated. Returning to the idea of reference and target systems, the efficiency of such free energy calculations can be greatly improved by sampling the reference ensemble sufficiently broadly to ascertain that meaningful information about the low-energy configurations of the target ensembles is acquired.

In 1977, Torrie and Valleau devised such an approach by introducing in their simulations non-Boltzmann weights, $\mathcal{V}_\xi(x)$, that are removed subsequently to yield an unbiased probability distribution [26]. If $\mathcal{P}(\xi)$ is the probability distribution resulting from these simulations, the actual free energy profile is not given by Eq. (4), but rather by:

$$\Delta A(\xi) = -\frac{1}{\beta}\ln\mathcal{P}(\xi) + \mathcal{V}_\xi(x) + \Delta A_0 \tag{5}$$

This approach, which became rapidly popular, is known as umbrella sampling (US) and has been applied successfully to a host of problems featuring a variety of reaction coordinates. The success of this method depends to a large extent on the initial choice of the external biases, $\mathcal{V}_\xi(x)$, that have to foresee the locations of maxima and minima in $\Delta A(\xi)$. Ideally, $\mathcal{V}_\xi(x)$ should be equal to $-\Delta A(\xi)$, thereby ensuring that the system would not "feel" the ruggedness of the free energy surface as it diffuses along ξ, and yield uniform probability distributions. This evidently would imply an *a priori* knowledge of the free energy behavior in the direction of the reaction coordinate, which is seldom the case. Whereas guessing the shape of $\mathcal{V}_\xi(x)$ is usually possible for related, well-documented systems, it becomes an intricate task in the event of qualitatively new problems.

8.2.2.1 **Umbrella Sampling or Stratification?**

To increase further the efficiency of US calculations, it is tempting to break down the reaction path into mutually overlapping "windows" or ranges of values of ξ – the vocabulary "window" bears evidently different meanings in US and FEP/TI methodologies, albeit the underlying idea of the approach is to express the total free energy difference as a sum of free energy differences between intermediate states. Overlap of adjacent windows is required for pasting the corresponding free energy profiles into one complete PMF over the entire reaction path. Such a chain of configurational energies can be obtained employing the so-called multistage sampling proposed in 1972 by Valleau and Card [27], which connects the reference and the target states when the low-energy regions of the latter overlap poorly. Sampling is generally confined in the window of interest by means of additional external biases consisting of substantial energy penalties. Individual free energy profiles are subsequently pasted together using, for instance, the self-consistent weighted histogram analysis method (WHAM) [28]. The temptation may be great to stratify further the free energy calculation by defining very narrow windows and assume Boltzmann sampling therein, thus obviating the need of guessing what $\mathcal{V}_\xi(x)$ should be. Whereas this scheme is expected to yield a reasonable local convergence, it is anticipated to suffer from large systematic errors arising from inaccurate matching of the many individual free energy profiles.

Returning to Eqs. (1) and (3), one might wonder legitimately why FEP and TI cannot be used for computing free energy profiles, the coupling parameter, λ, embracing a variety of possible reaction coordinates. In reality, a number of PMFs have been determined based on Eqs. (1) and (3), until the approach was eventually proven flawed [29, 30]. For instance, in the simple case of two tagged particles at infinite separation and moving toward each other, the flaw is rooted in the erroneous assumption that λ and the Cartesian coordinates, {x}, of the system in $\mathcal{H}(\mathbf{x}, \mathbf{p}x; \lambda)$ are independent. Such is obviously not the case, as {x} varies as λ is changed from an infinite to a relative separation.

8.2.2.2 Adaptive Biasing Force

To guarantee the independence of the Cartesian coordinates and the reaction coordinate, a transformation of the metric is required in the calculation of the first derivative of Eq. (4). After back-transformation into Cartesian coordinates, the derivative of the free energy with respect to ξ can be expressed as a sum of configurational averages at constant ξ [31]

$$\frac{\mathrm{d}A(\xi)}{\mathrm{d}\xi} = \left\langle \frac{\partial \mathcal{V}(\mathrm{x})}{\partial \xi} \right\rangle_{\xi} - \frac{1}{\beta} \left\langle \frac{\partial \ln|J|}{\partial \xi} \right\rangle_{\xi} = -\langle F_{\xi} \rangle_{\xi} \, . \tag{6}$$

Here, $|J|$ is the Jacobian of the transformation. In practice, the instantaneous component of the force along ξ, F_{ξ}, is accrued in bins of finite size, providing an estimate of $\mathrm{d}A(\xi)/\mathrm{d}\xi$. After a predefined number of observables of the instantaneous force are accumulated, the adaptive biasing force (ABF) [32] is applied to the system – that is, $-\langle F_{\xi}\rangle_{\xi}\nabla_{x}\xi$ – thereby yielding a Hamiltonian in which no average force is exerted along ξ. As a result, all values of the reaction coordinate, either contrained or unconstrained [33], are sampled with an equal probability which, in turn, improves dramatically the accuracy of the free energy estimates. Evolution of the system along ξ is fully reversible and governed mainly by its self-diffusion properties. It is apparent from the present description that this method is significantly more effective than US and its subsequent incarnations, because no *a priori* knowledge of the free energy hypersurface is required to define the necessary weights that guarantee uniform sampling along ξ. Often misconstrued, it should be underlined that, whereas ABF without a doubt improves sampling significantly along the reaction coordinate, the efficiency is plagued (as in any other free energy method), by any orthogonal, slowly relaxing degrees of freedom.

8.2.2.3 Non-Equilibrium Simulations for Equilibrium Free Energies

An alternative route to explore processes that span appreciable time scales, usually not amenable to MD simulations, consists of driving the system out of equilibrium. This can be achieved using the so-called steered MD approach, in which an external force is exerted onto the system to force it to drift along the desired direction of Cartesian space [34, 35]. Compared to US and ABF free energy calculations, SMD simulations apply either a constant or a time-varying force that yield significant deviations from equilibrium conditions. In constant-force (*cf*) SMD simulations, a constant force on the order of 10^{-12} to 10^{-9} N, is exerted onto a subset of atoms, in addition to the conservative forces of the potential energy function. This is distinct from constant-velocity (*cv*) SMD simulations, which mimic the action of a mobile cantilever acting on a substrate in atomic force microscopy (AFM) experiments. A subset of atoms are tethered harmonically to a given point in space that is moved at constant velocity along a well-delineated direction, which can be viewed as a reaction coordinate. In AFM experiments, the spring constant of the cantilever is equal to *ca.* 10^{-3} N m^{-1}, which, on the atomic scale, translates to rather large thermal fluctuations in the position of the attached substrate. In contrast, SMD utilize spring constants about two orders of magnitude stiffer, thereby providing a sharper spatial resolution and more detailed information about the interactions at play. Unfortunately, the time scales explored by *cv*-SMD simulations are substantially shorter than those characteristic of AFM experiments, *viz.* typically by a factor of 10^3 to 10^6. In order to cover large stretches of the reaction path, high pulling velocities are required of *ca.* $10^3\,\mathrm{m\,s^{-1}}$. This, in turn, generates a sizeable amount of irreversible work, w, often difficult to relate to equilibrium properties. It will be seen that meaningful information about the thermodynamics of the process may, nevertheless, be inferred from such non-equilibrium processes.

The second law of thermodynamics states that the average work, $\langle w \rangle$, involved in the irreversible transformation over a reaction coordinate, $\xi(t)$ – connecting a reference state, **0**, at $\xi(0)$, and a target state, **1**, at $\xi(\tau)$ – is greater than, or equal to the free energy difference for that particular transformation (*i.e.*, $\Delta A \leq \langle w \rangle$). The difference between the average work and the free energy accounts for dissipation effects in the course of the irreversible process. Jarzynski demonstrated, however, that even for non-equilibrium paths, the latter inequality can be turned into an equality [36]:

$$\langle \exp[-\beta w(\tau)] \rangle = \exp(-\beta \Delta A) \tag{7}$$

The left-hand side of Eq. (7) consists of a combination of an ensemble average over a set of initial conditions and a path average over MD trajectory realizations. The various initial conditions are selected according to the equilibrium Boltzmann probability in the reference state. The path average samples all possible realizations of dynamic paths obtained using an explicitly time-dependent Hamiltonian.

In sharp contrast to stochastic dynamics, deterministic dynamics implies that only one unique trajectory is generated for a given set of initial conditions, thereby obviating the need for path averaging. A fundamental difficulty concealed in Eq. (7) arises from the paucity of *meaningful* trajectories that correspond to small values of w, albeit contributing significantly to the left-hand term. Not too surprisingly, accurate estimates of ΔA generally require appreciable ensembles of appropriately sampled trajectories, and are, therefore, plagued by statistical errors. Yet, it can be shown that within the relevant pulling regime, employing stiff springs, w follows a Gaussian distribution and a truncation at the second order of the Jarzynski identity may be sufficient to yield accurate free energy differences [37, 38].

8.3
Point Mutations in Membrane Proteins

8.3.1
Why Have Free Energy Calculations Been Applied only Sparingly to Membrane Proteins?

A partial answer to this question can be found in the overwhelming length of the trajectory necessary to assume safely that the calculation has converged. Assessing the convergence properties and the error associated to a free energy calculation constitute a challenging task on account of the many possible sources of errors that are likely to affect the results differently. The systematic error arising from the choice of the force field parameters will undoubtedly affect the results of the simulation, but this contribution can be largely concealed by the statistical error due to insufficient sampling. Moreover, under the hypothetical assumption of an optimally designed potential energy function, quasi non-ergodicity scenarios that stem from slowly relaxing degrees of freedom constitute a common pitfall towards fully converged simulations [39]. Not too surprisingly, such scenarios are very common in membrane proteins, where the slowest motions can span the 10^{-6} to 10^{-3} s time scale. There are evidently other reasons accounting for the scarcity of free energy calculations in membrane proteins. Compared to the plethora of globular proteins, the three-dimensional structures of which have been made available, only about a hundred unique membrane proteins can be found in databases with a reasonable resolution. Furthermore, whereas MD simulations of globular proteins involve only aqueous environments, which may be modeled by means of continuum electrostatic approaches, computational investigation of membrane proteins requires the inclusion of fully hydrated lipid bilayers that are not described satisfactorily by implicit solvation models. The resulting large molecular assemblies, on the order of 10^5 to 10^6 atoms, cannot be examined by MD simulations beyond the 10^{-7} s time scale, which can be a serious limitation to achieve converged free energies. Free energy calculations have, nonetheless, been endeavored in integral membrane proteins to answer specific questions about their structure and function. In this section, a number of relevant applications of FEP and TI methodologies to "alchemical transformations" in these complex biological systems are reviewed.

One of the very first spectacular applications of thermodynamic perturbation theory to membrane proteins is the calculation of relative binding affinities of a series of monovalent cations towards gramicidin by Roux et al. [40]. The calculations were carried out in 1995, when MD simulations of lipid bilayers were still limited to subnanosecond time scales and the membrane environment had to be mimicked by an ensemble of neutral Lennard–Jones particles. This computational investigation suggested that doubly occupied states are more favorable for large cations, *viz.* Cs^+, than for small ones, *viz.* Li^+.

The translocation of water molecules from the bulk aqueous environment to specific binding sites of the bacteriorhodopsin transmembrane channel was investigated during the following year by the same group [41]. FEP and TI methods were employed to estimate the hydration free energy of well-delineated moieties, from which probabilities of occupancy of the bacteriorhodopsin channel were inferred. A possible continuous, 12 Å-long column of water molecules was investigated between the proton donor, the side chain of Asp^{96}, and its acceptor, the retinal Schiff base.

In the same year, Roux revisited the paradigmatic transmembrane channel of gramicidin, using a model formed by poly-L,D-alanine β-helices [42]. The affinity of potassium and chloride ions towards the model channel was probed, employing FEP calculations. These simulations revealed that the interior of the transmembrane protein interacted favorably with the cation, but unfavorably so with the anion. This result may be rationalized by the unbalanced stabilization effect of amino groups on Cl^- ions and of carbonyl groups on K^+ ions, the latter significantly more effective.

8.3.2 Gaining New Insights into Potassium Channels

Resolution of the pore domain of the KcsA potassium channel by MacKinnon and coworkers [43] imparted a new momentum in applications of MD simulations, in general, and free energy calculations, in particular, to large membrane proteins. In 2000, Bernèche and Roux delved into the ionization state of a key residue close to the selectivity filter of KscA, *viz.* Glu^{71}, the atomic coordinates of which were undetermined [44]. These authors proposed that this amino acid is protonated and forms a strong hydrogen bond with the nearby Asp^{80} residue, which does not induce any noteworthy deviation with respect to the original crystal structure. Two years later, they revisited this interaction and showed that the selectivity filter may be disrupted considerably when the Glu^{71} and Asp^{80} amino acids are ionized in each monomer of the homotetrameric structure of KcsA [45]. Employing a combination of finite-difference Poisson–Boltzmann electrostatics and FEP calculations in an explicit membrane environment, Bernèche and Roux evaluated the pK_a of these two residues, and showed that, in the case of Glu^{71}, it is shifted by about 10 pK_a units, thus supporting the hypothesis of a protonated state under normal conditions.

In pursuing their efforts to understand how ion selectivity is controlled in potassium channels, Roux and coworkers applied FEP calculations to KcsA, disrupting artificially the interaction of the cation K^+ with the different carbonyl moieties lining the narrow pore region of the channel [46]. By comparing simulations performed in a fully flexible channel and in a frozen one, nes light was shed on the dynamic nature of the pore region, which, combined with the intrinsic electrostatic properties of its constituent carbonyl groups, ensures ionic selectivity – in particular K^+ versus Na^+ ions. Interestingly enough, it was noted that removal of carbonyl–carbonyl interactions abolishes selectivity at the center of the narrow pore region, with which sodium ions may now interact favorably.

8.3.3
Tackling the Assisted Transport of Ammonium Using FEP

In order to investigate the binding affinity and selectivity of the bacterial ammonium transporter of *Escherichia coli* (AmtB), Luzhkov et al. made use of perturbation theory in an explicit membrane environment [47]. The group measured the binding free energy of ammonium to the carrier and concluded, in the light of the apparent pK_a that association favors ammonium rather than neutral ammonia. Point mutation of the highly conserved Asp160 amino acid into asparagine led to a dramatic reduction in the pK_a of the bound NH_4^+ and a loss of binding, which explains why the activity of AmtB is quenched under these circumstances [48]. In addition, investigation of putative intermediate states involved in the transfer of water, ammonia and ammonium suggests that the latter remains protonated until it reaches the interior of the channel.

8.3.4
How Relevant are Free Energy Calculations in Models of Membrane Proteins?

So far, we have seen that in the realm of membrane proteins, thermodynamic perturbation theory has been applied essentially to carriers and channels. Recently, however, Hénin et al. reported *in-silico* point mutations in a model of a G-protein-coupled receptor – the human receptor of cholecystokinin 1 (CCK1R) [49]. The challenge in these FEP calculations was twofold. From a technical perspective, all-atom MD simulations of membrane assemblies are computationally demanding on account of their size. This, in turn, constitutes an intrinsic limitation for accessing long, biologically relevant time scales. Whereas this aspect is not expected to represent an insurmountable obstacle in "alchemical transformations" involving rapidly relaxing degrees of freedom, it, does however preclude the capture of slow rearrangements in the membrane protein upon mutation. From a more fundamental point of view, endeavoring free energy calculations in a model of a receptor raises questions about the true relevance of costly simulations in the absence of well-resolved three-dimensional structures. In the particular case of CCK1R, site-directed mutagenesis experiments were performed to guide the construction of the *in-vacuo* model, fine-tuning the orientation of the transmembrane α-helices and refining the position of the agonist ligand – the nonapeptide CCK9 – in its designated binding pocket. Two criteria should be fulfilled prior to initiating the FEP calculations. A preliminary, sufficiently long MD simulation of the GPCR immersed in a fully hydrated lipid bilayer should be carried out to ascertain that the array of receptor–ligand interactions brought to light in experiment is preserved over time. In addition, it is crucial that the proposed "alchemical transformations" be applied to residues involved in direct, highly specific interactions. In this respect, point mutations in the agonist ligand CCK9 of its third, sulfated tyrosyl residue and its eight, aspartyl residue into alanine are pertinent, because these amino acids interact strongly with the receptor, in particular through the formation of salt bridges with arginine residues (see Fig. 8.3). Though the remark-

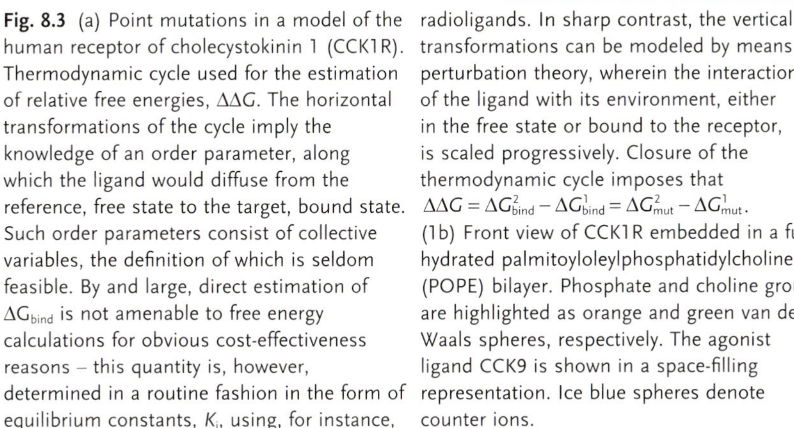

Fig. 8.3 (a) Point mutations in a model of the human receptor of cholecystokinin 1 (CCK1R). Thermodynamic cycle used for the estimation of relative free energies, $\Delta\Delta G$. The horizontal transformations of the cycle imply the knowledge of an order parameter, along which the ligand would diffuse from the reference, free state to the target, bound state. Such order parameters consist of collective variables, the definition of which is seldom feasible. By and large, direct estimation of ΔG_{bind} is not amenable to free energy calculations for obvious cost-effectiveness reasons – this quantity is, however, determined in a routine fashion in the form of equilibrium constants, K_i, using, for instance, radioligands. In sharp contrast, the vertical transformations can be modeled by means of perturbation theory, wherein the interaction of the ligand with its environment, either in the free state or bound to the receptor, is scaled progressively. Closure of the thermodynamic cycle imposes that $\Delta\Delta G = \Delta G^2_{bind} - \Delta G^1_{bind} = \Delta G^2_{mut} - \Delta G^1_{mut}$. (1b) Front view of CCK1R embedded in a fully hydrated palmitoyloleylphosphatidylcholine (POPE) bilayer. Phosphate and choline groups are highlighted as orange and green van der Waals spheres, respectively. The agonist ligand CCK9 is shown in a space-filling representation. Ice blue spheres denote counter ions.

able agreement between theory and experiment for the reproduction of the relative binding constants does not validate *per se* the present model, it suggests that carefully probed models can help provide a reliable structural basis for improving our understanding of the function of GPCRs and designing new, potent leads.

8.4 Assisted Transport Phenomena Across Membranes

8.4.1 Gramicidin: A Paradigm for Assisted Transport Across Membranes

One of the first free energy profiles determined to improve our understanding of how membrane proteins convey chemical species across the biological membrane was proposed as early as 1991 by Roux and Karplus [50, 51]. Turning to a simplistic representation of gramicidin, formed by a periodic poly-L,D-alanine, these authors established the PMF for sodium and potassium permeation. The free energy barrier of ionic conduction was found to be about fourfolds larger for Na^+ than for K^+. Breaking down the net free energy change into components, the role played

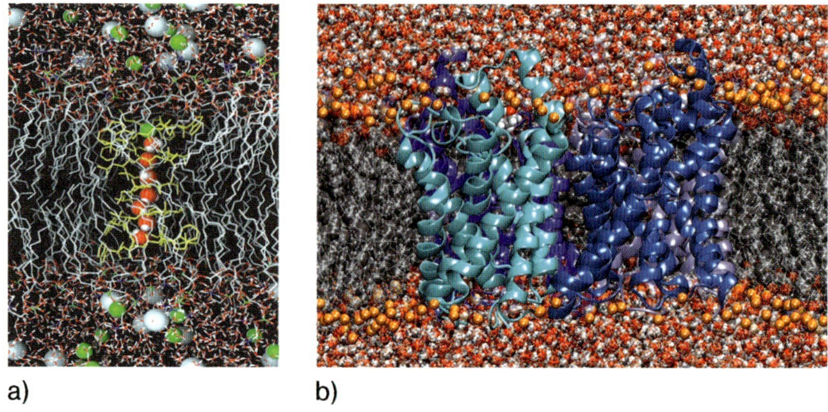

Fig. 8.4 Free energy calculations and assisted transport phenomena. (a) Permeation of potassium ions across gramicidin A embedded in a fully hydrated dimyristoylphosphatidylcholine (DMPC) bilayer [52]. K^+ ions are highlighted as green van der Waals spheres. Seven water molecules shown in a space-filling representation form the single-file water wire that spans the channel. (b) Assisted transport of glycerol in the *Escherichia coli* facilitator GlpF embedded in a fully hydrated palmitoylethanolaminephosphatidylcholine [53] (POPE) bilayer. The four monomeric channels forming the complete protein structure are displayed using different colors. In the cyan, left-hand-side channel, a glycerol molecule shown in a space-filling representation is found near the so-called periplasmic vestibule.

by hydration water molecules on the activation free energy was emphasized. Roux and Karplus further argued that the free energy behavior is primarily dictated by ancillary, weaker water–water, water–peptide and peptide–peptide interactions, rather than direct interactions involving the cations. Plasticity of the channel was hypothesized to be crucial for an immediate, fluid-like response to the presence of the cation.

In 2004, Roux and coworkers revisited the prototypical case of ionic conduction through gramicidin A by means of a 100-ns MD simulation [52] (Fig. 8.4). To gain insight into the mechanism for potassium ion permeation, a two-dimensional free energy surface was generated, as a function of both relative axial and radial coordinates. From the resulting free energy changes, fine-tuned to account for the absence of explicit polarization effects in the simulations, single-channel conductance was inferred and shown to be of unprecedented agreement with experiment. Deconvolution of the mean force acting along the order parameter reveals a noteworthy propensity of the water wire to stabilize K^+ ions and a particular role played by the orientation of the constituent water molecules on conductivity.

8.4.2
Free Energy Calculations and Potassium Channels

An understanding of how ion channel selectivity requires at its core the elucidation of how attractive ion–channel and repulsive ion–ion interactions balance out. By using US simulations, Bernèche and Roux [54] tackled in 2001 the mechanism of K^+ ion conduction in KcsA, for which the structure of the pore had been resolved two years previously [43]. The free energy calculations endeavored by Bernèche and Roux brought to light two prevailing states, wherein either two or three potassium ions occupy the selectivity filter. The highest free energy barrier towards ion permeation was estimated to be on the order of 2 to 3 kcal mol^{-1}, which suggests that conduction is limited by self-diffusion in the channel.

Using US simulations, Cohen and Schulten examined the mechanism of anionic conduction in the voltage-gated ClC channel [55]. In order to fulfill its regulatory function in vertebrates, this membrane protein selectively assists the transport of small anions, preventing permeation of other chemical species. Investigation of the conduction process in ClC was carried out by translocating a chlorine ion across the channel embedded in a realistic membrane environment. The suggested mechanism implies that a Cl^- ion bound to the center of the pore can be pushed out and replaced by a second ion entering the pore region. Accordingly, conduction through a two-ion process has a relatively low free energy barrier, in contrast to a single-ion process that exhibits a high barrier due to the considerable energy penalty imposed by extraction of a single anion from the pore. These simulations further shed light on the role played by Cl^- ions in blocking undesired chemical species at the narrowest region of the pore through disruption of the flow of water molecules across the channel.

In the effervescent research area of voltage-gated potassium channels, Treptow and Tarek provided recently new insights into the energetics of K^+ conduction through the activation gate and the role played by the permeation pathway, using equilibrium ABF free energy calculations [56]. These authors considered three distinct pathways, wherein the activation gate of the *Shaker* and the KV1.2 channels is in a closed, a partially opened and a fully opened conformation, and demonstrated that ion permeation is unfavorable if the constriction region of the channel is narrower than ca. 5 Å. The energy penalty is proposed to stem from an incomplete hydration of the potassium ions. Beyond this critical radial threshold, diffusion of fully hydrated cations is enabled, which is reflected in a flat free energy profile.

Also recently, Beckstein and Sansom employed US simulations to probe the hypothesis of a hydrophobic gating mechanism in the nicotinic acetylcholine receptor (nAChR) – a ligand-gated ion channel mediating synaptic neurotransmission [57]. PMF calculations indicated that the constriction region of the pore, a 3 Å-radius hydrophobic gate, is characterized by a free energy barrier of about 6 kcal mol^{-1} towards permeation of sodium ions. This result suggests that hydrophobic pores may act as functional barriers targeted at limiting ionic conduction.

8.4.3
Non-Equilibrium Simulations for Understanding Equilibrium Phenomena

As has been commented on above, the reconciliation of AFM experiments with SMD simulations is a cumbersome task, essentially because the latter are carried out much faster than the former. In 1999, Gullingsrud et al. investigated three distinct approaches for the post-treatment of SMD data and the reconstruction of PMFs, taking into account the non-equilibrium nature of SMD simulations [58]. The first method consists of a direct application of the WHAM equations mentioned previously. The second relies on the use of a van Kampen Ω-expansion in the form of a Gaussian drift, which proved to be an appropriate route for exploiting SMD information. The third scheme is a minimization of an Onsager–Machlup action and appeared to be particularly promising because of its inherent flexibility. Yet, it is fair to recognize that these approaches were rapidly superseded by the ground-shaking identity put forth by Chris Jarzynski [36]. Gerhard Hummer and Attila Szabo employed this relationship between free energy and irreversible work to reconstruct PMFs in the quasi – equilibrium regime of AFM experiments [59] – which obviates the need of a friction coefficient, as was the case, for instance, in the aforementioned Gaussian drift scheme. The theoretical framework for coupling the Jarzynski identity to SMD numerical experiments was further developed in 2003 by Park et al., and applied to the *in-vacuo* helix–coil transition of the prototypical deca-alanine peptide chain [37].

8.4.4
Deciphering Transport Mechanisms in Aquaporins.

In 2002, Jensen et al. published the first application of the Jarzynski identity to membrane proteins, dissecting the energetics of glycerol conduction through *E. coli* acquaglyceroporin GlpF [53], a 2.2-Å structure which had been resolved two years earlier by Stroud and coworkers [60]. By analyzing the time series derived from their SMD simulations, Jensen et al. reconstructed the PMF delineating glycerol conduction in the four channels of the homotetrameric membrane protein. Among the remarkable features brought to light in this study, two appreciable free energy barriers emerge in the region formed by the selectivity filter, prefaced by a rather deep minimum in the periplasmic vestibule, appropriate for efficient uptake of glycerol.

The NPA motif of GlpF – two loops of aspartate–proline–asparagine near the center of the channel – can be seen as an effective bulwark against proton transfer from the periplasm to the cytoplasm. At the qualitative level, an excess of protons attempting to cross the channel by hopping between water molecules is expected to run against the intrinsic dual orientation of the water file, induced by the electric field generated by the core of the membrane protein. Movement of excess protons through the channel as a single H_3O^+ moiety or a larger protein–water complex is hindered by a strictly conserved arginine adjacent to the channel. In order to disentangle the mechanism whereby proton exclusion is realized in aquaporins, Ilan et al. turned to a quantum mechanical–classical, multi-state valence-bond model to construct the PMF delineating proton transfer in GlpF [61]. Protons were pulled through the channel irreversibly and the equilibrium free energy change was recovered by means of the Jarzynski identity. Different valence-bond states were considered, corresponding to possible proton hydration motifs, viz. the Zundel,

$H_5O_2^+$, versus the Eigen, $H_9O_4^+$, cation. A free energy barrier of sufficient height to preclude proton transfer is found at the location of the NPA motif.

The unique role played by the NPA motif of GlpF was the object of a related investigation targeted at identifying the molecular mechanism responsible for preventing proton translocation across aquaporins. To reach this goal, Chakrabarti et al. determined the reversible work involved in the transfer of a proton across the water wire formed in the pore, and the subsequent reorganization of that water wire [62]. In the absence of protons, the latter step is impeded by the bipolar orientation of water around the NPA motif. On the other hand, proton transfer is strongly disfavored by an appreciable free energy barrier emerging at the location of the NPA sequence. As mentioned above, the electric field created by the core of the channel appears to act against proton translocation from the periplasm to the cytoplasm.

Aquaporins have been the theater of an active research aimed at understanding how nature controls, in a selective fashion, the passage of small molecules across the otherwise impervious biological membrane. By combining non-equilibrium MD simulations and the Jarzynski equality, Tajkhorshid and coworkers tried to illuminate the key features that distinguish *E. coli* AqpZ, a pure water channel, from GlpF, a channel that allows water and small, linear alcohols (*e.g.*, glycerol) to translocate from the periplasm to the cytoplasm [63]. Forced passage of glycerol through AqpZ yields a free energy barrier approximately threefold higher than that found in GlpF, which stems primarily from steric hindrances in the narrow region of the selectivity filter. The computed PMFs also highlight differences in the periplasmic vestibule of the channels, the deeper minimum characteristic of GlpF being hypothesized to enhance the capture of glycerol [64, 65].

Probing of the mechanism of ion conduction through the tetrameric pore of aquaporin-1 (Aqp1) represents yet another beautiful illustration where free energy calculations can help interpret inferences based on experimental data [66]. To address the much-debated issue that Aqp1 can function as an ion channel upon activation via the cyclic nucleotide cGMP, ion permeation through the tetrameric pore was simulated. It followed from the latter that the central pore is rapidly flooded with water molecules – a hydration process largely facilitated by subtle conformational changes in the pore-lining residues. The rather strong interaction of cGMP with the arginine-rich D-loop was proposed to act as the trigger that opens the cytoplasmic gate, a mechanism that is supported by site-directed mutagenesis.

8.4.5
Non-Equilibrium Simulations and Potassium Channels

As noted earlier in this section, the conduction of potassium ions by channels is an intricate problem that can be, at least in part, addressed by free energy calculations. Treptow and Tarek applied SMD simulations and the Jarzynski identity to investigate the influence of the sequence of amino acids of the channel on the ionic conduction properties of the selectivity filter [67]. In the light of the PMFs obtained for KcsA, KV1.2 and mutant potassium channels, these authors concluded that the free energy barrier towards K^+ permeation not only depends intrinsically on the amino acids lining the selectivity filter, but is also modulated by the distribution of charged residues along the channel sequence. This result demonstrates the long-range effect of charges on the free energy barrier.

8.5
Recognition and Association in Membrane Proteins

8.5.1
The "Two-Stage" Model

Modeling recognition and association phenomena in membranes constitute yet another computational challenge, rooted in the time scales spanned by the collective motions characteristic of lipid bilayers. These appreciable time scales, from the point of view of theoretical and computational biophysics, can be rationalized by the slow rearrangement of the lipid chains upon insertion and translation of protein segments. In this section, focus is centered primarily on the assembly of simple transmembrane segments into functional entities, rather than on the oligomerization of membrane proteins, the description of which is evidently not amenable to statistical simulations. Altough membrane proteins can be structurally complex, their transmembrane regions, as has been seen so far, are often quite simple, consisting in general of a bundle of α-helices or a barrel of β-strands. Popot and Engelman proposed, over 15 years ago, a rather convenient framework for understanding how transmembrane domains form. In their "two-stage" model [68], independently stable elements of secondary structure first insert into the lipid bilayer, prefacing lateral translation and formation of specific interactions, which result in higher-order, native structures. Whereas the underlying thermodynamics of the whole process is, by and large, well appreciated, detail of the interactions that drive recognition and association of transmembrane α-helices remains only fragmentary. Yet, new insight into recognition and association phenomena can be gained from simple, paradigmatic models consisting of a few independently folded transmembrane peptides. Turning to such prototypical systems is legitimate, granted that a number of membrane proteins retain their main functionalities, even though a large fraction of their structure has been removed [69]. The dimeric transmembrane region of glycophorin A (GpA), a glycoprotein ubiquitous to the human erythrocyte membrane, is a perfect example of such paradigmatic systems. The membrane-spanning domain of GpA is formed by 40 residues, among which 24 actually adopt an α-helical conformation. It has been hypothesized that helix–helix association is promoted by specific interactions of seven amino acids located on one face of each α-helix, which was further demonstrated by NMR spectroscopy in detergent micelles [70]. The latter investigation revealed that the two helical segments interact with a right–handed crossing angle of approximately 40° [70, 71].

8.5.2
Glycophorin A: A Paradigmatic System for Tackling Recognition and Association in Membranes

The inherent time scale limitations of MD simulations of membrane proteins in explicit lipid environments – which are intimately related to inadequate computational power – suggest from the onset that following chronologically the recognition and association of model peptides such as GpA and measuring accurately the associated free energy change is still out of reach. This explains for the most part why the prototypical transmembrane region of GpA has been studied either as a preassembled dimer in lipid bilayers [72] and in detergent micelles formed around it [73, 74], or as a dynamic dimerization process in a simpler dielectric medium [75]. However, none of these computational investigations delved, into the determination of the free energy profiles characterizing the reversible association of the α-helical segments. There are a number of reasons for this. First, the definition of a non-equivocal order parameter likely to capture the subtle rearrangement of the participating side chains as the membrane-spanning segments move back and forth from each other constitutes by itself a challenging problem. Second, the choice of an optimum free energy method for addressing the problem of reversible association of transmembrane α-helices is particularly intricate. Given the complexity of the problem, a number of conventional approaches are ruled out, such as US. The latter rather definitive assertion can be easily understood by realizing that for one particular inter-helical distance, the two transmembrane segments may be oriented quite differently, so that the forces exerted along the order parameter – whatever that order parameter might be – may also be very different. This, in turn, tends to imply that a unique definition of those external biases of Eq. (5), necessary to attain uniform probability distributions, is simply impossible. On the other hand, brute application of the perturbation theory embodied in Eq. (2) raises other conceptual difficulties, concealed in the erroneous assumption that the order parameter and the Cartesian coordinates are independent variables [76].

A significant step forward was made in 2005 by Hénin et al. towards understanding the recognition and association of transmembrane α-helices [77]. By using the recently developed ABF approach, these authors estimated the free energy of GpA α-helix dimerization in a membrane mimetic formed by a lamella of dodecane in equilibrium with two lamellae of water above and below it. Turning to rudimentary hydrocarbon media is fully justified by the aforementioned stringent limitations of MD simulations to capture the slow relaxation of the collective motions in lipid bilayers [78]. Furthermore, measurements of association constants, K_a, for the transmembrane α-helical segments, either by analytical ultracentrifugation [79] or fluorescence resonance energy transfer (FRET) [80, 81] is usually carried out in a variety of micellar environments. For this reason, direct comparison of experimentally and computationally determined 1-M standardized dimerization free energies [82] is difficult. It remains that ABF simulations illuminate distinct regimes in the sequence of events of the recognition and association processes. Deconvolution of

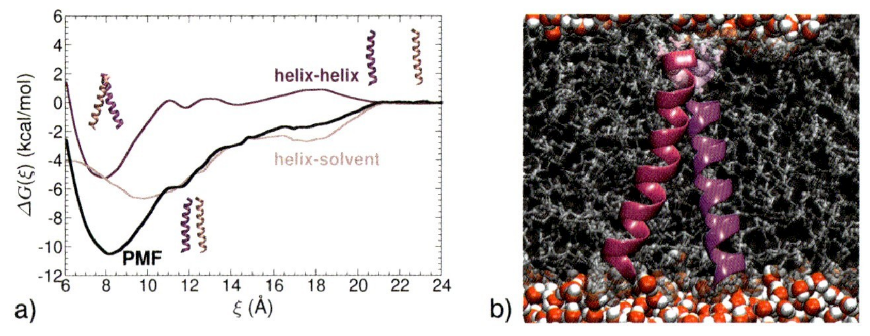

Fig. 8.5 Reversible α-helix dimerization in the transmembrane domain of glycophorin A. (a) Free energy profile delineating the reversible dissociation of the α-helix dimer. The order parameter, ξ, is defined as the distance separating the centers of mass of the two transmembrane segments. The deep single minimum of the PMF characterizes an ensemble of native-like structures, wherein the two α-helices interact with the signature crossing angle of ca. 40° (b).

the complete PMF into free energy components indicates that at large separations, the upright α-helices are pushed together by the solvent, which is conducive to the occurrence of preliminary inter-helical contacts. This important step precedes the formation of native, primarily van der Waals contacts, which is concomitant with the mutual tilt of the transmembrane segments characteristic of the dimeric structure (see Fig. 8.5).

8.6 Conclusions

Although the theoretical foundations of free energy calculations were first laid over 50 years ago, their application to membrane proteins has been much more recent. To a large extent, this situation is due to conceptual and technical considerations, rather than to methodological ones. This statement does not necessarily imply that free energy methods have not evolved during the past decades, but rather that the underlying theory was known long before being applied to non-trivial molecular systems of both chemical and biological relevance. For instance, the first spectacular application of the FEP machinery to a concrete problem of chemistry was the “alchemical transformation” of methanol into ethane by Jorgensen and Ravimohan [21], and this paved the road to the computation of differential free energies of solvation from first principles. Yet, Jorgensen and Ravimohan had to wait for 30 years since Robert Zwanzig established the perturbation formula in 1954 [11], to harness the computational power required in free energy calculations. Today, with 20 years of hindsight, point mutation of methanol in bulk water sounds like a simple task, but at the time it posed a daunting computational challenge. Today, with the continuously increasing performance of computers, and in particular, the availability of massively parallel machines, the size–scale limitations of statistical simulations have been pushed back to the extent that free energy calculations in membrane assemblies are now about as expensive as they were back in 1985 for a simple solute immersed in an aqueous medium.

In addition to the technical limitations rooted in the performance/price ratio of contemporary computers, the development of suitable potential energy functions for describing lipid environments has also been very recent [83, 84]. Progress in that direction was admittedly impeded by the limited access of research groups to large computer facilities. As underlined recently, only long MD simulations are capable of highlighting conceptual flaws both in the parametrization of the force field and in the strategy adopted to simulate lipid bilayers [85]. Another reason accounting for the rather embryonic stage of free energy calculations applied to membrane proteins lies in the paucity of well-resolved three-dimensional structures. This is reflected in the glaring imbalance in the number of simulations performed on globular and integral membrane proteins.

The bolstering results reported herein suggest, nevertheless, that free energy methods represent a powerful tool for fathoming, at the atomic level, the mechanisms whereby membrane proteins fulfill their cellular function. Recent developments on the methodological front have opened new vistas for investigating transport phenomena in membrane channels and carriers. Conjunction of SMD simulations and the Jarzynski identity [37], or use of the ABF approach [32, 33], provides promising perspectives for tackling the energetics that characterizes the assisted translocation of small molecules and ions across the biological membrane. These simulations are particularly appealing, because they complement experiment by offering a tangible link between the structural modifications that occur in response to an external perturbation and the associated free energy behavior. Cutting-edge applications of free energy calculations to membrane proteins are envisioned to increase significantly as new three-dimensional structures become available.

Summary

Free energy calculations aimed at understanding how membrane proteins function have recently emerged as a powerful complement to experiment, bridging structure and energetics. The recent application of these numerical simulations to large membrane assemblies are based on deeply intertwined factors that range from the performance of modern computer architectures to the development of suitable potential energy functions for modeling lipid bilayers. In this chapter, the theoretical foundations of free energy calculations are summarized, emphasizing how the methodology has evolved over the past decades, and which biophysical problems can be addressed by a given class of methods. Recent strategies for improving the reliability of the simulations, while making them more cost–effective are outlined. Selected applications of free energy calculations to transport and recognition–association processes in membrane proteins are reviewed. Though still in a nascent stage, these calculations are anticipated to play rapidly a pivotal, predictive role in deciphering the molecular mechanisms underlying membrane protein function.

Acknowledgments

The authors gratefully acknowledge Drs. François Dehez, Andrew Pohorille, Benoît Roux, Mark Sansom and Mounir Tarek for their valuable input and insightful discussions.

References

1 Allen, M. P.; Tildesley, D. J., *Computer Simulation of Liquids*, Clarendon Press: Oxford, **1987**.

2 Frenkel, D.; Smit, B., *Understanding molecular simulations: From algorithms to applications*, Academic Press: San Diego, **1996**.

3 Tajkhorshid, E.; Nollert, P.; Jensen, M. Ø.; Miercke, L. J. W.; O'Connell, J.; Stroud, R. M.; Schulten, K., Control of the selectivity of the aquaporin water channel family by global orientational tuning, *Science* **2002**, *296*, 525–530.

4 Aksimentiev, A.; Schulten, K., Imaging α-hemolysin with molecular dynamics: Ionic conductance, osmotic permeability, and the electrostatic potential map, *Biophys. J.* **2005**, *88*, 3745–3761.

5 Sotomayor, M.; Vasquez, V.; Perozo, E.; Schulten, K., Ion conduction through MscS as determined by electrophysiology and simulation, *Biophys. J.* **2007**, *92*, 886–902.

6 Shelley, J. C.; Shelley, M. Y.; Reeder, R. C.; Bandyopadhyay, S.; Klein, M. L., A coarse grain model for phospholipid simulations, *J. Phys. Chem. B.* **2001**, *105*, 4464–4470.

7 Nielsen, S. O.; Lopez, C. F.; Srinivas, G.; Klein, M. L., Coarse grain models and the computer simulation of soft materials, *J. Phys.: Condens. Matter* **2004**, *16*, R481–R512.

8 Shih, A. Y.; Freddolino, P. L.; Arkhipov, A.; Schulten, K., Assembly of lipoprotein particles revealed by coarse-grained molecular dynamics simulations, *J. Struct. Biol.* **2006**, *157*, 579–592.

9 Kirkwood, J. G., Statistical mechanics of fluid mixtures, *J. Chem. Phys.* **1935**, *3*, 300–313.

10 Kirkwood, J. G., in *Theory of liquids*, Alder, B. J., Ed., Gordon and Breach: New York, **1968**.

11 Zwanzig, R. W., High-temperature equation of state by a perturbation method. I. Nonpolar gases, *J. Chem. Phys.* **1954**, *22*, 1420–1426.

12 Widom, B., Some topics in the theory of fluids, *J. Chem. Phys.* **1963**, *39*, 2808–2812.

13 Torrie, G. M.; Valleau, J. P., Nonphysical sampling distributions in Monte Carlo free energy estimation: Umbrella sampling, *J. Comput. Phys.* **1977**, *23*, 187–199.

14 Bennett, C. H., Efficient estimation of free energy differences from Monte Carlo data, *J. Comput. Phys.* **1976**, *22*, 245–268.

15 Chipot, C. Calculating free energy differences from perturbation theory. In: *Free energy calculations. Theory and applications in chemistry and biology*, Chipot, C.; Pohorille, A., Eds. Springer Verlag, Berlin, Heidelberg, **2006**.

16 Lu, N.; Kofke, D. A.; Woolf, T. B., Staging is more important than perturbation method for computation of enthalpy and entropy changes in complex systems, *J. Phys. Chem. B* **2003**, *107*, 5598–5611.

17 Lu, N.; Kofke, D. A.; Woolf, T. B., Improving the efficiency and reliability of free energy perturbation calculations using overlap sampling methods, *J. Comput. Chem.* **2004**, *25*, 28–39.

18 Mark, A. E. Free Energy Perturbation Calculations. In: *Encyclopedia of computational chemistry*, Schleyer, P. v. R.;

Allinger, N. L.; Clark, T.; Gasteiger, J.; Kollman, P. A.; Schaefer III, H. F.; Schreiner, P. R., Eds., vol. *2*. Wiley & Sons, Chichester, **1998**, pp. 1070–1083.

19 Beutler, T. C.; Mark, A. E.; van Schaik, R. C.; Gerber, P. R.; van Gunsteren,W. F., Avoiding singularities and neumerical instabilities in free energy calculations based on molecular simulations, *Chem. Phys. Lett.* **1994**, *222*, 529–539.

20 Tembe, B. L.; McCammon, J. A., Ligand–receptor interactions, *Comp. Chem.* **1984**, *8*, 281–283.

21 Jorgensen, W. L.; Ravimohan, C., Monte Carlo simulation of differences in free energies of hydration, *J. Chem. Phys.* **1985**, *83*, 3050–3054.

22 Bash, P. A.; Singh, U. C.; Brown, F. K.; Langridge, R.; Kollman, P. A., Calculation of the relative change in binding free energy of a protein–inhibitor complex, *Science* **1987**, *235*, 574–576.

23 Bash, P. A.; Singh, U. C.; Langridge, R.; Kollman, P. A., Free energy calculations by computer simulation, *Science* **1987**, *236*, 564–568.

24 Straatsma, T. P.; Berendsen, H. J. C., Free energy of ionic hydration: Analysis of a thermodynamic integration technique to evaluate free energy differences by molecular dynamics simulations, *J. Chem. Phys.* **1988**, *89*, 5876–5886.

25 Chandler, D., *Introduction to modern statistical mechanics*, Oxford University Press, **1987**.

26 Torrie, G. M.; Valleau, J. P., Monte Carlo study of phase separating liquid mixture by umbrella sampling, *J. Chem. Phys.* **1977**, *66*, 1402–1408.

27 Valleau, J. P.; Card, D. N., Monte Carlo estimation of the free energy by multistage sampling, *J. Chem. Phys.* **1972**, *57*, 5457–5462.

28 Kumar, S.; Bouzida, D.; Swendsen, R. H.; Kollman, P. A.; Rosenberg, J. M., The weighted histogram analysis method for free energy calculations on biomolecules. I. The method, *J. Comput. Chem.* **1992**, *13*, 1011–1021.

29 Carter, A.; Cicotti, G.; Hynes, J. T.; Kapral, R., Constrained reaction coordinate dynamics for the simulation of rare events, *Chem. Phys. Lett.* **1989**, *156*, 472–477.

30 den Otter, W. K.; Briels, W. J., The calculation of free-energy differences by constrained molecular dynamics simulations, *J. Chem. Phys.* **1998**, *109*, 4139–4146.

31 den Otter, W. K., Thermodynamic integration of the free energy along a reaction coordinate in Cartesian coordinates, *J. Chem. Phys.* **2000**, *112*, 7283–7292.

32 Darve, E.; Pohorille, A., Calculating free energies using average force, *J. Chem. Phys.* **2001**, *115*, 9169–9183.

33 Hénin, J.; Chipot, C., Overcoming free energy barriers using unconstrained molecular dynamics simulations, *J. Chem. Phys.* **2004**, *121*, 2904–2914.

34 Isralewitz, B.; Gao, M.; Schulten, K., Steered molecular dynamics and mechanical functions of proteins, *Curr. Opin. Struct. Biol.* **2001**, *11*, 224–230.

35 Isralewitz, B.; Baudry, J.; Gullingsrud, J.; Kosztin, D.; Schulten, K., Steered molecular dynamics investigations of protein function, *J. Mol. Graph. Model.* **2001**, *19*, 13–25.

36 Jarzynski, C., Nonequilibrium equality for free energy differences, *Phys. Rev. Lett.* **1997**, *78*, 2690–2693.

37 Park, S.; Khalili-Araghi, F.; Tajkhorshid, E.; Schulten, K., Free energy calculation from steered molecular dynamics simulations using Jarzynski's equality, *J. Chem. Phys.* **2003**, *119*, 3559–3566.

38 Park, S.; Schulten, K., Calculating potentials of mean force from steered molecular dynamics simulations, *J. Chem. Phys.* **2004**, *120*, 5946–5961.

39 Chipot, C.; Pohorille, A., Eds., *Free energy calculations. Theory and applications in chemistry and biology*, Springer-Verlag: Berlin, Heidelberg, **2007**.

40 Roux, B.; Prod'hom, B.; Karplus, M., Ion transport in the gramicidin channel: molecular dynamics study of single and double occupancy, *Biophys. J.* **1995**, *68*, 876–892.

41 Roux, B.; Nina, M.; Pomès, R.; Smith, J. C., Thermodynamic stability of water molecules in the Bacteriorhodopsin proton channel: A molecular dynamics and free energy perturbation study, *Biophys. J.* **1996**, *71*, 670–681.

42 Roux, B., Valence selectivity of the gramicidin channel: a molecular dynamics free energy perturbation study, *Biophys. J.* **1996**, *71*, 3177–3185.

43 Doyle, D. A.; Cabral, J. Morais; Pfuetzner, R. A.; Kuo, A.; Gulbis, J. M.; Cohen, S. L.; Chait, B. T.; MacKinnon, R., The structure of the potassium channel: molecular basis of K^+ conduction and selectivity, *Science* **1998**, *280*, 69–77.

44 Bernèche, S.; Roux, B., Molecular dynamics of the KcsA K^+ channel in a bilayer membrane, *Biophys. J.* **2000**, *78*, 2900–2917.

45 Bernèche, S.; Roux, B., The ionization state and the conformation of Glu-71 in the KcsA K^+ channel, *Biophys. J.* **2002**, *82*, 772–780.

46 Noskov, S. Y.; Bernèche, S.; Roux, B., Control of ion selectivity in potassium channels by electrostatic and dynamic properties of carbonyl ligands, *Nature* **2001**, *431*, 830–834.

47 Luzhkov, V. B.; Almlöf, M.; Nervall, M.; Åqvist, J., Computational Study of the Binding Affinity and Selectivity of the Bacterial Ammonium Transporter AmtB, *Biochemistry* **2006**, *45*, 10807–10814.

48 Javelle, A.; Severi, E.; Thornton, J.; Merrick, M., Ammonium sensing in *Escherichia coli*. Role of the ammonium transporter AmtB and AmtB-GlnK complex formation, *J. Biol. Chem.* **2004**, *279*, 8530–8538.

49 Hénin, J.; Maigret, B.; Tarek, M.; Escrieut, C.; Fourmy, D.; Chipot, C., Probing a model of a GPCR/ligand complex in an explicit membrane environment. The human cholecystokinin-1 receptor, *Biophys. J.* **2006**, *90*, 1232–1240.

50 Roux, B.; Karplus, M., Ion transport in a model gramicidin channel. Structure and thermodynamics, *Biophys. J.* **1991**, *59*, 961–981.

51 Roux, B.; Karplus, M., Molecular dynamics simulations of the gramicidin channel, *Annu. Rev. Biophys. Biomolec. Struct.* **1994**, *23*, 731–761.

52 Allen, T. W.; Andersen, O. S.; Roux, B., Energetics of ion conduction through the gramicidin channel, *Proc. Natl. Acad. Sci. USA* **2004**, *101*, 117–122.

53 Jensen, M. Ø.; Park, S.; Tajkhorshid, E.; Schulten, K., Energetics of glycerol conduction through aquaglyceroporin GlpF, *Proc. Natl. Acad. Sci. USA* **2002**, *99*, 6731–6736.

54 Bernèche, S.; Roux, B., Energetics of ion conduction through the K^+ channel, *Nature* **2001**, *414*, 73–77.

55 Cohen, J.; Schulten, K., Mechanism of anionic conduction across ClC, *Biophys. J.* **2004**, *86*, 836–845.

56 Treptow, W.; Tarek, M., Molecular restraints in the permeation pathway of ion channels, *Biophys. J.* **2006**, 91, L26–L28.

57 Beckstein, O.; Sansom, M., A hydrophobic gate in an ion channel: the closed state of the nicotinic acetylcholine receptor, *Phys. Biol.* **2006**, *3*, 147–159.

58 Gullingsrud, J.; Braun, R.; Schulten, K., Reconstructing potentials of mean force through time series analysis of steered molecular dynamics simulations, *J. Comput. Phys.* **1999**, *151*, 190–211.

59 Hummer, G.; Szabo, A., Free energy reconstruction from nonequilibrium single-molecule pulling experiments, *Proc. Natl. Acad. Sci. USA* **2001**, *98*, 3658–3661.

60 Fu, D., Libson, A.; Miercke, L. J.;Weitzman, C.; Nollert, P.; Krucinski, J.; Stroud, R. M., Structure of a glycerol-conducting channel and the basis for its selectivity, *Science* **2000**, *290*, 481–486.

61 Ilan, B.; Tajkhorshid, E.; Schulten, K.; Voth, G. A., The mechanism of proton exclusion in aquaporin channels, *Proteins: Struct. Func. Bioinfo.* **2004**, *55*, 223–228.

62 Chakrabarti, N.; Tajkhorshid, E.; Roux, B.; Pomès, R., Molecular basis of proton blockage in aquaporins, *Structure* **2004**, *12*, 65–74.

63 Wang, Y.; Schulten, K.; Tajkhorshid, E., What makes an aquaporin a glycerol channel? A comparative study of AqpZ and GlpF, *Structure* **2005**, *13*, 1107–1118.

64 Jensen, M. O.; Tajkhorshid, E.; Schulten, K., The mechanism of glycerol conduction in aquaglyceroporins, *Structure* **2001**, *9*, 1083–1093.

65 Lu, D. Y.; Grayson, P.; Schulten, K., Glycerol conductance and physical asymmetry of the *Escherichia coli* glycerol facilitator GlpF, *Biophys. J.* **2003**, *85*, 2977–2987.

66 Yu, J.; Yool, A. J.; Schulten, K.; Tajkhorshid, E., Mechanism of gating and ion conductivity of a possible tetrameric pore in aquaporin-1, *Structure* **2006**, *14*, 1411–1423.

67 Treptow, W.; Tarek, M., K^+ conduction in the selectivity filter of potassium channels is monitored by the charge distribution along their sequence, *Biophys. J.* **2006**, *91*, L81–L83.

68 Popot, J. L.; Engelman, D. M., Membrane protein folding and oligomerization: The two-stage model, *Biochemistry* **1990**, *29*, 4031–4037.

69 Montal, M., Molecular mimicry in channel-protein structure, *Curr. Opin. Struct. Biol.* **1995**, *5*, 501–506.

70 MacKenzie, K. R.; Prestegard, J. H.; Engelman, D. M., A transmembrane helix dimer: Structure and implications, *Science* **1997**, *276*, 131–133.

71 Smith, S. O.; Song, D.; Shekar, S.; Groesbeek, M.; Ziliox, M.; Aimoto, S., Structure of the transmembrane dimer interface of glycophorin A in membrane bilayers, *Biochemistry* **2001**, *40*, 6553–6558.

72 Petrache, H. I.; Grossfield, A.; MacKenzie, K. R.; Engelman, D. M.; Woolf, T. B., Modulation of glycophorin A transmembrane helix interactions by lipid bilayers: Molecular dynamics calculations, *J. Mol. Biol.* **2000**, *302*, 727–746.

73 Braun, R.; Engelman, D. M.; Schulten, K., Molecular dynamics simulation of micelle formation around dimeric glycophorin A transmembrane helices, *Biophys. J.* **2004**, *87*, 754–763.

74 Bond, P. J.; Cuthbertson, J. M.; Deol, S. S.; Sansom, M. S. P., MD simulations of spontaneous membrane protein/ detergent micelle formation, *J. Am. Chem. Soc.* **2004**, *126*, 15948–15949.

75 Im, W.; Feig, M.; Brooks III, C. L., An implicit membrane generalized born theory for the study of structure, stability, and interactions of membrane proteins, *Biophys. J.* **2003**, *85*, 2900–2918.

76 Pohorille, A.; Wilson, M. A.; Chipot, C., Membrane peptides and their role in protobiological evolution, *Orig. Life and Evol. Biosph.* **2003**, *33*, 173–197.

77 Hénin, J.; Pohorille, A.; Chipot, C., Insights into the recognition and association of transmembrane α-helices. The free energy of α-helix dimerization in glycophorin A, *J. Am. Chem. Soc.* **2005**, *127*, 8478–8484.

78 Stockner, T.; Ash, W. L.; MacCallum, J. L.; Tieleman, D. P., Direct simulation of transmembrane helix association: Role of asparagines, *Biophys. J.* **2004**, *87*, 1650–1656.

79 Fleming, K. G.; Ackerman, A. L.; Engelman, D. M., The effect of point mutations on the free energy of transmembrane α-helix dimerization, *J. Mol. Biol.* **1997**, *272*, 266–275.

80 Fisher, L. E.; Engelman, D. M.; Sturgis, J. N., Detergents modulate dimerization, but not helicity, of the glycophorin A transmembrane domain, *J. Mol. Biol.* **1999**, *293*, 639–651.

81 Fisher, L. E.; Engelman, D. M.; Sturgis, J. N., Effects of detergents on the association of the glycophorin A transmembrane helix, *Biophys. J.* **2003**, *85*, 3097–3105.

82 Fleming, K. G., Standardizing the free energy change of transmembrane helix–helix interactions, *J. Mol. Biol.* **2002**, *323*, 563–571.

83 Tieleman, D. P.; Marrink, S. J.; Berendsen, H. J. C., A computer perspective of membranes: Molecular dynamics studies of lipid bilayer systems, *Biochim. Biophys. Acta* **1997**, *1331*, 235–270.

84 Chipot, C.; Klein, M. L.; Tarek, M. Modeling lipid membranes. In: *The handbook of materials modeling. Methods and models of materials modeling*, Catlow, R.; Shercliff, H.; Yip, S., Eds., vol. *1*. Kluwer Academic Publishers, Dordrecht, **2005**, pp. 929–958.

85 Anézo, C.; de Vries, A. H.; Höltje, H. D.; Tieleman, P.; Marrink, S. J., Methodological issues in lipid bilayer simulations, *J. Phys. Chem. B* **2003**, *107*, 9424–9433.

9
Neutrons to Study the Structure and Dynamics of Membrane Proteins

Kathleen Wood and Giuseppe Zaccai

9.1
General Introduction

Neutrons represent a unique probe with which to study matter, as their wavelengths and energies make them useful for investigations of both molecular structure and dynamics. In this chapter, the particularities of neutrons are discussed which make them suitable for the study of the structural and dynamic properties of membrane proteins. A diagram representing time and space domains accessible by neutron scattering is provided in Fig. 9.1. To highlight the power of the technique the chapter includes examples which focus, whenever possible, on the Purple Membrane (PM), for which a wealth of experimental data currently exists.

9.2
Introduction to Neutrons

9.2.1
Production and Properties of the Neutron

The neutron is an elementary particle, which was first discovered in 1932 by J. Chadwick (Chadwick, 1932a, 1932b). Its importance as a probe for the study of condensed matter was recognized by the award of a Nobel Prize in 1994 to Shull and Brockhouse, who were pioneers in the field. As these authors noted in their Nobel citation, neutrons enable us to know "... where atoms are and what atoms do". Some basic properties of the neutron are listed in Table 9.1.

Neutrons are produced in reactors through the fission of uranium-235, or in spallation sources when an accelerated proton beam hits a heavy metal target. As the neutron mass is relatively large, once produced, the kinetic energy of the neutron can be changed by collision with atomic nuclei in a process called *moderation*: the neutrons are bought to equilibrium at the temperature of the atomic nuclei. The energy of the neutrons is therefore a Maxwellian distribution, corresponding to the

Biophysical Analysis of Membrane Proteins. Investigating Structure and Function. Edited by Eva Pebay-Peyroula

ISBN: 978-3-527-31677-9

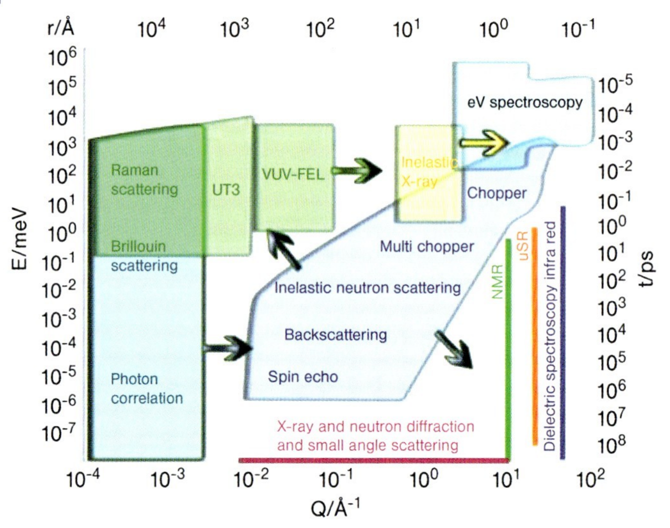

Fig. 9.1 The complementarity of neutrons with other biophysical methods: comparison of time and space ranges accessible to different techniques. Chopper, multi chopper, backscattering and spin echo denote different types of neutron spectrometers. (Adapted from www.neutron-eu.net.)

Table 9.1 Properties of the neutron.

Charge	0
Mass	$1.00866\,u = 1.675 \times 10^{-27}\,kg$
Spin	1/2
Energy (E), temperature (T), frequency (ω), speed (v)	$E = k_B T = \hbar\omega = 1/2\, mv^2$
Momentum (**p**), wavevector (**k**)	$\mathbf{p} = m\mathbf{v} = \hbar\mathbf{k}$
Wavelength (λ)	$\lvert\mathbf{k}\rvert = 2\pi/\lambda$

For a neutron energy of 25 meV, $T = 293\,K$, $\lambda \sim 1.8\,Å$, $k \sim 3.5\,Å^{-1}$, $v \sim 2.2\,km\,s^{-1}$

thermal energy of the particles with which they are moderated. As shown in Table 9.1, a neutron moderated at room temperature has an energy of 25 meV and travels with a velocity of around $2\,km\,s^{-1}$, a relatively slow speed, it is useful in the design of certain spectrometers (see below).

9.2.2
Interaction Between Neutrons and Matter

The neutron has a mass similar to that of the proton or hydrogen atom, no charge, and a spin of 1/2 (see Table 9.1). Unlike X-rays, the neutron interacts negligibly with an atom's electronic cloud and predominantly with the nucleus; it is for this reason that there is an important isotope effect in neutron scattering. The range of interaction between a neutron and a nucleus is of the order of 10^{-15} m. The nucleus can therefore be considered point-like as the neutron wavelength is five orders of magnitude larger, implying that the scattered waves are isotropic (the scattering does not decrease with the angle, as for X-ray scattering). Scattering by a nucleus is characterized by a single parameter b, the scattering length. The scattering cross-section is simply $4\pi b^2$. A list of scattering lengths and cross sections for atoms useful in biological studies is provided in Table 9.2 (the terms "incoherent" and "coherent" are explained later in the chapter). In contrast to X-ray and electron scattering, there is no "pattern" to these values: the heavier elements do not dominate the scattering, and the scattering power of different isotopes may be very different. The case of hydrogen (^1H) and its heavier isotope deuterium (^2H or D) is of particular interest, for which both coherent and incoherent cross sections are extremely different. The negative sign of b_{coh} in the case of hydrogen is associated with a change in phase of the scattered neutron wave compared to the scattering by other nuclei.

The weakness of the interaction, as shown by the small values of scattering cross sections (naming the unit of these as "barn" was clearly ironic), means that the total scattering from a sample can simply be considered as the sum of the scattering from all atoms.

A diagram of the scattering process is shown in Fig. 9.2. During a scattering event, a neutron can exchange both momentum and energy with the sample. $\mathbf{k}_i$ is

Table 9.2 Neutron scattering cross sections useful for studying biological samples. (From Sears, 1984.)

	Scattering length (10^{-15} m)		*Scattering cross-section (barns) (1 barn = 10^{-28} m^2)*	
	Coherent	*Incoherent*	*Coherent*	*Incoherent*
^1H	−3.74	25.2	1.76	79.9
^2H = D	6.67	4.03	5.60	2.04
C	6.65	0	5.55	0
N	9.37	1.98	11.0	0.49
O	5.81	0	4.23	0
S	2.80	0	0.99	0

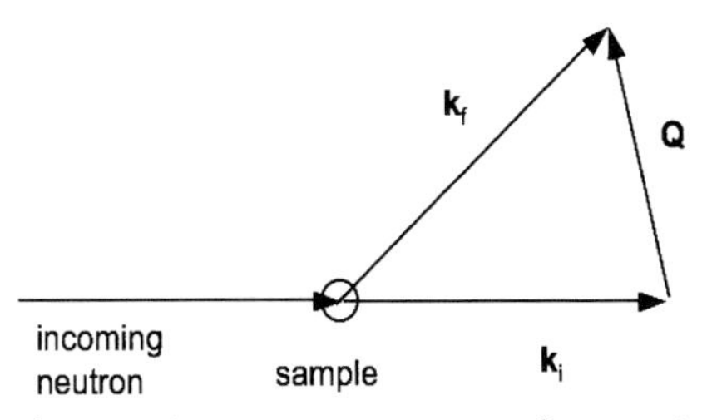

Fig. 9.2 Schematic representation of a scattering event, conservation of momentum and energy imply: $\mathbf{Q} = (\mathbf{k}_f - \mathbf{k}_i)$ and $\hbar\omega = E_i - E_f$.

the incoming wavevector (its modulus k_i is equal to $2\pi/\lambda$, where λ is the neutron wavelength), and $\mathbf{k}_f$ the final wavevector after the scattering process. $\mathbf{Q}$ is defined as the difference between $\mathbf{k}_f$ and $\mathbf{k}_i$. If there is no energy exchange, the modulus k_i is equal to k_f, and the process is defined as elastic.

9.2.3 Scattering Law

In a neutron scattering experiment, the number of scattered neutrons is measured in a certain solid angle $d\Omega_f$, in the direction of the wavevector $\mathbf{k}_f$ with a final energy between E_f and $E_f + dE_f$ and with an incident wavevector $\mathbf{k}_i$. This quantity is called the "double-differential cross section", $d^2\sigma/d\Omega_f dE_f$.

It can be shown that for a system of N identical atoms of equal scattering length b:

$$\frac{d^2\sigma}{d\Omega_f dE_f} = N\frac{k_f}{k_i} b^2 S(\mathbf{Q},\omega) \tag{1}$$

where $S(\mathbf{Q},\omega)$ is called the neutron scattering function.

$$S(\mathbf{Q},\omega) = \frac{1}{2\pi\hbar N}\sum_{jk}\int_{-\infty}^{\infty} dt < e^{-i\mathbf{Q}\cdot\mathbf{r}_j(0)} e^{i\mathbf{Q}\cdot\mathbf{r}_k(t)} > e^{-i\omega t} \tag{2}$$

Here, $\mathbf{r}_j(0)$ and $\mathbf{r}_k(t)$ are the positions of atoms j at time zero and k at time t respectively. Measuring $S(\mathbf{Q},\omega)$ provides the correlations of the positions of the atoms at different times, and measuring this function allows the microscopic properties of the system – both spatial and temporal – to be studied.

9.2.4 Coherent and Incoherent scattering

Mathematically, it is simple to divide Eq. (2) into two parts, the so-called "coherent" and "incoherent" parts.

$$\frac{d^2\sigma}{d\Omega_f dE_f} = \frac{d^2\sigma_{inc}}{d\Omega_f dE_f} + \frac{d^2\sigma_{coh}}{d\Omega_f dE_f} \tag{3}$$

Coherent scattering provides information about cooperative effects, either structural (if a sample is crystalline, Bragg peaks will be measured), or dynamic (in a crystalline sample, certain modes of propagation of waves can be seen). Incoherent scattering arises because, in standard experiments without polarization, neither the neutrons nor the nuclei in the sample have homogeneous spin values. The incoherent part can be thought of as representing the correlation of a single atom with itself in time: it contains information about single-particle or "self" dynamics. Each process is characterized by a specific cross-section (Table 9.2).

The cross-sections listed in Table 9.2 make neutron scattering interesting for the study of membrane proteins for several reasons:

- The coherent scattering is analyzed in membrane diffraction, crystallography, reflectivity and small-angle neutron scattering (SANS) measurements to provide information at different levels of resolution on the structure of the sample. The coherent scattering cross sections of hydrogen and deuterium are of opposite sign, and of about the same amplitude: if one sample can be labeled with a deuterated component then, by comparing the scattered intensities from the same hydrogenated sample, the structure of the label can be identified, even at low resolution.
- The difference in contrast between H_2O and D_2O mixes can be used to "match out" certain components of a system at low resolution.
- The incoherent scattering is usually analyzed in neutron-scattering experiments in which the energy transfer is measured. The incoherent cross-section from hydrogen is an order of magnitude greater than all others commonly found in biological samples. Hydrogen atoms are fairly evenly distributed throughout biological macromolecules, and their dynamics reflects the groups to which they are attached; using energy resolved incoherent neutron scattering measurements therefore allows investigations to be made of the global dynamics of macromolecules. As incoherent scattering from deuterium is – in contrast – extremely low, it is possible, by labeling, to concentrate on dynamics of different parts of the sample.

Although their energies and wavelengths make neutrons an ideal probe for condensed matter studies, the low flux available means that experiments are only feasible with large-scale instruments. A typical flux on a small-angle scattering instrument is around 10^7 neutrons $cm^{-2}\,s^{-1}$, whereas on a spectrometer it is about 10^6 neutrons $cm^{-2}\,s^{-1}$ [at the Institut Laue Langevin (ILL) in Grenoble, France which, at the time of writing, is the most intense source in the world]. Coupled with the low-scattering cross-sections, the low flux implies that experiments are relatively long and can require large amounts of sample. Although less than 1 mg is necessary for a SANS experiment, at least 100 mg are required for a dynamics experiment, with typical crystal sizes ranging from 0.01 to 1.0 mm^3. The low flux is compensated to some extent by large beam dimensions (of the order of centimeters). The following examples of studies should provide convincing evidence that the effort of obtaining suitable samples is well justified. In order to reveal the full power of neutron techniques, however, hydrogen–deuterium (H–D) labeling is necessary, and some methods for labeling membrane proteins are described later in the chapter.

Further information relating to neutron scattering, together with links to the websites of all major neutron scattering centers are available at:

- European Neutron Portal: http://www.neutron-eu.net/
- Neutron Scattering Web: http://www.neutron.anl.gov/

As labeling is now recognized as being of primary importance for biological samples, specialist laboratories are under development worldwide. For example, in Grenoble, the ILL-EMBL deuteration laboratory is a user-facility available to all European laboratories; information on this facility is available at: http://www.ill.fr/YellowBook/deuteration/index.htm. Similar initiatives are being started at several other facilities.

9.2.5 Instruments

Neutron diffractometers and SANS instruments essentially count neutrons as a function of the scattering angle, and display few particularities compared to X-ray diffraction. For example, at the ILL small-angle measurements are performed on D11 and D22, membrane diffraction on D16, and reflectometry on D17.

Neutron spectroscopy instruments use certain specific properties of the neutron, and measure not only the scattering angle but also changes in energy. Neutron detectors are devices that can only "count" neutrons but do not resolve their energy. As the incident and scattered energies must be known, the beam must first be "prepared" for the measurement by selecting a particular wavelength or energy, and the energy of the scattered neutron measured. Different methods are available to determine neutron energies, the two main techniques being time-of-flight and diffraction from crystals:

- As mentioned earlier, a thermal neutron travels relatively slowly, such that, in 1935, J. R. Dunning showed that they could be selected depending on their velocity by using "choppers", which are rotating disks with slits. A first chopper will send a pulse of neutrons of different speeds, while a second chopper rotating at a given speed will only allow neutrons of a certain wavelength (lambda) to pass through. (In reality, additional choppers are required to remove contamination from neutrons of other wavelengths). On a time-of-flight spectrometer, the choppers are used to select a particular energy, and the time is then measured between when the neutrons interact with the sample and when they hit the detectors. By knowing the time-of-flight and the distance, it is possible to calculate the neutrons' outgoing energy. Examples of these types of spectrometer at the ILL include the IN5 and IN6.

- Crystals can be used to select a particular wavelength and therefore energy of neutrons, both to monochromate the incoming beam and to analyze the scattered intensity. The so-called "backscattering" spectrometers which use this method are based on the fact that Bragg's law is extremely wavelength selective for diffraction when the scattering angle is equal to 180°. Generally, in backscattering spectrometers, it is the incoming energy which is varied, either by heating the monochromating crystal or by moving it on a Doppler drive. Examples of these types of spectrometer at the ILL include the IN10, IN13, and IN16.

9.3 Introduction to Bacteriorhodopsin and the Purple Membrane

Bacteriorhodopsin (BR), and the Purple Membrane (PM) in which it is found, have provided biophysicists with exceptional models for the study of biological membranes. PM can be purified in relatively large quantities necessary for neutron scattering from the plasma membrane of the archaeon *Halobacterium salinarum*, a halophilic organism that thrives in salt-saturated brines exposed to bright sunlight. PM is naturally organized as a highly ordered two-dimensional (2-D) lattice, composed of 75% (wt.%) of a single protein, BR, and 25% of various lipid species. It remains the natural biological membrane with the best-characterised structure (here, the term "structure" refers to all of the elements that make it a functional unit, BR, lipids, and hydration water).

Bacteriorhodopsin itself has become a paradigm for membrane proteins in general. It folds into a seven-transmembrane helical structure with short interconnecting loops (see Fig. 9.3a), and has a molecular mass of approximately 26 kDa. The early structural studies of PM revealed the strength of electron microscopy for the study of membranes with 2-D organization. The structure (as ascertained by Unwin and Henderson, 1975) which achieved a resolution of 7 Å in the plane and 14 Å in the membrane normal, showed the helical nature of BR and its arrangement in trimers within the PM. The projected structure published in 1975 is shown in Fig. 9.3b. The first high-resolution structure of BR was obtained using cryoelectron microscopy (Henderson et al., 1990), and since then the structure has been resolved to increasingly better resolution using X-ray crystallography (Belrhali et al., 1999; Luecke et al., 1999; Lanyi and Schobert, 2006).

The transmembrane helices of BR surround a retinal molecule (shown in red in Fig. 9.3a) which is covalently bound via a Schiff base to Lys^{216}. Bacteriorhodopsin is a nano-machine which functions as a proton pump. Upon absorption of a photon of "green" light, a change in isomerization of the retinal molecule is induced. The protein then undergoes a photocycle, which results in a proton being transported across the cell membrane, from the cytoplasmic to the extracellular side, thereby converting light energy into an electrochemical proton gradient,

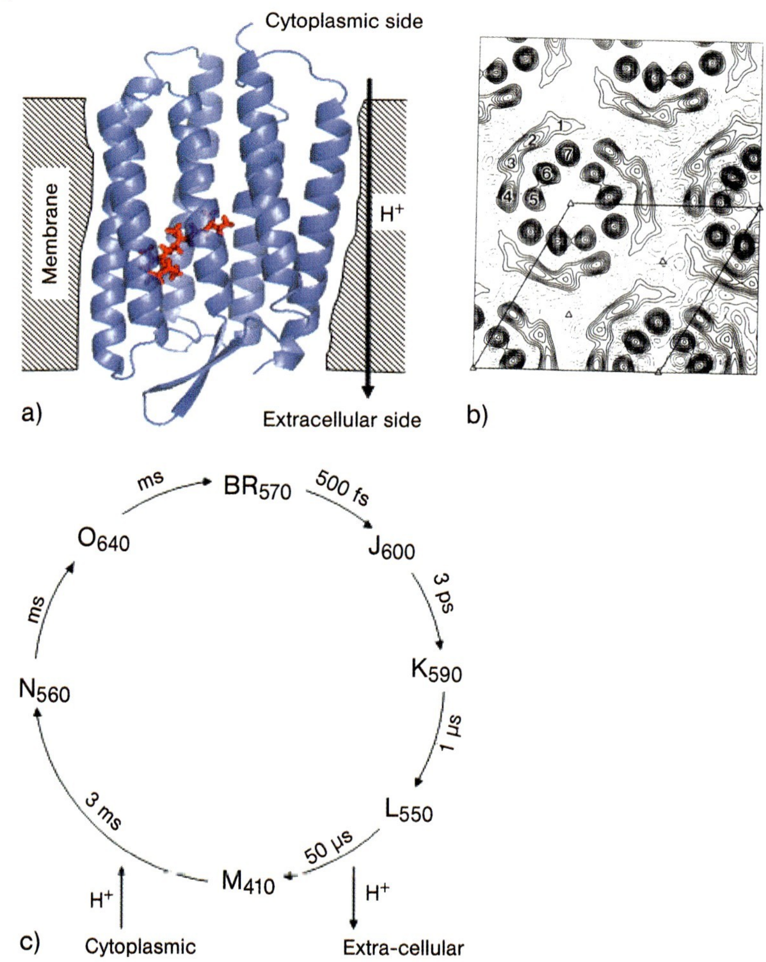

Fig. 9.3 (a) Bacteriorhodopsin (BR) in a schematic bilayer. The structure is taken from Belrhali et al. (1999), PDB entry 1QHJ, with the retinal molecule displayed in red. Fig. made with Pymol (DeLano, 2002). (b) Projection of purple membrane (PM) "in-plane" from the electron microscopy study of Unwin and Henderson (1975). (c) The photocycle of BR. The capital letters indicate the different intermediates, and their indices are the absorption maxima in nanometers. From Oesterhelt (1998). Approximate times for the formation of the different intermediates after absorption of a photon are given.

which is used for ATP production. The photocycle passes through several intermediates, which are characterized by their absorption maxima. An overview of the BR photocycle is provided by Oesterhelt (1998), whilst a schematic representation is shown in Fig. 9.3c. Several photo-intermediates in which structural changes are seen compared to the ground-state structure have also been resolved by X-ray crystallography (Schobert et al., 2002; Lanyi and Schobert, 2003, 2006; Kouyama et al., 2004; Takeda et al., 2004).

Early neutron experiments suggested that, in order to transport protons, conformational flexibility is essential to BR (Zaccai, 1987), and the PM is a particularly well-suited model both for structural and dynamic studies and their relation to biological activity.

9.4 Methods for Labeling

9.4.1 Biosynthetic Labeling

Although the electron microscopy studies of the PM were pioneering in the membrane structure field, neutron diffraction coupled to H–D labeling enabled the electron density map to be interpreted. In addition to providing complementarity to electron microscopy during the initial investigations, neutrons have also revealed information not accessible by other techniques. The elegant labeling techniques which have been crucial to the neutron studies are briefly outlined at this point (Zaccai, 2000b).

It was found that, by growing *H. salinarum* on a standard medium (natural abundance in hydrogen and deuterium), and by supplementing the organism with a given deuterated amino acid, it was possible to obtain extremely high levels of incorporation of the deuterated amino acid, with little spread to other regions of the membrane (Engelman and Zaccai, 1980). For dynamic studies on specific regions of the membrane, it was shown that the reverse was also possible, namely to grow the archaea in a deuterated medium with certain amino acids hydrogenated (Reat et al., 1998).

It has been shown possible to label specifically the retinal (Jubb et al., 1984), and the leucine, valine, phenylalanine, tryptophan and methionine residues (Engelman and Zaccai, 1980; Popot et al., 1989; Reat et al., 1998). Although not yet achieved, the labeling of other residues is also thought possible: amino acids expected not to label well include asparagine, glutamine and glutamic acid, as these are located at the earlier stages of the metabolic pathways and may be scrambled by *H. salinarum*, with the label not being directly incorporated.

Although during the 1990s several groups labored intensely to grow crystals of BR in order to obtain high-resolution structures of the protein, neutrons still offered unique possibilities for studying the structure of the natural lipids of the membrane. By growing the cells on a standard medium and by adding deuterated glucose, deuteration was found to occur exclusively within the sugar moieties of one of the major glycolipids (Weik et al., 1998a).

9.4.2 Reconstitution

In 1986, Popot and coworkers (Popot et al., 1986) demonstrated a second approach to labeling, by first disassociating BR from its lipids and then reconstituting it. It was shown to be possible to reassemble different parts of the membrane from differently labeled cultures. The lipids and retinal are separated from BR using gel chromatography, after which, by using a specific ratio of *H. salinarum* lipids, retinal and BR, these components could be precipitated with potassium dodecylsulfate to reform crystalline patches. Popot's group subsequently showed that the crystalline patches formed had the same lattice, with unchanged unit cell dimensions, and with a similar long-range order to the naturally forming membrane patches.

Also in 1986, Trewhella and colleagues (Trewhella et al., 1986) used the reconstitution technique described, coupled with H–D labeling, to perform structural investigations. Chymotrypsin was used to cleave BR into two fragments before reconstituting it; these two fragments consisted of five and two transmembrane helices. By comparing the diffraction from a native membrane with that of a membrane where two of the supposed helices (at this time the helical structure was not universally accepted) were deuterated, it was possible to localize the fragments in a 2-D map. Samatey et al. (1994) subsequently used both biosynthetic labeling and reconstitution to establish the rotational orientation of the helices of BR.

9.5 Neutrons for Structural Studies of Membrane Proteins

9.5.1 Neutron Diffraction

Neutron diffraction measurements on PM furnished structural information on all three of the membrane's components, namely BR, lipids, and water. Most such investigations have included H–D labeling, either by biosynthetic incorporation of the label, or by using H_2O–D_2O exchange.

In a typical experiment, the data are collected from labeled and unlabeled PM, after which standard crystallographic difference Fourier method or model fit approaches are used to locate the labels in 2-D projections.

9.5.1.1 Bacteriorhodopsin

Two-dimensional neutron diffraction powder patterns, collected from stacked, oriented PM with different deuterated amino acids showed at an early stage, before little was known about membrane proteins, that BR was an "inside-out" protein (Engelman and Zaccai, 1980), with hydrophobic residues pointing into the lipid environment. The technique was used to identify the position and orientation of transmembrane helices when the BR sequence became available, and the distribution of other amino acids within the structure (Popot et al., 1989; Samatey et al., 1994).

An early study using deuterium labeling (Jubb et al., 1984) enabled the authors to position the retinal in the 2-D electron microscopy map. Later, a study was performed showing that, during the photocycle, the retinal is tilted compared to its ground state (Hauss et al., 1994); this involved using retinal molecules with two different deuteration states, and trapping BR in an intermediate structure.

Neutron diffraction has also been used to study changes in the structure of BR at different stages of the photocycle. For example, Dencher et al. (1989) used a low temperature and guanidine hydrochloride to trap BR in its M intermediate, and concluded that structural changes had occurred within the protein. Others (Weik et al., 1998b) used a mutant rather than trapping procedures to perform a similar study.

9.5.1.2 **Lipids**

Studies of the glycolipids in PM provided the first structural localization of glycolipids with respect to a transmembrane protein in a natural membrane environment (Weik et al., 1998a). The difference Fourier maps between labeled (natural membrane with one type of glycolipid deuterated) and unlabeled samples are shown in Fig. 9.4. The localization of two lipid molecules in the membrane, called peak 1 and peak 2, is clearly visible. (The third difference feature seen in Fig. 9.4 is actually weak: the real density is obtained by dividing by three, as it is on a threefold axis.) An analysis of these results suggested the existence of stacking interactions between aromatic residues and the sugar moieties, and provided insight into the possible role of glycolipids in the formation of the 2-D crystal patches. The neutron diffraction identification of the lipid position also permitted assignment of the corresponding electron density in the crystal structures of BR.

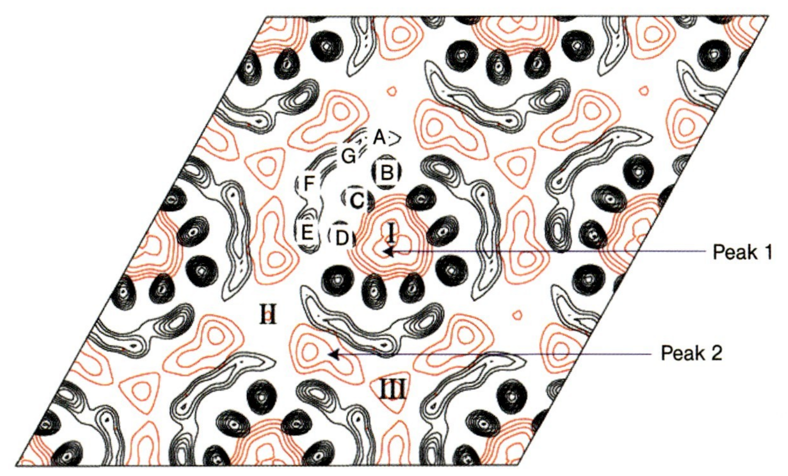

Fig. 9.4 Difference Fourier map between labeled (deuterated glycolipids) and unlabeled purple membrane (PM) samples. Positive difference contour levels are red which correspond to the deuterium label. The native membrane is represented in black. (From Weik et al., 1998.)

9.5.1.3 Water

Neutrons remain an exceptional probe for studying the structure of hydration water. By using H_2O–D_2O exchange, unique information has been obtained on the hydration of PM. In contrast to X-rays, neutron scattering at low resolution enables information to be acquired on the extent and location of hydration, even when the molecules are disordered. In the case of PM, Zaccai and Gilmore established at an early stage that hydration occurred around the lipid head groups, and that there were no pockets in the membrane that would allow for a bulk water channel (Zaccai and Gilmore, 1979). This was followed by several other hydration studies. For example, it was found that in contrast to the BR interior, the hydration around the lipid head groups was modulated by the relative humidity in the membrane atmosphere (Rogan and Zaccai, 1981). A more precise study of water molecules located several in the proton pathway, which was a step towards the elucidation of its mechanism (Papadopoulos et al., 1990). Neutron diffraction experiments on PM as a function of relative humidity and temperature helped relate the hydration structure to photocycle activity, and also suggested the importance of dynamics for function (Zaccai, 1987).

The studies conducted by Dencher et al. (1989) and by Weik et al. (1998b) also employed H_2O–D_2O exchange to study hydration structure as well as structural changes. The results of both studies demonstrated minimal change in hydration between the ground and M states, thereby supporting the idea that, contrary to certain models, structural changes are not correlated with major hydration changes within BR.

9.5.2 Low-Resolution Studies

9.5.2.1 Small-Angle Neutron Scattering of Membrane Proteins in D-Vesicles

In a small-angle scattering experiment, as low resolution is examined, the measured intensity depends on the contrast between the scattering particle and its solvent. In the case of neutron scattering, it is possible to change the contrast using different proportions of D_2O in the solvent, or by labeling different parts of the diffusing particle. The neutron-scattering densities of different components of membranes, as a function of the percentage of D_2O in their solvent, are illustrated graphically in Fig. 9.5. The contrast here is the difference between scattering density of the particle and its solvent. For example, the data in Fig. 9.5 show that at 0% D_2O (100% H_2O), the contrast between protein and water is approximately $2.5 \times 10^{10}\,cm^{-2}$, whereas the contrast from a deuterated protein at 0% D_2O is much larger (approximately $7 \times 10^{10}\,cm^{-2}$).

At certain points on the graph, the scattering lengths for some components are equal to the scattering length of the solvent; therefore, there is no contrast between the two, and no scattering, and this permits the "matching-out" of different components. For example, for a sample containing a membrane protein embedded in a lipid bilayer, at 40% D_2O, the protein will be effectively "invisible", and it is possible to examine the bilayer. Contrast variation studies using SANS can be

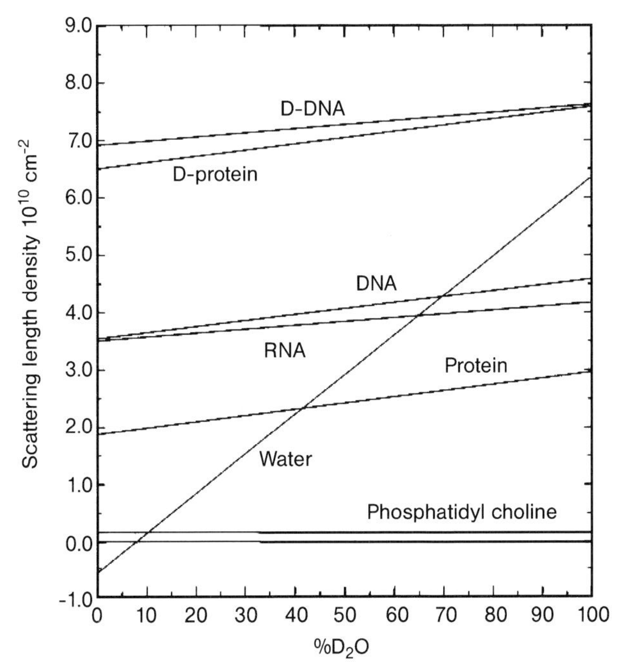

Fig. 9.5 Neutron-scattering densities of various biological molecules as a function of the percent of D_2O in their solvent.

particularly useful in determining how the complexes of different components are formed.

Hunt et al. (1997) subsequently showed that contrast matching could be used to study BR in deuterated phospholipid vesicles. In this type of experiment, the first step is to determine at what point the vesicles are effectively "matched out". A component can be matched out if it can be considered to be homogeneous in scattering density at the resolution of the study. This is the case for deuterated phospholipid molecules, but not for the unlabeled molecule in which the head groups have higher scattering density than the fatty acid chains. The matching point is found by measuring the scattering from the phospholipids only at different concentrations of H_2O–D_2O, and interpolating the measurements to determine the value of the intensity at $\mathbf{Q}=0$. This type of measurement for the phospholipid vesicles, as used by Hunt and coworkers, is shown in Fig. 9.6a. Here, it can be seen that, for the type of phospholipid used in this study, when there is 94% D_2O in the solvent the vesicles are not seen. The scattering profile of phospholipids with BR incorporated is then measured at 94% D_2O. Two important pieces of information can be gained very simply from this measurement:

- From the value of the intensity at $Q=0$, the molecular mass of the protein can be estimated.

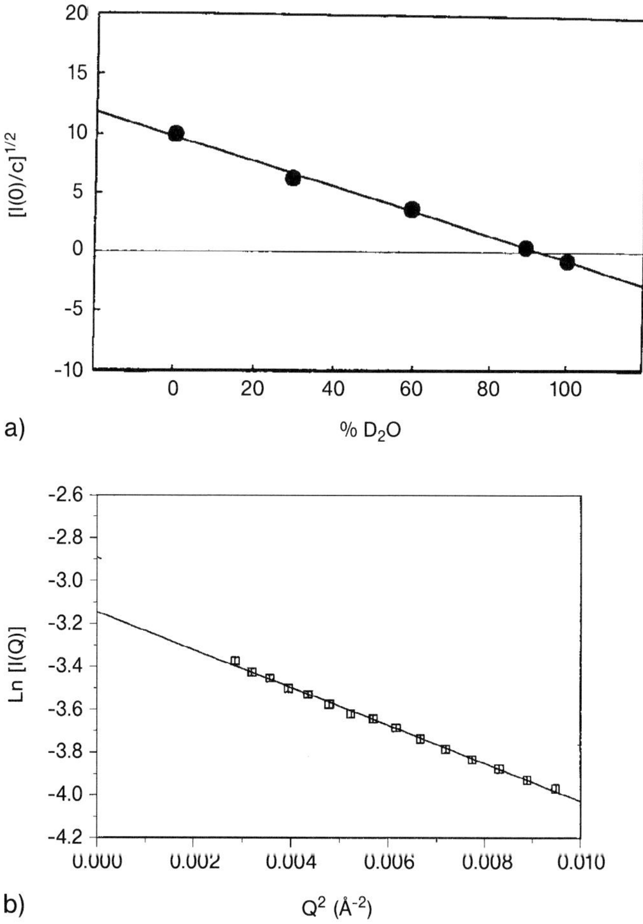

Fig. 9.6 (a) Estimated value of scattered intensity at $\mathbf{Q}=0$ for phospholipid vesicles as a function of percentage of D_2O solvent content. (b) Logarithm of the scattered intensity for bacteriorhodopsin (BR) in vesicles plotted as a function of Q_2 at 94% D_2O. (From Hunt et al., 1997.)

- By plotting the logarithm of the intensity as a function of wavevector squared (Fig. 9.6b), it is possible to determine a radius of gyration of BR, equal to three times the negative slope.

The values obtained provide information on the aggregation state of the protein in the vesicle, and also its global shape. The study described here proved that it was possible to perform SANS measurements on membrane proteins and to determine structural parameters; however, this type of experiment remains challenging.

SANS represents an extremely useful technique to study structural changes when a component of the solvent is changed (e.g., studying the effect of a change in pH, salt concentration, etc.). Gilbert and co-workers (Gilbert et al., 1999) used contrast variation and SANS in combination to study the effect of the addition of the pore-forming toxin pneumolysin on the thickness of vesicles. As the vesicles may be considered as models for the cell membrane, the results were interpreted in the context of understanding how the toxin causes perforation of the membrane.

9.5.2.2 Low-Resolution Single-Crystal Studies

The efforts which are being made to grow membrane-protein crystals for X-ray diffraction studies are yielding fascinating results that are rich in atomic detail. A neutron crystallographic study can complement such studies by providing information on a component which is not seen due to disorder, although this has not yet been achieved for BR crystals. At the time when the first membrane protein crystals were produced using detergents, Roth and coworkers (Roth et al., 1989) performed a neutron crystallographic study of a bacterial photosynthetic reaction center to determine the detergent structure, as only one molecule of detergent was sufficiently well-ordered to be seen in the X-ray structure.

The achievements of Roth et al. (1989) were made possible with H_2O/D_2O contrast variation studies, similar to those of SANS. The crystals are measured at several H_2O/D_2O contents, with the contrast between the detergent, reaction center and the solvent differing. A particular phasing protocol is then followed, to calculate a low-resolution structure. At around 40% D_2O the average protein is hardly visible, and the detergent is most clearly seen. The packing of the detergent in the crystal (Fig. 9.7a) was found to be concentrated in rings around the transmembrane helices of the reaction center (Fig. 9.7b).

An understanding of the protein–detergent interaction can provide a model for the interaction between the lipid bilayer and the protein *in vivo*. It is also of great importance for understanding the process of crystallization of membrane proteins. A similar type of investigation was also performed on several other crystals. By studying OmpF porin structures, Pebay-Peyroula et al. (1995) again displayed the detergent covering the hydrophobic moieties of the protein, and found rings of aromatic residues delimiting the detergent-binding zone, thus supporting the idea of the importance of these residues in the stability of membrane proteins in their natural lipid bilayers. The ability to observe the distribution of detergent in two different crystalline forms of OmpF (tetragonal and trigonal) pointed to different interactions dictating the crystallization process, gaining insight into the crystallization processes of membrane proteins (Penel et al., 1998). A schematic representation of the distribution is shown in Fig. 9.8.

9.5.2.3 Reflectivity

Just as when light traverses media of different refractive indices, a neutron beam is in part refracted and in part reflected at an interface between media of different

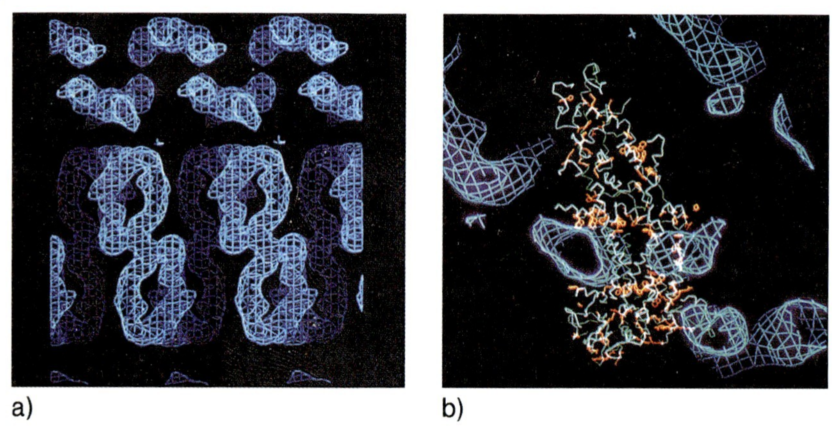

Fig. 9.7 (a) View of the packing of detergent rings in a crystal of bacterial photosynthetic reaction center. (b) The reaction center molecule (the backbone is represented in green, and Trp, Arg and His residues in red) surrounded by the detergent, represented in blue. (From Roth et al., 1989.)

scattering length density. In the case of a single layer in the beam (Fig. 9.9a), an interference pattern is observed between the neutrons reflected from the first and second borders of the layer. An example of an interference pattern for a free-standing film is shown in Fig. 9.9b. The interference pattern contains information about the thickness of the layer and its scattering density distribution.

The fact that reflectivity contains information about layered samples, and that only extremely small amounts of sample are required, has very quickly made the technique popular not only for the study of membranes but also for monitoring the interaction of proteins with membranes.

Two types of scattering can be distinguished in a reflectivity measurement, namely specular and off-specular. When the incident and scattered beam angles are identical and the beams are in the same plane (as in Fig. 9.9a), the scattering is termed "specular". If the layer is homogeneous in scattering density along the surface, then this is the only type of scattering that will occur. If, however, in the plane of the layer there are differences in scattering density (e.g., a membrane in which proteins are imbedded), then a second type of scattering occurs, for which the angles are different; this is termed "off-specular" scattering. Although this contains information about the in-plane structure (which may be of great interest in the study of membrane proteins), the difficulties associated with sample preparation and data interpretation means that analyses of this type of scattering remain a challenge for future membrane studies.

The analysis of specular scattering provides information about the thickness of the different components of the layer (which have different scattering length densities). A single reflectivity curve can by fitted by a family of scattering density profiles. An example of a bilayer membrane is shown in Fig. 9.9c (from Krueger

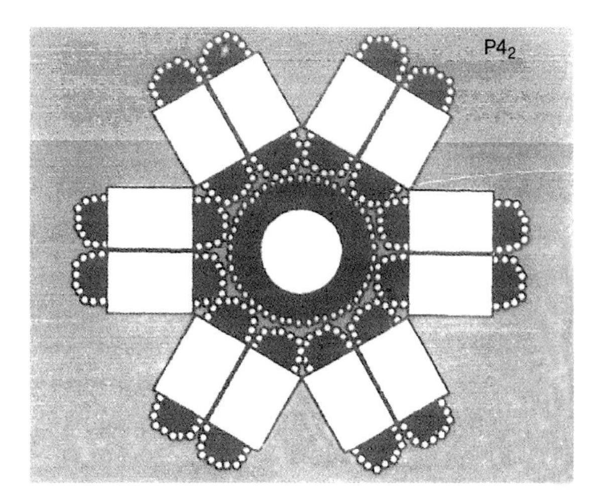

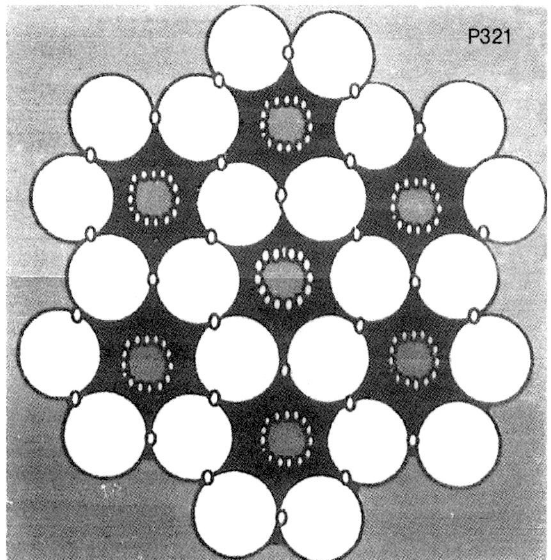

Fig. 9.8 Schematic representation of two different crystal forms of OmpF porin from *Escherichia coli*. In the $P4_2$ form, the crystal is formed by hydrophilic protein–protein and detergent–detergent interactions. In the P321 form, the crystal is stabilized by hydrophobic protein–protein interactions. (From Penel et al., 1998.)

et al., 2001), and a model of the bilayer can then be built up with a resolution of the order of Ångstroms. Such studies are becoming increasingly biomimetic, with improvements in the way that the bilayer is prepared (Vacklin et al., 2005a). Apart from determining the structure of membranes, the specular reflection technique is also very effective for studying protein–membrane interactions. For example,

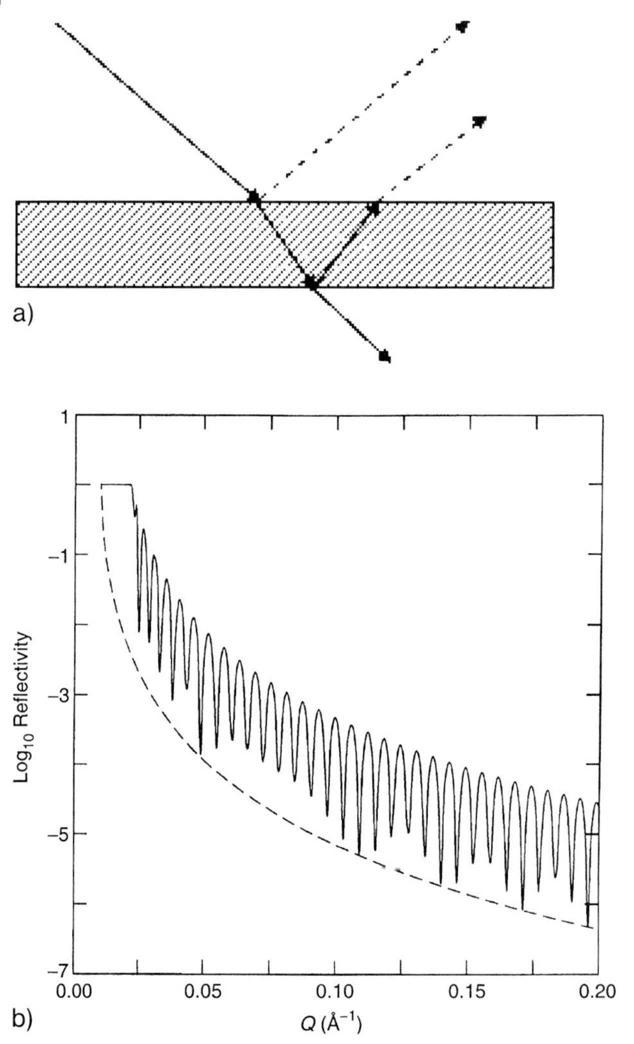

Fig. 9.9 (a) Schematic representation of a reflectivity experiment on a single thin film, represented by the hashed box; incident, reflected and refracted beams are shown as arrows. An interference pattern is observed between the beams represented by dashed arrows. (b) Specular neutron reflectivity for a free standing film of 100 nm-thick nickel (From Majkrzak et al., 2006.) (c) Family of scattering length density profiles obtained from fits to the neutron reflectivity data shown in the inset, all the profiles fit the data equally well. The different chemical groups of the sample are indicated. (From Krueger et al., 2001.)

Johnson et al. (2005) used a combination of X-ray and neutron reflectivity to reveal structural information for the binding of a neural cell adhesion molecule to a membrane. Likewise, a study of venom phospholipidase A_2 revealed how the originally water-soluble protein became adsorbed to the membrane (Vacklin et al., 2005b).

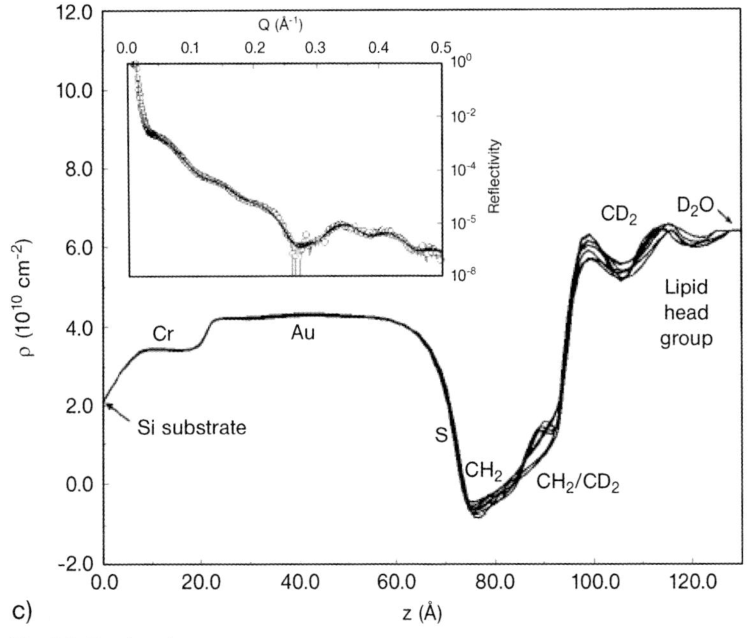

Fig. 9.9 *Continued*

The reflectivity technique is developing quickly and exciting results are expected within the next few years.

9.6 Neutrons for Dynamical Studies of Membrane Proteins

9.6.1 Energy-Resolved Experiments

As mentioned earlier the measurement of scattered neutrons as a function of momentum and energy transfer provides information on the molecular dynamics of a sample. The discussion here focuses on incoherent scattering measurements. As the incoherent cross section of hydrogen is an order of magnitude larger than all others, the scattering is dominated by hydrogens, which reflect the dynamics of the groups to which they are attached.

A schematic representation of scattered spectra is given in Fig. 9.10; the comprehension of this figure is important, as it incorporates all of the key elements of energy-resolved neutron scattering. In the figure, the scattering is represented at two temperatures, T_1 and T_2, where T_1 is the lower temperature. As can be seen, compared to diffraction an extra dimension of intensity is measured: the intensity is recorded both as a function of the wavevector $\mathbf{Q}$ and of the energy transfer, E.

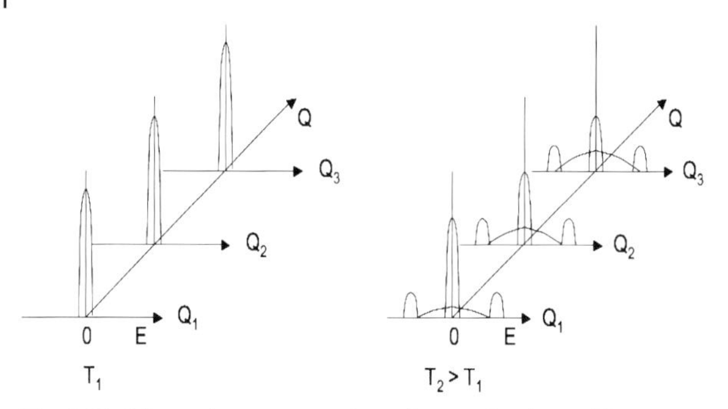

Fig. 9.10 Schematic representation of scattering intensities as a function of wave-vector (**Q**) and energy transfer (E) at two temperatures, T_1 and T_2.

If a neutron interacts with an immobile atom, its momentum changes, but its energy is conserved. At very low temperature (e.g., 20 K), all atomic motion can be considered to be frozen out, with the atoms immobile (this corresponds to T_1 in Fig. 9.10). All of the neutrons scattered by the sample are scattered elastically, ($|\mathbf{k}_i| = |\mathbf{k}_f|$), and an "elastic peak", centered at zero energy transfer, is measured at all **Q**-values. For an ideal spectrometer, the elastic peak is a Dirac function, whereas for a real spectrometer the elastic peak has a certain form and width, which is referred to as the instrument's *resolution*.

At higher temperatures, the atoms begin to move so that the elastic peak begins to diminish in height, and two other types of scattering are observed: quasi-elastic and inelastic, which increase as a function of temperature (temperature T_2 in Fig. 9.10). The total scattering is generally normalized to unity, and the sum of elastic, quasi-elastic and inelastic is equal to one at all temperatures.

Quasielastic broadening is due to diffusive processes within the sample. The quasielastic peak is centered at zero energy transfer, and its width is inversely proportional to the relaxation time. Different models are used to interpret this broadening.

Inelastic scattering arises from the vibrations of individual atoms at defined frequencies. Details of the information that can be gathered from these three types of scattering in the case of the PM model will be discussed after some brief points have been made with regards to the accessible time and space scales.

9.6.1.1 **Time and Space Scales**

The narrower the energy resolution, the more easily it will be possible to distinguish quasi-elastic broadening from an elastic peak. Just as **Q**-space corresponds in reciprocal fashion to real space, energy corresponds in reciprocal fashion to time. The energy and **Q**-values accessible to a spectrometer correspond to a window in space and time. A time-of flight spectrometer typically has an energy resolution (the width of the measured elastic peak at half-maximum) that can be varied between 50 and 250 μeV, with a trade-off for better energy resolution always being lower flux. This corresponds to times of 2 to 10 ps. Backscattering spectrometers have narrower energy resolutions (~1 μeV) corresponding to longer times (~1 ns).

It is worth noting that these are the time scales also accessible to molecular dynamics simulations, and the combination of these two approaches has been particularly fruitful, both for the development of atomic force fields used for molecular dynamics simulations, and for the interpretation of neutron scattering data. A typical study using this combination of techniques was conducted by Hayward et al. (2003).

9.6.2
Elastic Scattering and Atomic Mean Square Displacements

Performing full-energy scans with sufficient statistical accuracy to be able to interpret the observed scattering through complete models describing the atomic motion can often be prohibitive in terms of measuring time for biological samples. A relatively rapid method of characterizing a sample is, therefore, simply to measure the elastic intensity as a function of temperature. During such as experiment the scattered neutrons' energy is recorded, but only at the elastic peak, thereby significantly reducing the experimental time. It is possible to extract, from the way that the logarithm of the elastic intensity decreases as a function of $\mathbf{Q}^2$, an atomic mean square displacement (MSD); this corresponds to the mean fluctuation that an atom undergoes in the time and space window as defined by the instrument's energy resolution and Q-range.

Following the dynamics–function hypothesis proposal by Zaccai (1987), MSD-values were first calculated for dry PM (0.02 g D_2O g^{-1} BR) and for wet PM (0.55 g g^{-1}) (Ferrand et al., 1993); the MSD-values are shown in Fig. 9.11. At low temperature, both the dry and wet samples behave like harmonic solids, with their MSDs as a function of temperature staying linear. However, at about 230 K, the MSDs become very different, with the wet sample now exploring a much larger conformational space, whilst the dry sample does not. On a functional level, the photocycle is slowed down by several orders of magnitude in the dry sample. It was postulated, therefore, that the membrane requires these larger conformational motions in order to function correctly.

Using labeling techniques, a similar type of dynamic study was then further extended to concentrate on different parts of the membrane. Reat et al. (1998) produced a completely deuterated PM in which the retinal, tryptophan and methionine residues were all hydrogenated. The tryptophan and methionine residues are mainly located in the retinal-binding pocket. The MSDs of the two samples are shown in Fig. 9.12 (both are hydrated to 0.45 g D_2O g^{-1} PM). An investigation of the dynamics of such a sample furthered the understanding of how the pump functions, with the core of BR being found significantly stiffer than the remainder of the membrane. The result was correlated to the function, with the "valve" of the pump (corresponding to the Schiff base proton changing its accessibility from one side of the membrane to the other) being held in a stiff local environment while the remainder of the protein was softer.

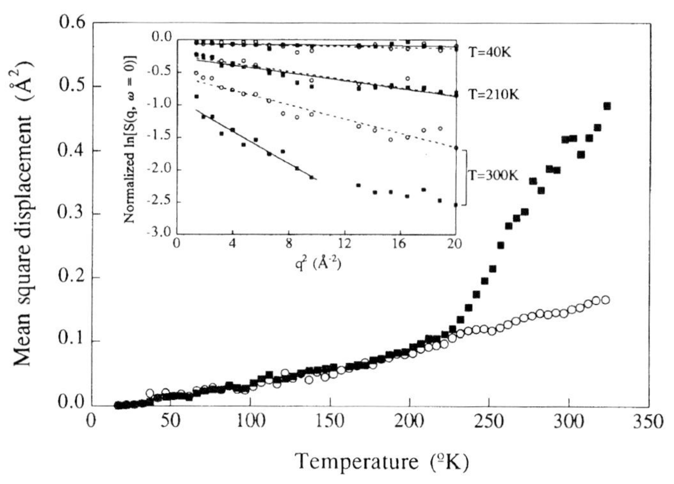

Fig. 9.11 Mean square displacements in dry (○) and hydrated (■) purple membrane (PM) as a function of temperature. The inset corresponds to examples of the fits of the normalized intensity as a function of $\mathbf{Q}^2$. Measured on the IN13 backscattering spectrometer at the ILL; energy resolution ~10 μeV, $1<\mathbf{Q}^2<8\,\text{Å}^2$. (From Ferrand et al., 1993.)

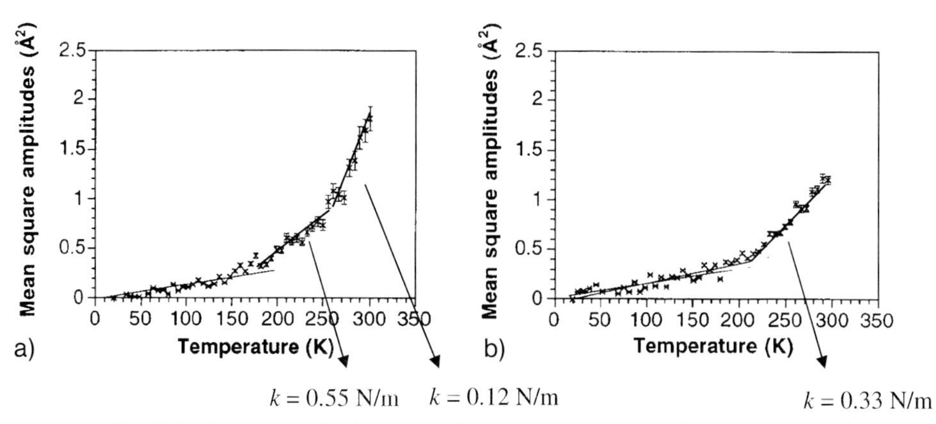

Fig. 9.12 Mean square displacements of a completely deuterated membrane (A) and a deuterated membrane with hydrogenated retinal, tryptophan and methionine residues (B). Measured on the IN10 backscattering spectrometer at the ILL, energy resolution ~1 μeV, $0.2<\mathbf{Q}^2<3\,\text{Å}^2$. (From Reat et al., 1998.) The definition of the mean square displacement in this study differs by a factor 2 compared to the data shown in Fig. 9.11.

Subsequently, Zaccai (2000a) introduced a "force constant analysis" to describe this "stiffness" and "softness" of proteins in a more quantitative manner. The force constant is proportional to the inverse of the slope of the MSD as a function of temperature, and is expressed in Newtons m^{-1}. Examples of the force constant analysis in the case of the labeled PM from the study by Reat and colleagues (Reat et al., 1998) are illustrated graphically in Fig. 9.12, where the force constant is denoted *k*. It should be noted that in this figure the force constant at high temperature is much lower for the unlabeled sample than for the labeled sample, which translates into a “stiffer” potential that governs atomic motions.

9.6.3 Quasi-Elastic Scattering

Quasi-elastic scattering is rich in information and, in the case of simple systems, it can be modeled to establish microscopic properties. In the case of biological membranes, fitting models to this scattering is of course more limited in its extension because of their complexity. Nonetheless, Fitter and coworkers have shown in several studies the validity of this approach, having successfully modeled the diffusive behavior of the hydrogen atoms in PM, by fitting the quasielastic data with several Lorentzian functions (Fitter et al., 1996, 1997). The model involved dividing the hydrogen atoms into several classes, methyl group rotation (three-site jump) and two-site jumps. These authors also found diffusive motions to be highly hydration-dependent, and examined the differences in these motions both parallel and perpendicular to the plane of the membrane.

9.6.4 Inelastic Scattering

Peaks can be identified in a neutron-scattering spectrum at specific energies, referred to as “inelastic scattering”. Such peaks are due to energy exchange with atomic vibrations at precise frequencies within the sample. Notably, inelastic measurements have been used in the development of molecular dynamics simulations.

Ferrand and coworkers (Ferrand et al., 1993) measured incoherent inelastic spectra of dry and hydrated PM; the spectra are illustrated in Fig. 9.13. The measurements show that, in the wet sample and at higher temperatures, the protein vibrations shift to lower frequencies, corresponding to larger time scales.

9.6.5 Other Types of Measurement

In the present chapter, attention has been focused on the better established techniques, it should be noted that two other types of measurement are used:

- All studies presented so far have been based on measuring the “self” dynamics of atoms through the study of incoherent neutron scattering. Neutrons represent a powerful means of studying the collective dynamics (phonons) of an ordered sample using coherent neutron scattering. These types of measurement have been

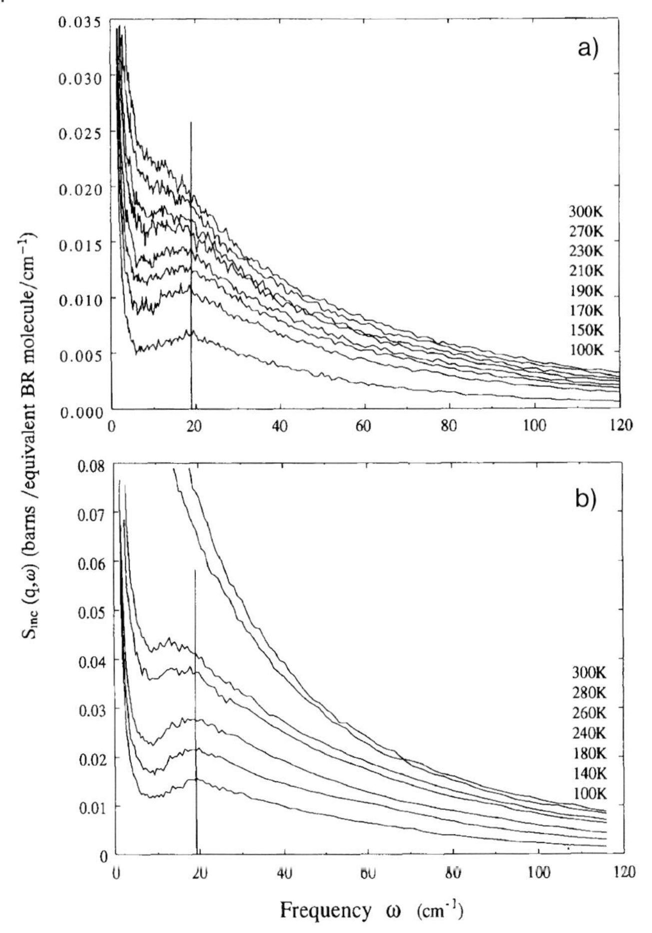

Fig. 9.13 Temperature dependence of the incoherent inelastic scattering, in dry (a) and wet (b) purple membrane (PM). The intensity in this part of the spectra increases with increasing temperature (the lowest intensity corresponds to 100 K). (From Ferrand et al., 1993.)

fundamental in condensed matter studies using so-called "triple-axis" spectrometer. A lack of suitable samples has meant the technique has only recently been applied to biological samples: Rheinstadter and coworkers (Rheinstadter et al., 2004) studied the dispersion relation of phonons in lipid samples, and further reports of this type of study are expected.

- The studies cited throughout this chapter have been performed on neutron instruments that explore movements on the ps–ns timescale (time-of-flight and backscattering spectrometers). Currently, a different type of spectrometer – neutron spin echo – is being used to study biological macromolecules on a timescale which reaches ~100 ns. A first report in which dispersion in aligned bilayers was examined has been published (Rheinstadter et al., 2006).

As mentioned above, neutrons are of particular interest in the study of water. Here, although focus has centered on the structure of hydration water, neutrons have also been instrumental in understanding the microscopic dynamic properties of the liquid without which, as most people agree, life would not be possible. By studying PM in D_2O and H_2O and correctly subtracting the signal from the membrane, it has been possible to deduce the diffusion coefficient for water in contact with PM (Lechner et al., 1994). Other investigations were related to the behavior of water confined near PM (Lechner et al., 1998; Weik et al., 2005).

9.7 Take-Home Message

Although neutron scattering often requires strong efforts in sample preparation, it is ultimately able to provide (often) unique information that is not accessible by other means:

- In studies of low-resolution structures it is possible, by using deuterium labeling, to identify disordered components such as detergent/lipid and water molecules. As membrane proteins cannot generally function in the absence of either element, an understanding of the interaction between a protein and its environment can be considered as important as knowing the protein structure in atomic detail.
- In aiming to understand the complex dynamic picture of membrane proteins, it will surely be necessary to combine many techniques. For this, neutron scattering can play a role, as the time and space scales accessed are often not available with other methods. Moreover, neutron scattering is especially well-suited for combination with molecular dynamics simulations.

The future of neutron scattering is bright, with the construction of a new generation of sources in America and Japan underway, and a European source currently under discussion.

References

Belrhali, H., Nollert, P., Royant, A., Menzel, C., Rosenbusch, J. P., Landau, E. M., and Pebay-Peyroula, E. (1999) Protein, lipid and water organization in bacteriorhodopsin crystals: a molecular view of the purple membrane at 1.9 Å resolution. *Structure, 7,* 909–917.

Chadwick, J. (1932a) The existence of a neutron. *Proc. Roy. Soc. A, 136,* 692–708.

Chadwick, J. (1932b) Possible existence of a neutron. *Nature, 129,* 312.

DeLano, W. L. (2002) The PyMOL Molecular Graphics System. http://www.pymol.org. DeLano Scientific, San Carlos, CA, USA.

Dencher, N. A., Dresselhaus, D., Zaccai, G., and Buldt, G. (1989) Structural changes in bacteriorhodopsin during proton translocation revealed by neutron diffraction. *Proc. Natl. Acad. Sci. USA, 86,* 7876–7879.

Engelman, D. M. and Zaccai, G. (1980) Bacteriorhodopsin is an inside-out protein. *Proc. Natl. Acad. Sci. USA, 77,* 5894–5898.

Ferrand, M., Dianoux, A. J., Petry, W., and Zaccai, G. (1993) Thermal motions and function of bacteriorhodopsin in purple membranes: effects of temperature and hydration studied by neutron scattering. *Proc. Natl. Acad. Sci. USA, 90,* 9668–9672.

Fitter, J., Lechner, R. E., Buldt, G., and Dencher, N. A. (1996) Internal molecular motions of bacteriorhodopsin: hydration-induced flexibility studied by quasielastic incoherent neutron scattering using oriented purple membranes. *Proc. Natl. Acad. Sci. USA, 93,* 7600–7605.

Fitter, J., Lechner, R. E., and Dencher, N. A. (1997) Picosecond molecular motions in bacteriorhodopsin from neutron scattering. *Biophys. J., 73,* 2126–2137.

Gilbert, R. J., Heenan, R. K., Timmins, P. A., Gingles, N. A., Mitchell, T. J., Rowe, A. J., Rossjohn, J., Parker, M. W., Andrew, P. W., and Byron, O. (1999) Studies on the structure and mechanism of a bacterial protein toxin by analytical ultracentrifugation and small-angle neutron scattering. *J. Mol. Biol., 293,* 1145–1160.

Hauss, T., Buldt, G., Heyn, M. P., and Dencher, N. A. (1994) Light-induced isomerization causes an increase in the chromophore tilt in the M intermediate of bacteriorhodopsin: a neutron diffraction study. *Proc. Natl. Acad. Sci. USA, 91,* 11854–11858.

Hayward, J. A., Finney, J. L., Daniel, R. M., and Smith, J. C. (2003) Molecular dynamics decomposition of temperature-dependent elastic neutron scattering by a protein solution. *Biophys. J., 85,* 679–685.

Henderson, R., Baldwin, J. M., Ceska, T. A., Zemlin, F., Beckmann, E., and Downing, K. (1990) Model for the structure of bacteriorhodopsin based on high-resolution electron cryo-microscopy. *J. Mol. Biol., 213,* 889–929.

Hunt, J. F., McCrea, P. D., Zaccai, G., and Engelman, D. M. (1997) Assessment of the aggregation state of integral membrane proteins in reconstituted phospholipid vesicles using small angle neutron scattering. *J. Mol. Biol., 273,* 1004–1019.

Johnson, C. P., Fragneto, G., Konovalov, O., Dubosclard, V., Legrand, J.-F., and Leckband, D. E. (2005) Structural studies of the neural-cell-adhesion molecule by x-ray and neutron reflectivity. *Biochemistry, 44,* 546–554.

Jubb, J. S., Worcester, D. L., Crespi, H. L., and Zaccai, G. (1984) Retinal location in purple membrane of *Halobacterium halobium*: a neutron diffraction study of membranes labelled in vivo with deuterated retinal. *EMBO J., 3,* 1455–1461.

Kouyama, T., Nishikawa, T., Tokuhisa, T., and Okumura, H. (2004) Crystal structure of the L intermediate of bacteriorhodopsin: evidence for vertical translocation of a water molecule during the proton pumping cycle. *J. Mol. Biol., 335,* 531–546.

Krueger, S., Meuse, C. W., Majkrzak, C. F., Dura, J. A., Berk, N. F., Tarek, M., and Plant, A. L. (2001) Investigation of hybrid bilayer membranes with neutron reflectometry: probing the interactions of melittin. *Langmuir, 17,* 511–521.

Lanyi, J. K. and Schobert, B. (2003) Mechanism of proton transport in bacteriorhodopsin from crystallographic structures of the K, L, M1, M2, and M2′ intermediates of the photocycle. *J. Mol. Biol.*, *328*, 439–450.

Lanyi, J. K. and Schobert, B. (2006) Structural changes in the L photointermediate of bacteriorhodopsin. *J. Mol. Biol.*, *365*, 1379–1392.

Lechner, R. E., Dencher, N. A., Fitter, J., Buldt, G., and Belushkin, A. V. (1994) Proton diffusion on purple membrane studied by neutron scattering. *Biophys. Chem.*, *49*, 91–99.

Lechner, R. E., Fitter, J., Dencher, N. A., and Hauss, T. (1998) Dehydration of biological membranes by cooling: an investigation on the purple membrane. *J. Mol. Biol.*, *277*, 593–603.

Luecke, H., Schobert, B., Richter, H. T., Cartailler, J. P., and Lanyi, J. K. (1999) Structure of bacteriorhodopsin at 1.55 Å resolution. *J. Mol. Biol.*, *291*, 899–911.

Majkrzak, C., Berk, N., Krueger, S., and Perez-Salas, U. (2006) in *Neutron Scattering in Biology* (Springer-Verlag, Berlin Heidelberg), pp. 225–263.

Oesterhelt, D. (1998) The structure and mechanism of the family of retinal proteins from halophilic archaea. *Curr. Opin. Struct. Biol.*, *8*, 489–500.

Papadopoulos, G., Dencher, N. A., Zaccai, G., and Buldt, G. (1990) Water molecules and exchangeable hydrogen ions at the active centre of bacteriorhodopsin localized by neutron diffraction. Elements of the proton pathway? *J. Mol. Biol.*, *214*, 15–19.

Pebay-Peyroula, E., Garavito, R. M., Rosenbusch, J. P., Zulauf, M., and Timmins, P. A. (1995) Detergent structure in tetragonal crystals of OmpF porin. *Structure*, *3*, 1051–1059.

Penel, S., Pebay-Peyroula, E., Rosenbusch, J., Rummel, G., Schirmer, T., and Timmins, P. A. (1998) Detergent binding in trigonal crystals of OmpF porin from *Escherichia coli*. *Biochimie*, *80*, 543–551.

Popot, J. L., Trewhella, J., and Engelman, D. M. (1986) Reformation of crystalline purple membrane from purified bacteriorhodopsin fragments. *EMBO J.*, *5*, 3039–3044.

Popot, J. L., Engelman, D. M., Gurel, O., and Zaccai, G. (1989) Tertiary structure of bacteriorhodopsin. Positions and orientations of helices A and B in the structural map determined by neutron diffraction. *J. Mol. Biol.*, *210*, 829–847.

Reat, V., Patzelt, H., Ferrand, M., Pfister, C., Oesterhelt, D., and Zaccai, G. (1998) Dynamics of different functional parts of bacteriorhodopsin: H-2H labeling and neutron scattering. *Proc. Natl. Acad. Sci. USA*, *95*, 4970–4975.

Rheinstadter, M. C., Ollinger, C., Fragneto, G., Demmel, F., and Salditt, T. (2004) Collective dynamics of lipid membranes studied by inelastic neutron scattering. *Phys. Rev. Lett.*, *93*, 108107.

Rheinstadter, M. C., Haussler, W., and Salditt, T. (2006) Dispersion relation of lipid membrane shape fluctuations by neutron spin-echo spectrometry. *Phys. Rev. Lett.*, *97*, 048103.

Rogan, P. K. and Zaccai, G. (1981) Hydration in purple membrane as a function of relative humidity. *J. Mol. Biol.*, *145*, 281–284.

Roth, M., Lewit-Bentley, A., Michel, H., Deisenhofer, J., Huber, R., and Oesterhelt, D. (1989) Detergent structure in crystals of a bacterial photosynthetic reaction centre. *Nature*, *340*, 659–662.

Samatey, F. A., Zaccai, G., Engelman, D. M., Etchebest, C., and Popot, J. L. (1994) Rotational orientation of transmembrane alpha-helices in bacteriorhodopsin. A neutron diffraction study. *J. Mol. Biol.*, *236*, 1093–1104.

Schobert, B., Cupp-Vickery, J., Hornak, V., Smith, S., and Lanyi, J. (2002) Crystallographic structure of the K intermediate of bacteriorhodopsin: conservation of free energy after photoisomerization of the retinal. *J. Mol. Biol.*, *321*, 715–726.

Sears, V. F. (1984) *Thermal neutron scattering lengths and cross sections for condensed-matter research* (Atomic Energy Canada Ltd, Chalk River, Ontario, Canada).

Takeda, K., Matsui, Y., Kamiya, N., Adachi, S., Okumura, H., and Kouyama, T. (2004) Crystal structure of the M intermediate of bacteriorhodopsin: allosteric structural changes mediated by sliding movement of a transmembrane helix. *J. Mol. Biol.*, *341*, 1023–1037.

Trewhella, J., Popot, J. L., Zaccai, G., and Engelman, D. M. (1986) Localization of two chymotryptic fragments in the structure of renatured bacteriorhodopsin by neutron diffraction. *EMBO J.*, *5*, 3045–3049.

Unwin, P. N. T. and Henderson, R. (1975) Molecular structure determination by electron microscopy of unstained crystalline specimens. *J. Mol. Biol.*, *94*, 425–440.

Vacklin, H. P., Tiberg, F., Fragneto, G., and Thomas, R. K. (2005a) Composition of supported model membranes determined by neutron reflection. *Langmuir*, *21*, 2827–2837.

Vacklin, H. P., Tiberg, F., Fragneto, G., and Thomas, R. K. (2005b) Phospholipidase A2 hydrolysis of supported phospholipid bilayers: a neutron reflectivity and ellipsometry study. *Biochemistry*, *44*, 2811–2821.

Weik, M., Patzelt, H., Zaccai, G., and Oesterhelt, D. (1998a) Localization of glycolipids in membranes by in vivo labeling and neutron diffraction. *Mol. Cell*, *1*, 411–419.

Weik, M., Zaccai, G., Dencher, N. A., Oesterhelt, D., and Hauss, T. (1998b) Structure and hydration of the M-state of the bacteriorhodopsin mutant D96N studied by neutron diffraction. *J. Mol. Biol.*, *275*, 625–634.

Weik, M., Lehnert, U., and Zaccai, G. (2005) Liquid-like water confined in stacks of biological membranes at 200 k and its relation to protein dynamics. *Biophys. J.*, *89*, 3639–3646.

Zaccai, G. (1987) Structure and hydration of purple membranes in different conditions. *J. Mol. Biol.*, *194*, 569–572.

Zaccai, G. (2000a) How soft is a protein? A protein dynamics force constant measured by neutron scattering. *Science*, *288*, 1604–1607.

Zaccai, G. (2000b) Moist and soft, dry and stiff: a review of neutron experiments on hydration-dynamics-activity relations in the purple membrane of Halobacterium salinarum. *Biophys. Chem.*, *86*, 249–257.

Zaccai, G. and Gilmore, D. J. (1979) Areas of hydration in the purple membrane of *Halobacterium halobium*: a neutron diffraction study. *J. Mol. Biol.*, *132*, 181–191.

Part V
Spectroscopies

10
Circular Dichroism: Folding and Conformational Changes of Membrane Proteins

Nadège Jamin and Jean-Jacques Lacapère

10.1
Introduction

Circular dichroism (CD) is an absorption spectroscopy method that utilizes the differential absorption of right- and left-circularly polarized light by chiral molecules. Most biological molecules, including proteins, are chiral molecules and therefore can be studied by using CD. The theory of CD and its applications to biochemistry have been described and discussed in many excellent books and review articles (e.g., Refs. [1–3]). In this chapter, attention will be focused on the applications of CD to membrane proteins.

The chromophores of interest in proteins are:

- below 240 nm, the peptide groups in the far-UV
- between 260 and 320 nm, in the near UV, the aromatic residues (Phe, Tyr and Trp)
- the disulfide groups (about 260 nm)
- above 350 nm, extrinsic chromophores such as heme groups, chlorophyll, and flavins.

As a consequence, a variety of types of information are obtained from CD data obtained from the different spectral regions analyzed, and especially of the secondary structure composition from the far-UV region, the tertiary structure fingerprint from the near-UV region, and information regarding the environments of the bound cofactors or ligands. In contrast to nuclear magnetic resonance (NMR) or X-ray crystallography – both of which provide high-resolution data – CD is a low-resolution spectroscopy that describes overall structural features. Other spectroscopic techniques, such as infrared (IR) or Raman also provide structural information (see Chapters 11 and 12).

In general, membrane proteins (MPs) solubilized in detergent micelles do not present any particular difficulties, and CD spectra are recorded as for soluble proteins. In the presence of lipids, however, various complications caused by artifacts such as light scattering and absorption flattening can be observed, especially when

Biophysical Analysis of Membrane Proteins. Investigating Structure and Function. Edited by Eva Pebay-Peyroula

ISBN: 978-3-527-31677-9

the proteins are present as large membrane fragments or are inserted into liposomes. The first-mentioned artifact of light scattering arises when the size of the membrane fragments is comparable to the wavelength of far-UV light. If both right- and left-circular polarized light are scattered to different degrees, then the differential scattering will appear as CD [4]. This effect can be overcome by moving the sample cell close to the detector collecting much of the scattered light.

The second artifact, of absorption flattening, is caused by the inhomogeneous distribution of the protein in the detergent micelles or lipid vesicles. For example, the protein may be concentrated in the micelles or vesicles, and absent from the surrounding solvent or specific/unspecific aggregation of proteins, often as a result of the sample preparation method. This inhomogeneous distribution gives rise to an apparent reduction of the CD in regions of high absorption. The flattening effect can be eliminated by incorporating the membrane protein in small unilamellar vesicles at a lipid/protein ratio sufficiently high that, on average, there is only one protein molecule per vesicle [5, 6]. However, this environment may differ greatly from the native environment, and therefore it is important for CD recording and analysis to establish an environment where the protein is both fully functional and in a dispersed state.

10.2 Secondary Structure Composition

One of the most prevalent applications of CD spectroscopy is in determining the secondary structure contents of the protein by analysis of the far-UV spectra. In this region, the CD bands arise mainly from the peptide groups, and are sensitive to the backbone conformations; this is illustrated in Fig. 10.1, which shows the different shapes and magnitudes of CD spectra corresponding to different secondary structures such as α-helix, β-sheet, or unordered structure. The determination of the secondary structure contents can be made with simple and quick experiments, using a relatively small amount of protein. It serves as a common check of the folded state of the protein, although it cannot be concluded from this analysis whether the protein has a native tertiary fold or is a molten globule. Indeed, this state reflects a compact protein form which conserves most of its secondary structure, but lacks some or all of the tertiary interactions typical of native globular proteins. Information regarding tertiary interactions may be provided by an analysis of the near-UV spectra (see below) or by using others techniques (see also Chapters 11 and 12).

The current methods of analysis of protein secondary structure from CD data are empirical. In fact, they compare the spectrum of the protein with an unknown secondary structure content with the spectra of a set of reference proteins with known three-dimensional (3-D) structure obtained via X-ray crystallography. These analysis are based on the following assumptions:

- the solution and solid-state structure are equivalent
- the protein CD spectrum is expressed as a linear combination of individual secondary structure component

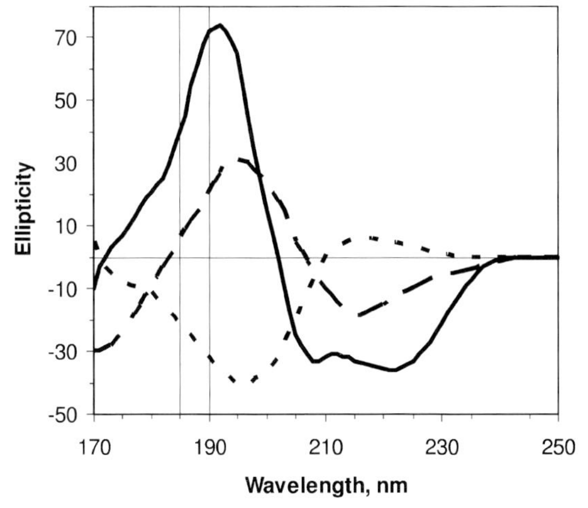

Fig. 10.1 Far-UV spectra associated with various types of secondary structure. Solid line, α-helix; long dashed line, β-sheet; short dashed line, unordered structure. The vertical dotted lines represent the low wavelength limit obtained for conventional CD spectra. Lower limits are only reached with synchrotron radiation circular dichroism (SRCD).

spectra and the contribution from the tertiary structure is negligible

- the geometric variability of the secondary structural elements is not taken into account
- the contributions from aromatic residues and disulfide bonds are negligible.

The assignment of secondary structures from X-ray structures is not straightforward, however, as it involves several parameters leading thus to different secondary structure contents depending on the algorithm used [7]. The secondary structure contents of membrane protein as defined in protein database (PDB) files might provide significantly different values from that obtained using different algorithms [8]. As an example, Fig. 10.2 shows clearly a difference of at least 5% between the secondary structure content of the various structure of sarcoplasmic reticulum Ca-ATPase as described in the PDB file (Fig. 10.2, left panel) or as obtained using the DSSP (Definition of Secondary Structure of Proteins) algorithm [9] (Fig. 10.2, right panel). The DSSP algorithm is used to define the secondary structure of the reference proteins sets. Depending on the reference proteins sets, five to six classes of secondary structure are defined: six classes including distorted and regular helix, distorted and regular sheet, turn and unordered [10]; six classes including helix, 3_{10}-helix, sheet, turn, polyproline-II helix, unordered; or five classes including helix, sheet, polyproline-II helix, turn, unordered.

The methods used to fit the experimental data with the data from the reference proteins data sets differ either in the algorithms or the implementation of variable

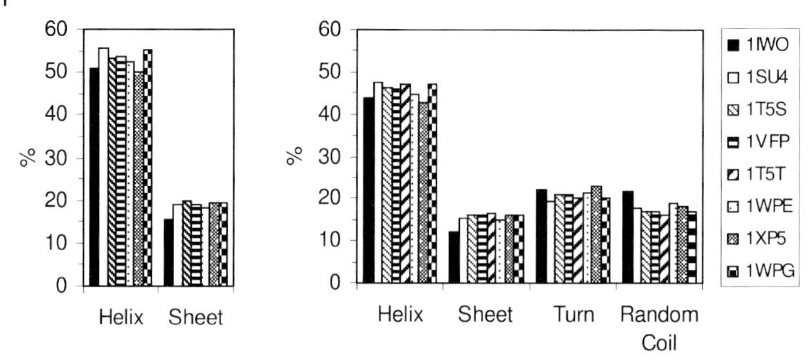

Fig. 10.2 Secondary structure of the various atomic structures of the sarcoplasmic reticulum Ca-ATPase. Left: α-helical and β-sheet structures descriptions from the Protein Database (PDB) files. Right: Secondary structure assignment from atomic coordinates using the DSSP program. (PDB files 1IWO, 1SU4, 1T5S and 1VFP, 1T5T and 1WPE, 1XP5 and 1WPG correspond to calcium-free, calcium-bound, AMPPCP-bound, ADP-bound, F4-bound reaction intermediates of Ca-ATPase, respectively.)

selection, or both. The most commonly used methods are implemented in the computer programs CONTINLL, SELCON3 and CDSSTR [11, 12].

The accuracy of this analysis will depend on the one hand, on the number and variety of proteins encompassing different secondary structures and folds and constituting these reference data sets and, on the other hand, on the correct magnitude of the spectra which relies on the correct determination of the protein concentration, the exact cell pathlength and the calibration of the dichrograph.

The use of reference data sets including only soluble protein for the analysis of CD spectra of membrane proteins has been questioned [7, 13]. Recently, Sreerama et al. (2004) [14] have shown that the inclusion of 13 membrane proteins in the available soluble protein reference sets increases the performance of the analysis both for soluble and membrane proteins, which suggests that the important point is the number and variety of proteins of the database. Today, only the reference databases, SMP50 and SMP56, as found in the CDPro program package [12], include spectroscopic data from membrane proteins, a total of 13 membrane proteins over 50 and 56 proteins. The data sets SMP50 and SMP56 differ from each other by the low wavelength cut-offs (185 and 190 nm, respectively). The set of the 13 membrane proteins is not optimal as it is dominated by α-helical proteins. Indeed, only four over 13 membrane proteins (porins) are mainly β-structures (50 to 60% β-sheet, less than 7% β-helix) and the nine other membrane proteins are mainly α-structures (44–76% α-helix, less than 7% β-sheet except for Ca-ATPase). Therefore, increasing the number and variety of membrane proteins in the database is necessary for improving the accuracy of this method to membrane protein. A new reference database, which includes CD data from 74 soluble proteins, has been very recently proposed covering both secondary structure and fold space and also using spectral data from synchrotron radiation circular dichroism (SRCD) spectroscopy – that is, spectral data down to 170 to 175 nm [15]. In practice, it is recommended to use different algorithms and reference data sets as provided by

CDPro or DICHROWEB packages [12, 16] to find a reliable estimate of the secondary structure contents.

The parameters that affect the magnitude of the spectra are the protein concentration, the cell pathlength and, the calibration of the dichrograph.

The inaccurate determination of protein concentration is the major factor that contributes to poor estimates of secondary structure. For example, during a recent data analysis it was found that a 10% error in concentration of mouse translocator protein (previously named peripheral-type benzodiazepine receptor; PBR) (mTSPO) in dodecyl-phosphocholine (DPC) micelles led to an error of 10% of α-helical structure over $50 \pm 5\%$ α-helix, and that this error must be added to the error due to the use of limited reference proteins data sets. It is recommended that the protein concentration be determined in the most accurately possible manner (±5% or less), using one or several methods [17]. Indeed, membrane proteins preparations usually contain detergents and lipids that may affect classical protein concentration determination using colorimetric measurements. A very commonly used method for proteins containing Tyr and Trp measures absorbance at 280 nm and calculates the protein concentration with an absorption coefficient determined from the amino acid composition [18].

The pathlength of cells above 1 mm (used for conventional circular dichroism) are usually accurately determined by the manufacturer, but for micrometer-length cells (as used for SRCD) the pathlength should be determined [19].

Misalignment or miscalibration of the monochromator of the dichrograph represent another source of magnitude errors, and modified procedures for calibration have recently been proposed [19, 20].

Due to the inaccuracy of the secondary structure estimation derived from CD spectra in the far-UV region, the most common application of far-UV CD is determination of the quality of the refolding process of native extracted or recombinant expressed protein, the detection of conformational changes as a result of environmental changes, ligand binding, or mutations. For example, the CD spectra and channel activities of different preparations of the voltage-dependent anion-selective channels (VDAC) have been compared [21]. Purified VDAC from *Neurospora crassa* (ncVDAC) or *Saccharomyces cerevisiae* mitochondria (scVDAC) exhibit different CD spectrum in non-denaturing detergent [1% lauryldimethylamine oxide (LDAO), pH 7]. The CD spectrum of ncVDAC had characteristics of higher β-sheet content than the CD spectrum of scVDAC. Moreover, bacterially expressed His6-containing scVDAC displayed CD spectrum intermediate between that of mitochondrial ncVDAc and scVDAC. The reasons for such differences are still unexplained.

The detergent environment of a membrane protein is an important factor for its folding and secondary structure. As an example, Fig. 10.3 shows the spectrum of recombinant mTSPO solubilized in various detergent micelles. Recombinant mTSPO, which is mostly found in the inclusion bodies of bacteria, is obtained by sodium dodecylsulfate (SDS) solubilization [22]. The samples of recombinant mTSPO solubilized in the other detergent micelles (Fig. 10.3) were obtained by detergent exchange during purification process. The overall shape of the CD

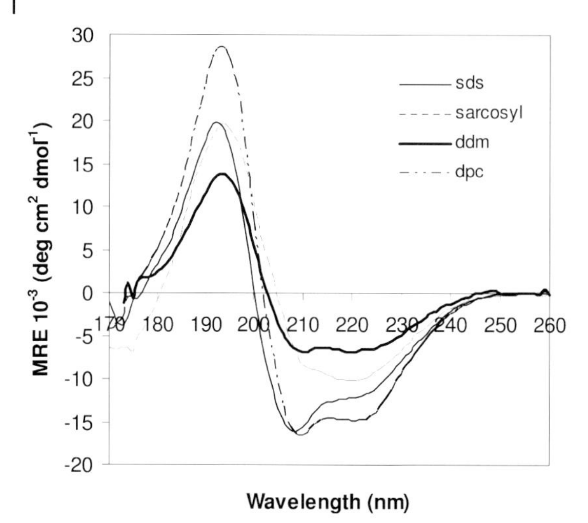

Fig. 10.3 Far-UV circular dichroism (CD) spectra of recombinant mTSPO solubilized in detergent micelles of sodium dodecylsulfate (sds), sarcosyl, dodecylmaltoside (ddm), and dodecyl-phosphocholine (dpc). Spectra were recorded on station CD12.1 at the SRS (Daresbury). The temperature was 20 °C, the cell path length 10 or 20 μm, the protein concentrations ranged from 1.7 to 5.1 mg mL^{-1}, and the detergent concentrations from 1 to 5%. In order to avoid beam light damage, three spectra and three baselines (one scan each) using three independent loaded cells were recorded. Spectra were averaged, baseline-corrected and smoothed using a Savitsky–Golay filter included in CDtool package [44].

spectra of the recombinant protein in the different detergents suggests a predominant α-helical structure with difference in α-helicity depending on the detergent used. Such variations of membrane proteins CD spectra depending on the type of micelles have already been observed for the small multidrug transporters, SugE and TehA [23]. The analysis of the CD data has been performed using three algorithms, SELCON3, CONTINLL and CDSSTR, as proposed by the CDPro and DICHROWEB packages and six protein reference datasets. The results of such an analysis are reported Table 10.1 for the samples in DPC micelles. The nrmsd (normalized root mean square deviation) is a fit parameter which measures the difference between the experimental and calculated spectra. Nrmsd values lower than 0.1 indicate close agreement between experimental and calculated spectra. The data in Table 10.1 indicate similar values of nrmsd for a given program using different reference proteins sets. Despite SELCON3 yielding similar secondary structure contents as CONTINLL and CDSSTR, the calculated and experimental spectra show higher deviation. Similar estimations of secondary structures contents are obtained with the different programs. Therefore, in the absence of structural information, a reliable estimate could be given by averaging solutions from different reference sets and different programs. The analysis of the spectra yielded an estimate of 50 ± 5% (DPC), 43 ± 4% (SDS), 28 ± 4% (sarcosyl) and 26 ± 7% (dodecylmaltoside, DDM) α-helical structure (Table 10.2). It should be mentioned

Tab. 10.1 Secondary structure fractions of mTSPO in DPC detergent micelles determined from CD spectrum.

Database	*Program*	*α-helix/*	*β-strand/%*	*Turn/%*	*Unordered/%*	*nrmsd*[(a)] *CDPro*	*nrmsd*[(a)] *DICHROWEB*
SP29	CONTINLL	48	11	20	21	0.03	(0.08)
	CDSSTR	50	10	19	21	0.03	(0.02)
	SELCON3	49	12	19	20	(b)	0.07
SP37	CONTINLL	46	13	19	22	0.03	(0.06)
	CDSSTR	49	13	17	21	0.03	(0.02)
	SELCON3	48	12	18	22	(b)	0.07
SP43	CONTINLL	51	7	17	25	0.01	(b)
	CDSSTR	53	10	14	23	0.02	(0.02)
	SELCON3	50	10	15	25	(b)	0.05
SP175	CONTINLL	45	13	11	31	(b)	0.07
	CDSSTR	46	13	11	30	(b)	0.02
SMP50	CONTIN	51	7	17	25	0.02	(b)
	CDSSTR	52	9	17	22	0.03	(b)
SMP56	CONTIN	53	6	15	26	0.02	(b)
	CDSSTR	55	7	14	24	0.03	(b)
Mean value		50 ± 5	9 ± 4	16 ± 4	25 ± 5		

The analysis of CD data was performed using three algorithms: CONTINLL, CDSSTR and SELCON3 included either in the CDPro and DICHROWEB packages. The reference proteins datasets used were SP29 (29 soluble proteins, 178–260 nm), SP37 (37 soluble proteins, 185–240 nm), SP43 (43 soluble proteins, 190–240 nm), SP175 (74 soluble proteins, 175–240 nm), SMP50 (SP37 soluble proteins, 13 membrane proteins, 185–240 nm), and SMP56 (SP43 soluble proteins, 13 membrane proteins, 190–240 nm).

Notes: (i) Both packages gave very similar estimations of secondary structure even if not giving identical nrmsd values; (ii) graphical representations are often useful to check the quality of the fit. Values of secondary structure shown in this table are related to nrmsd values not between parentheses.

Mean values were obtained by averaging all data shown in the table; error values were calculated using minima and maxima.

(a) nrmsd: Normalized root-mean-square deviation; this is used as a measure of the goodness of fit between the experimental spectrum and the calculated secondary structure.

(b) Analysis not performed.

DPC, dodecyl-phosphocholine; mTSPO, mouse translocator protein.

that this analysis yields similar estimations of the secondary structures (α-helix, β-sheet, turns and unordered structures) in sarcosyl and DDM, although the spectra are quite different. This may actually reflect the difficulties of the different algorithms in estimating the content of β- and other structures as a result of the spectral variability of these structures and/or the lack of representative data in the reference sets.

Although conformational changes of membrane proteins are functionally important, they may not involve secondary structure changes. CD experiments performed with Ca-ATPase in different conformational states did not reveal any major secondary structural changes [24, 25]. In fact, the recent published crystal structures of Ca-ATPase indicate that, although important conformational changes

Tab. 10.2 Secondary structure fractions of mTSPO in various detergent micelles determined from CD spectra.

Detergent	*% α-helix*	*% β-strand*	*% Turn*	*% Unordered*
SDS	43 ± 4	10 ± 4	19 ± 6	28 ± 7
DPC	50 ± 5	9 ± 4	16 ± 4	25 ± 5
Sarcosyl	28 ± 4	25 ± 6	18 ± 6	29 ± 5
DDM	26 ± 7	27 ± 9	17 ± 5	30 ± 9

The analysis of CD spectra was performed as shown in Table 10.1 and the data averaged. DDM, dodecylmaltoside; DPC, dodecyl-phosphocholine; mTSPO, mouse translocator protein; Sarcosyl, sodium dodecanoyl sarcosine; SDS, sodium dodecylsulfate.

occur, these involve essentially large domain movements [26], with almost no secondary structural changes (see Fig. 10.2). Indeed, careful analysis of secondary structure composition of the various atomic structures of the Ca-ATPase available in the PDB using the DSSP program shows that maximal changes are lower than 5%.

10.3 Tertiary Structure Fingerprint

The near-UV CD spectra arise from the contributions of all the aromatic side-chains (Tyr, Trp, Phe) as well as disulfide bonds (near 260 nm) and non-protein cofactors or ligands if present. Trp give rise to a peak near 290 nm and fine-structured bands between 290 and 310 nm, Tyr a peak between 275 and 282 nm, and Phe a sharp fine structure between 255 and 270 nm [27].

The interpretation of the spectral bands in the near-UV is complicated by the numerous factors than can affect their shape. These include the number of each aromatic side chain, their mobility and environment (hydrogen bonds, polar groups, polarizability), and the presence of nearby aromatic residues (exciton coupling for aromatic side chains separated by a distance less than 1 nm). An increased flexibility of the protein will give rise to a signal of lower intensity. For example, a molten globule state will be characterized by very weak near-UV CD signals as a result of the high mobility of aromatic side chains. Proteins with a great number of aromatic residues can display a small near-UV CD signal as a result of the addition of both negative and positive contributions of the numerous aromatic side chains. At present, the development of CD theory is insufficient to allow any detailed interpretation of the near-UV spectra in structural terms, and so the analysis of the near-UV CD spectra remains qualitative.

In order to assess the contribution of the different side chains, mutation experiments that do not affect the tertiary fold could be performed – for example, Trp into Phe, and Cys into Ser mutations [27]. By using mutagenesis, Baneres et al.

[28] proposed the assignment of a disulfide bond band in the near-UV CD spectra of the recombinant 5-HT_4 receptor, a G-protein-coupled receptor (GPCR). The CD spectrum in the near-UV included a broad negative band centered at 265 nm (see Fig. 1. in Ref. [28]). These authors succeeded in assigning this band to the disulfide bond (Cys93–Cys184) as the mutation of either each cysteine or both by a serine led to a large decrease of a band around 265 nm, without any major other alteration of the CD spectrum.

Large variations of the near-UV CD spectra can be observed as a consequence of changes in the environment of the aromatic side chains. An example of such a variation is given by the present authors' study on the mTSPO. The mTSPO is composed of a large number of aromatic residues: 6 Phe, 10 Tyr, and 12 Trp. The near-UV CD spectra of mTSPO solubilized in four different detergent micelles SDS, Sarcosyl, DDM and DPC, are reported in Fig. 10.4. In the presence of SDS micelles, the CD spectrum shows a large positive band from about 255 to 305 nm, whereas in the presence of DDM or DPC micelles the magnitude of the CD spectra is smaller and its shape different, with minima centered at about 270 nm and shoulder bands around 285 to 295 nm. These spectral variations suggest different environments of Phe and Tyr residues (between about 260 and 280 nm) as well as of the Trp residues (between about 290 and 305 nm). Therefore, the four different detergent micelles induce very large spectral variations as a result of difference in the environments of all type of aromatic side chains: Phe, Tyr and Trp.

Using near-UV CD and NMR spectroscopies, Taylor et al. [29] demonstrated that the integral membrane colicin E1 immunity protein (ImmE1) has a compact

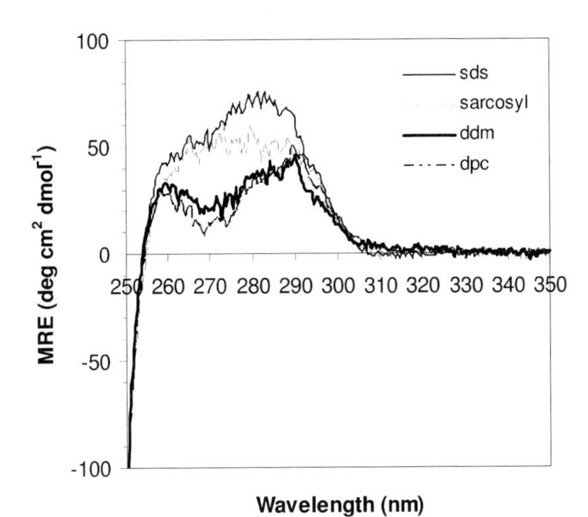

Fig. 10.4 Near-UV circular dichroism (CD) spectra of recombinant mTSPO solubilized in detergent micelles of sodium dodecylsulfate (sds), sarcosyl, dodecylmaltoside (ddm) and dodecyl-phosphocholine (dpc). Spectra were recorded with a JobinYvon CD6 spectrophotometer. The temperature was 20 °C, the cell path length 5 mm, and the protein and detergent concentrations were to 1.0 mg mL^{-1} and 1%, respectively. Sixteen spectra were recorded, averaged, and baseline-corrected for each detergent.

folded conformation in a chloroform/methanol/water solvent system. Mutants of Cys residues predicted to be located in the transmembrane domains were constructed by substitution with either Ala or Trp. All of the Cys to Trp mutants showed a positive CD signal at 290 nm, implying that the Trp residues were in a constrained environment. Some of the double mutants Cys to Ala and Cys to Trp showed a decrease of the signal in the region 270 to 285 nm, suggesting that some of the Tyr side chains had lost their constrained environment. Moreover, the *in-vivo* activity of all these mutants was tested and revealed a correlation between the amplitude of the CD signal in the region 270 to 285 nm and the *in-vivo* activity of ImmE1: the largest amplitude was correlated with the greatest activity.

It should be noted that the signals in the near-UV region are smaller than in the far-UV region. Therefore, recording near-UV CD spectra is more demanding in terms of the amount of purified membrane protein required; hence, such studies have been less frequently performed.

10.4 Extrinsic Chromophores

CD in the near-UV, visible and near-IR can provide information about the environments of the protein cofactors or bound ligands. Generally, the free cofactor or ligand has very weak or no CD spectrum, and acquires chirality upon binding to the protein. For example, the chromophore of bacteriorhodopsin (BR), retinal, acquires chirality due to its asymmetric environment, whereas it is an optically inactive molecule. Although the origin of this CD effect has been the subject of numerous studies, the situation remains not fully understood [30]. The CD of retinal has been used to follow the stability of the BR complexes within the crystal [31]. Another example concerns the pigments rings present in the light-harvesting complexes which exhibit visible and near-IR CD. In this case, the sign of the CD bands and their amplitudes depends on the orientation of the pigments within the protein complex, and has been recently modeled [32, 33]. The shape of these bands is used to follow the selective release, removal and reconstitution of the light-harvesting pigments from the complexes [34].

10.5 Conformational Changes upon Ligand Binding

Conformational changes upon ligand binding are an essential part of the mechanism of the action and regulation of membrane proteins. These changes can be probed by variations of the far-and/or near-UV CD spectrum. Selected examples of such conformational changes are presented below.

The sodium channel is a multimolecular complex that possesses a large α-subunit, composed of approximately 2000 amino acids, forming the channel's pore and gating machinery and containing the principal drug-binding sites. This α-subunit is composed of four highly homologous domains, each with six predicted transmembrane domains assembled in a clockwise pattern around the central pore. The sodium channel may have different conformational states depending on its activation level. Neurotoxins and drugs that modify channel function have been shown to bind preferentially to specific conformational states. CD was used to examine the conformational changes associated with the binding of two ligands of the inactivated and activated forms of the open states, lamotrigine and batrachotoxin, respectively [35]. The sodium channel from *Electrophorus electricus* was purified from electric eel membranes and solubilized in mixed lipid/detergent micelles for CD experiments. Analysis of the far-UV spectrum revealed a 55% α-helical secondary structure – a higher content than was predicted (40%). The addition of a large excess of lamotrigine in order to shift the equilibrium towards the conformation of the open, inactivated state led to an increase of ~9% in helix content, as shown by the CD spectrum. The addition of an amount of batrachotoxin sufficient to cause steady-state activation of the sodium channel, led to a 6% increase in helical content. The differences between the CD spectra in the presence of lamotrigine and batrachotoxin were greater than the standard deviations of the measurements, and thus reflected different channel conformations upon binding of these two molecules. Furthermore, the addition of both lamotrigine and batrachotoxin caused a 13% increase in α-helical content, which suggested that the addition of both molecules which bind to distinct and overlapping sites causes a further increase of helical structure.

Ligand binding may not induce changes in the secondary structure (see the example above of Ca-ATPase with various bound ligands), but does cause changes in the tertiary structure. If these conformational changes were to involve direct or indirect alterations in the environments of aromatic residues or disulfide bonds, they may be studied by near-UV spectra. For example, changes in the tertiary structure of a GPCR in the presence of agonists and antagonists binding has been monitored by near-UV CD. It has been suggested that GPCRs exist in at least three conformational states: a Rg state corresponding to a totally inactive state; an R state (without) ligand that can activate G-proteins; and a R* state which corresponds to an active state stabilized by an agonist. The interactions between a GPCR (the serotonin 5-HT4a5 receptor) and agonists and antagonists have been studied using near-UV CD [28]. The differences between signals of bound receptor and the sums of the free receptor and free ligand signals report on the induced conformational changes of the receptor upon ligand binding. The difference spectra upon binding to a full agonist, an inverse agonist, a neutral antagonist and a partial agonist, essentially included a broad band centered at 265 nm of different sign and intensity depending on the presence of the ligand. This band was identified as being due to the unique disulfide bridge that connects the extracellular tip of the transmembrane domain 3 and the extracellular loop e2. The sign of the band at 265 nm is related to the handedness of the disulfide bridge, and therefore to the conformation of the e2 loop. The difference spectra upon binding the different ligands therefore reflect different conformations of the e2 loop upon receptor activation.

10.6 Folding/Unfolding

The process of recombinant membrane protein folding is poorly understood as a result of the technical difficulties encountered when monitoring this process, and of the few examples. The key questions here are related to secondary and tertiary structures, such as the relationship between the formation of these structures and the membrane insertion, or the relationship between the formation of secondary and tertiary structures. At present, no method is available for determining at high resolution a tertiary structure under the conditions of folding. However, numerous experimental techniques have been used to monitor the folding kinetics of membrane proteins *in vitro*, including far-UV CD coupled with stopped-flow to acquire secondary structural information about the transient forms observed during the folding process [36].

Bacteriorhodopsin is one of the most studied membrane proteins, and serves as a model for α-helical membrane protein folding (for a review, see Ref. [37]). Bacteriorhodopsin is a transmembrane protein that binds a retinal chromophore covalently. CD spectra indicate that the native protein has an α-helical content of about 74%, equivalent to about seven transmembrane helices, while the unfolded protein has an α-helical content of about 42%, equivalent to about four transmembrane helices [38]. Therefore, about 70 amino acids (equivalent to three transmembrane helices) change conformation during this refolding process. To trigger the folding reaction, the partially denatured membrane protein is mixed with the renaturing detergents or lipids, thus diluting the denaturant. As a result of this mixing, some difficulties must be overcome in using CD, notably changes in the size and shape of the micelles and vesicles that can occur over time. Moreover, it is necessary to follow conformational changes not only at a single wavelength but over a large spectral range, because of the pitfalls in measuring membrane protein CD. The changes in far-UV spectra during refolding have been monitored at 195 and 224 nm [38] to account for changes in both the content of α-helical and disordered structures. Far-UV CD has been used in combination with time-resolved fluorescence and absorption spectroscopy to follow the refolding kinetics. The refolding of the SDS-denatured state of bacteriorhodopsin (bO) in DMPC/DHPC micelles in the presence of retinal (R) occurs according to the simplified following reaction scheme in which retinal binds after a rate-limiting step. bO is the denatured SDS state (that cannot bind retinal), and I1 and I2 are intermediates that form prior to retinal binding (I_R and bR forms).

$$\mathrm{bO} \rightleftharpoons \mathrm{I}_1 \rightleftharpoons \mathrm{I}_2 \overset{\mathrm{R}}{\rightleftharpoons} \mathrm{I}_\mathrm{R} \longrightarrow \mathrm{bR}$$

The CD data revealed monoexponential kinetics with an apparent rate constant of $0.077\,\mathrm{s}^{-1}$ at pH 8 at both 195 and 224 nm. The CD amplitude variations of the 195- and 224-nm bands during this slow phase account for about 30 amino acids (at least one transmembrane helix) undergoing a conformational change from disordered to α-helical conformation. The remaining 40 amino acids fold to

α helices during the 20-s experimental dead time. A comparison with fluorescence data indicated that this slow phase corresponds to formation of the I_2 intermediate which binds retinal. The rate-limiting folding of I_2 has been suggested to reflect either the folding and insertion of parts of helices F and G, or the packing of the core regions of most of the transmembrane helices, followed by formation of the helix ends [37]. More recent time-resolved site-specific fluorescent labeling studies [39] discarded the first suggestion and proposed that I_2 formation involves insertion of helix D as well as helix formation at the ends and helix packing.

10.7 Conclusion and Perspectives

Circular dichroism is a convenient spectroscopic method for following the membrane protein folding and conformational changes that occur during the activation and regulation of these proteins. Conventional dichrographs enable rapid and easy measurements of secondary structure contents as well as tertiary structure fingerprints, as long as the membrane proteins possess aromatic residues or disulfide bonds. It must be mentioned at this point that the amount of proteins required to perform such experiments is relatively high (0.1–1 $mg\,mL^{-1}$ and 0.5–2 $mg\,mL^{-1}$ for far-UV and near-UV CD experiments, respectively), especially when knowing the difficulties encountered in producing recombinant membrane proteins. For far-UV experiments, conventional instruments cannot record spectra below 185 to 190 nm as a result of the high absorbance of the buffer, solvent or sample and the low intensity of the light sources. As a result of additional bands observed below 190 nm for the different types of secondary structure, the analysis of the secondary structure contents should be more accurate using data below 190 nm [40]. Circular dichroism using the synchrotron radiation (SRCD) as a light source allows CD measurements of protein samples in aqueous solution as low as 160 nm due to the higher intensity of the light source [41]. The other advantages of using SRCD are due to the high signal-to-noise ratio of the SRCD spectra, which not only makes the monitoring of small conformational changes possible but also speeds up the recording. However, the light beam may induce denaturation due to local heating or degradation due to free-radical formation [42, 43]. Moreover, in order to record spectra down to 160 nm, small pathlength cells (10 μm) are used, such that concentrated samples are needed (ca. 10 $mg\,mL^{-1}$). This raises the question of the sample preparation for such experiments, as such high concentrations will not only increase the amounts of detergent and lipid required to maintain the membrane protein in a functional state but also increase the protein–protein interactions, thereby inducing aggregation.

Far-UV CD is a suitable technique to follow the folding/unfolding of membrane proteins. However, a purified membrane protein exhibiting native secondary structure might not have native tertiary fold. Whilst near-UV experiments can provide structural information about the tertiary fold, more theoretical investigations are needed in order to interpret the spectra. Ligand binding can be monitored using far-UV spectra when secondary structure changes are involved, or near-UV CD spectra when the environments of the aromatic residues or disulfide bonds are affected by the binding.

Acknowledgments

The authors' studies reported in this chapter were supported by the Institut National de la Santé et de la Recherche Médicale (INSERM), the Centre National de la Recherche Scientifique (CNRS), the Agence Nationale de la Recherche (ANR) grant 06-BLAN-0190-01 (to JJL) and the Commissariat à l'Energie Atomique (CEA). The circular dichroism experiments performed at the CCLR Daresbury Laboratory (UK) were supported by Grant 43056 from EU FP6 transnational access arrangement for travel and the use of beamline CD12.1. The authors thank J.-C. Robert (INSERM U773) for the preparation of TSPO samples, Prof. V. Papadopolos for donating the cDNA and expression system for TSPO, S. Miron (INSERM U350) for helpful discussions, and M.A. Ostuni, J.M. Neumann for their careful reading of the manuscript.

References

1 G. D. Fasman (Ed.) Circular dichroism and the conformational analysis of biomolecules, Plenum Press, New York, **1996**.

2 R. W. Woody. Circular dichroism. *Methods Enzymol.* **1995**, *246*, 34–71.

3 N. Sreerama and R. W. Woody. Computation and analysis of protein circular dichroism spectra. *Methods in Enzymol.* **2004**, *383*, 318–351.

4 C. Bustamante, I. Tinoco, Jr. and M. F. Maestre. Circular differential scattering can be an important part of the circular dichroism of macromolecules. *Proc. Natl. Acad. Sci. USA* **1983**, *80*, 3568–3572.

5 B. A. Wallace and D. Mao. Circular dichroism analyses of membrane proteins: an examination of differential light scattering and absorption flattening effects in large membrane vesicles and membrane sheets. *Anal. Biochem.* **1984**, *142*, 317–328.

6 N. A. Swords and B. A. Wallace. Circular dichroism analyses of membrane proteins: examination of environmental effects on bacteriorhodopsin spectra. *Biochem. J.* **1993** *289*, 215–219.

7 B. A. Wallace, J. G. Lees, A. J. W. Orry, A. Lobley and R. W. Janes. Analyses of circular dichroism spectra of membrane proteins. *Prot. Sci.* **2003**, *12*, 875–884.

8 J. Martin, G. Letellier, A. Marin, J. F. Taly, A. G. de Brevern and J. F. Gibrat. Protein secondary structure assignment revisited: a detailed analysis of different assignment methods. *BMC Struct Biol.* **2005**, *5*, 17.

9 W. Kabsch and C. Sander. Dictionary of protein secondary structure: pattern recognition of hydrogen-bonded and geometrical features. *Biopolymers* **1983**, *22*, 2577–2637.

10 N. Sreerama, S. Y. Venyaminov and R. W. Woody. Estimation of the number of α-helical and β-strand segments in proteins using circular dichroism. *Prot. Sci.* **1999**, *8*, 370–380.

11 S. Y. Venyaminov and J. T. Yang. Determination of protein secondary structure. In: *Circular Dichroism and the conformational analysis of biomolecules.* G. D. Fasman (Ed.), Plenum Press, New York, **1996**, pp. 69–107.

12 N. Sreerama and R. W. Woody. Estimation of protein secondary structure from circular dichroism spectra: comparison of

CONTIN, SELCON and CDSSTR method with an expanded reference set. *Anal. Biochem.* **2000**, *287*, 252–260.

13 P. Park, A. Perczel and G. D. Fasman. Differentiation between transmembrane and peripheral helices by the deconvolution of circular dichroism spectra of membrane proteins. *Prot. Sci.* **1992**, *1*, 1032–1049.

14 N. Sreerama and R. W. Woody. On the analysis of membrane protein circular dichroism spectra. *Prot. Sci.* **2004**, *13*, 100–112.

15 J. G. Lees, A. J. Miles, F. Wien and B. A. Wallace. A reference database for circular dichroism spectroscopy covering fold and secondary structure space. *Bioinformatics* **2006**, *22*, 1955–1962.

16 L. Whitmore and B. A. Wallace. DICHROWEB, an online server for protein secondary structure analyses from circular dichroism spectroscopic data. *Nucleic Acids Res.* **2004**, *32*, 668–673.

17 S. M. Kelly, T. J. Jess and N. C. Price. How to study proteins by circular dichroism. *Biochim. Biophys. Acta* **2005**, *1751*, 119–139.

18 C. N. Pace, F. Vajdos, L. Fee, G. Grimsley and T. Gray. How to measure and to predict the molar absorption coefficient of a protein. *Prot. Sci.* **1995**, *4*, 2411–2423.

19 A. J. Miles, F. Wien, J. G. Lees and B. A. Wallace. Calibration and standardisation of synchrotron radiation and conventional circular dichroism spectrometers. Part 2: Factors affecting magnitude and wavelength. *Spectroscopy* **2005**, *19*, 43–51.

20 A. J. Miles, F. Wien, J. G. Lees, A. Rodger, R. W. Janes and B. A. Wallace. Calibration and standardisation of synchrotron radiation circular dichroism and conventional circular dichroism spectrophotometers. *Spectroscopy* **2003**, *17*, 653–661.

21 D. A. Koppel, K. W. Kinnally, P. Masters, M. Forte, E. Blachly-Dyson and C. A. Mannella. Bacterial expression and characterization of the mitochondrial outer membrane channel. *J. Biol. Chem.* **1998**, *273*, 13794–13800.

22 J. J. Lacapere, F. Delavoie, H. Li, G. Peranzi, J. Maccario, J. Papadopoulos and B. Vidic. Structural and functional study of reconstituted peripheral benzodiazepine receptor. *Biochem. Biophys. Res. Commun.*, **2001**, *284*, 536–541.

23 C. Klammt, F. Löhr, B. Schäfer, W. Haase, V. Dötsch, H. Rüterjans, C. Glaubitz and F. Bernhard. High level cell-free expression and cell specific labeling of integral membrane proteins. *Eur. J. Biochem.* **2004**, *271*, 568–580.

24 J.-L. Girardet and Y. Dupont. Ellipticity changes of the sarcoplasmic reticulum Ca(2+)-ATPase induced by cation binding and phosphorylation. *FEBS Lett.* **1992**, *296*, 103–106.

25 P. Csermely, C. Katopis, B. A. Wallace and A. Martonosi. The E1–E2 transition of Ca2+-transporting ATPase in sarcoplasmic reticulum occurs without major changes in secondary structure. A circular-dichroism study. *Biochem J.* **1987**, *241*, 663–669.

26 N. Reuter, K. Hinsen, D. L. Stokes and J.-J. Lacapere. Transconformations of the SERCA1 Ca-ATPase: a normal mode study. *Biophys J.* **2003**, *85*, 2186–2197.

27 R. W. Woody and A. K. Dunker. Aromatic and cystine side chain circular dichroism in proteins. In: *Circular Dichroism and the conformational analysis of biomolecules.* G. D. Fasman (Ed.), Plenum Press, New York, **1996**, pp. 109–157.

28 J.-L. Banères, D. Mesnier, A. Martin, L. Joubert, A. Dumuis and J. Bockaert. Molecular characterization of a purified 5-HT4 receptor. *J. Biol. Chem.* **2005**, *280*, 20253–20260.

29 R. M. Taylor, S. D. Zakharov, J. B. Heymann, M. E. Girvin and W. A. Cramer. Folded state of the integral membrane colicin E1 immunity protein in solvents of mixed polarity. *Biochemistry* **2000**, *39*, 12131–12139.

30 E. Karnaukhova, K. L. Schey and R. K. Crouch. Circular dichroism and cross-linking studies of bacteriorhodopsin mutants. *Amino Acids* **2006**, *30*, 17–23.

31 Y. A. Isembarger and M. P. Krebs. Thermodynamic stability of the bacteriorhodopsin lattice as measured by lipid dilution. *Biochemistry* **2001**, *40*, 11923–11931.

32 S. Georgakopoulou, R. N. Frese, E. Johnson, C. Koolhaas, R. J. Cogdell, R. Van Grondelle and G. Van der Zwan. Absorption and CD spectroscopy and modelling of various LH2 complexes from purple bacteria. *Biophys. J.* **2002**, *82*, 2184–2197.

33 S. Georgakopoulou, R. Van Grondelle and G. Van der Zwan. Explaining the visible and near-infrared circular dichroism spectra of light-harvesting 1 complexes from purple bacteria: a modelling study. *J. Phys. Chem.* **2006**, *110*, 3344–3353.

34 N. J. Fraser, P. J. Dominy, B. Ucker, I. Simonin, H. Scheer and R. J. Cogdell. Selective release, removal, and reconstitution of bacteriophyll a molecules into the B800 sites of LH2 complexes from *Rhodopseudomonas acidophila* 10050. *Biochemistry* **1999**, *38*, 9684–9692.

35 N. B. Cronin, A. O'Reilly, H. Duclohier and B. A. Wallace. Binding of the anticonvulsant drug lamotrigine and the neurotoxin batrachotoxin to voltage-gated sodium channels induces conformational changes associated with block and steady state activation. *J. Biol. Chem.* **2003**, *278*, 10675–10682.

36 P. J. Booth, R. H. Templer, W. Meijberg, S. J. Allen, A. R. Curran and M. Lorch. In vitro studies of membrane protein folding. *CRC Biochem. Mol. Biol.* **2001**, *36*, 501–603.

37 P. J. Booth. Unravelling the folding of bacteriorhodopsin. *Biochem. Biophys. Acta* **2000**, *1460*, 4–14.

38 M. L. Riley, B. A. Wallace, S. L. Flitsch and P. J. Booth. Slow α-helix formation during folding of a membrane protein. *Biochemistry* **1997**, *36*, 192–196.

39 E. L. R. Compton, N. A. Farmer, M. Lorch, J. M. Mason, K. M. Moreton and P. J. Booth. Kinetics of an individual transmembrane helix during bacteriorhodopsin folding. *J. Mol. Biol.* **2006**, *357*, 325–338.

40 A. Toumadje, S. W. Alcorn and W. C. Johnson, Jr. Extending CD spectra of proteins below 168 nm improves the analysis for secondary structure. *Anal. Biochem.* **1992**, *200*, 321–331.

41 K. Gekko and K. Matsuo. Vacuum-ultraviolet circular dichroism analysis of biomolecules. *Chirality* **2006**, *18*, 329–334.

42 F. Wien, A. J. Miles, J. G. Lees, S. Vrønning-Hoffmann and B. A. Wallace. VUV irradiation effects on proteins in high-flux synchrotron radiation circular dichroism spectroscopy. *J. Synchrotron Rad.* **2005**, *12*, 517–523.

43 D. T. Clarke and G. Jones. CD12: a new high-flux beamline for ultra-violet and vacuum-ultraviolet circular dichroism on the SRS, Daresbury. *J. Synchrotron Rad.* **2004**, *11*, 142–149.

44 J. G. Lees, B. R. Smith, F. Wien, A. J. Miles and B. A. Wallace. CDtool-an integrated software package for circular dichroism spectroscopic data processing, analysis, and archiving. *Anal. Biochem.* **2004**, *332*, 285–289.

11
Membrane Protein Structure and Conformational Change Probed using Fourier Transform Infrared Spectroscopy

John E. Baenziger and Corrie J. B. daCosta

11.1
Introduction

The difficulties associated with the application of X-ray crystallography and nuclear magnetic resonance (NMR) spectroscopy to biological membranes still hinder the structural characterization of integral membrane proteins. These difficulties have limited the number of high-resolution structures that have been solved. Even when high-resolution structures are available, they are usually obtained from proteins solubilized in a detergent-environment, which can compromise the protein structure and/or lock a protein into a single conformational state. Non-invasive physical methods that probe both integral membrane protein structure and conformational change, while the proteins reside in their natural membrane environment, are thus essential for a complete understanding of membrane protein function.

Fourier transform infrared (FTIR) spectroscopy is one physical method that is readily applicable to the structural characterization of integral membrane proteins in their natural membrane environment. The technique requires only small amounts of protein (a few micrograms can suffice) and, in most cases, both data acquisition and interpretation are relatively rapid and simple such that surprisingly detailed insights into membrane protein structure can be obtained.

In this chapter, the application of FTIR spectroscopy to the structural characterization of integral membrane proteins is discussed, with attention focused mainly on data acquisition using the attenuated total reflectance (ATR) technique. First, the ATR methodology is described, after which details are provided as to how the technique can be used to probe both the biophysical features of membrane proteins and changes in both side chain and polypeptide backbone structure that occur during conformational change. Next, nature of the structural information that can be obtained from FTIR spectra is described. Finally, examples are provided, primarily from the authors' own research, on the nicotinic acetylcholine receptor (nAChR), and how this information can be used to probe features of membrane protein structure and function.

Biophysical Analysis of Membrane Proteins. Investigating Structure and Function. Edited by Eva Pebay-Peyroula

ISBN: 978-3-527-31677-9

11.2
FTIR Spectroscopy

FTIR spectroscopy is a technique that probes the frequencies of molecular vibrations [1]. A non-linear molecule with N atoms exhibits ($3N - 6$) fundamental vibrations. Each fundamental vibration involves every atom in the molecule vibrating at the same frequency although, in reality, the vibrations tend to be localized to a specific region or functional group, such as the C=O bond of a polypeptide backbone. If the fundamental vibration leads to a change in dipole moment (i.e., the vibration has a transition dipole), then the molecule will absorb electromagnetic radiation with a frequency that matches the vibrational frequency of the fundamental mode. The frequencies of molecular vibrations have the same energies as radiation in the infrared region of the electromagnetic spectrum.

The frequency and intensity of infrared light that is absorbed upon excitation of a vibrational mode are both governed by many factors, including the force constant of the bonded atoms, the masses of the vibrating atoms, and the orientation of the transition dipole relative to the electric field vector of the infrared light. Bond force constants are sensitive to the chemistry of the bonded atoms (i.e., double versus single bonds, protonated versus deprotonated, etc.) and their local environments (hydrogen bonding versus apolar, etc.). Infrared spectroscopy can thus provide insight into the local chemistry and environments of functional groups within a molecule, including the individual amino acid residues within a protein. Unfortunately, even a small protein of 20 kDa exhibits approximately 10^4 fundamental vibrations. The large number of vibrations leads to extensive band overlap and thus spectra with broad featureless peaks from which it is difficult to abstract residue-specific information. Difference methods have been developed that allow this residue-specific information to be probed and thus residue-specific changes in structure and environment that occur during protein conformational change to be examined. Difference spectroscopy has been used extensively to examine conformational change in light-activated proteins, such as bacteriorhodopsin (see below). In this chapter, attention is focused on a general approach for probing residue-specific vibrational information during membrane protein conformational change, which is the use of attenuated total reflectance to examine membrane receptor–ligand interactions.

11.2.1
Attenuated Total Reflectance FTIR Spectroscopy

In the classical FTIR experiment, an aqueous biological sample is sandwiched between two optically transparent windows separated by a thin Teflon spacer, and the transmittance/absorbance of infrared light measured (Fig. 11.1a and b). This approach is straightforward, but its application to membrane proteins is limited because the strong absorption bands of water (^1H_2O) overlap with many important protein vibrations and often saturate the absorption spectrum. In order to minimize this band overlap, most protein FTIR spectra are recorded from samples in

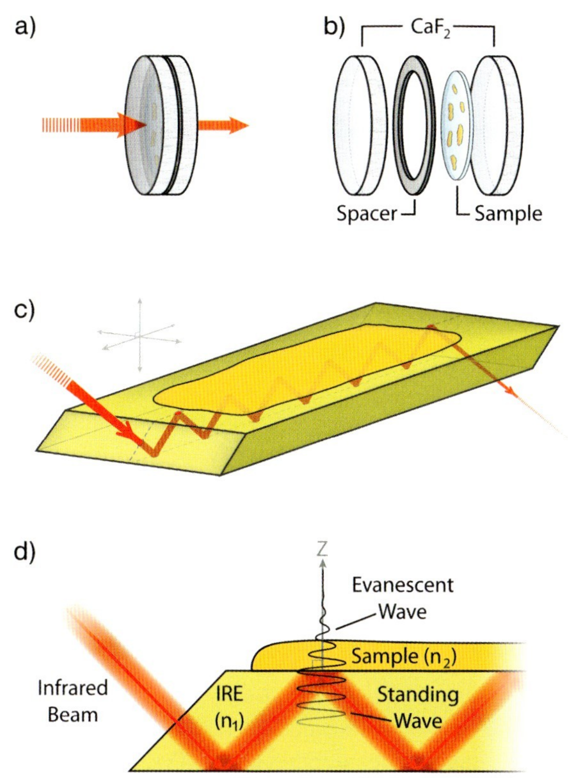

Fig. 11.1 Schematic diagram depicting the experimental set-up for acquiring FTIR spectra using the transmission (a and b) and attenuated total reflectance (ATR) (c and d) approaches. During transmission measurements, infrared radiation (red) passes through an aqueous sample which is sandwiched (with a Teflon spacer) between two CaF_2 windows. For ATR measurements, samples (with refractive index, n_2) are deposited on the surface of an internal reflection element (IRE, with a refractive index, n_1). Infrared radiation enters the IRE, where it undergoes total internal reflection, creating a standing wave within the IRE that is normal to the IRE surface. This standing wave extends beyond the surface of the IRE (evanescent wave), decaying exponentially as it moves away from the IRE surface.

^2H_2O, as opposed to ^1H_2O, buffer. Even with spectra recorded in ^2H_2O, however, very thin Teflon spacers (~10 μm) and relatively high concentrations of protein are still required to maximize the sample absorption relative to that of ^2H_2O.

The ATR technique represents an alternative for the acquisition of FTIR spectra of membrane proteins [2–4]. In the ATR experiment, the infrared light is focused on an optically transparent material, called an internal reflection element (IRE) (Fig. 11.1c). The light passes transversely through the IRE to strike the internal surface at an angle θ above the critical angle, $\theta_c = \sin^{-1} n_{21}$, where it undergoes total internal reflection ($n_{21} = n_2/n_1$, n_1 is the refractive index of the IRE, and n_2 is the refractive index of the sample on the IRE surface). Superimposition of the incoming and reflected waves at each reflection point yields a standing electromagnetic field within the IRE that is normal to the reflecting surface (Fig. 11.1d). Significantly, this electromagnetic field (E_o) penetrates beyond the IRE surface (the evanescent wave) with its strength (E) decaying exponentially according to:

$$E = E_o e^{-z/d_p}$$

where z is the distance perpendicular to the IRE surface, and d_p is the penetration depth of the infrared light defined as follows:

$$d_p = \frac{\lambda/n_1}{2\pi\sqrt{\sin^2\theta - n_{21}^2}}$$

The penetration depth is the distance from the IRE surface at which the intensity of the evanescent wave has diminished to roughly one-third its value at the IRE surface (E_o). For typical IREs, the penetration depth is between 0.5 and 2.0 μm, depending mainly on the refractive index of the IRE and the angle of incidence of the infrared light. This penetration depth is sufficient to allow the infrared light to interact strongly with membranes deposited on the surface of the IRE [3, 4].

There are several advantages to the acquisition of infrared spectra using the ATR approach. One advantage, relative to spectra recorded using transmission measurements, is that the high concentration of membranes on the ATR surface leads to a stronger sample absorption relative to that of the buffer. The low penetration depth of the IR light beyond the IRE also avoids saturation of the absorption signal from the bulk aqueous solution. In fact, excess buffer beyond the penetration depth of the evanescent wave is not probed in an ATR experiment. Membrane protein spectra can thus be recorded in the presence of excess buffer and/or while flowing buffer over the deposited membrane surface [5, 6]. The latter allows spectra to be recorded of the same protein sample under different buffer conditions, such as in the presence and absence of a specific ligand or increasing pH, simply by changing the buffer in the external sample compartment. As discussed in Section 11.2.2, measuring the differences between spectra of a protein under two different buffer conditions allows probing of the subtle vibrational alterations that occur in individual amino acid residues during membrane protein conformational change (see section 11.2.2).

Another advantage of the ATR technique is that it is ideally suited to exploit the sensitivity of the absorbance intensity of a given vibration to the orientation of its transition dipole relative to the electric vector of the incident radiation. By using polarized infrared light to record spectra from membrane samples oriented on the surface of a planar IRE, it is possible to define the orientations of functional groups relative to the bilayer normal [4]. Changes in functional group orientation, such as the tilting of transmembrane α-helices, during protein conformational change, can thus be monitored (see section 11.2.3).

Finally, as there are many different IRE materials, IRE geometries, and commercially available ATR accessories, it is easy to customize an ATR experiment for almost any application. For example, microsampling accessories allows the rapid recording of spectra from tiny aliquots of a solution (5 μL or less). The present authors have used microsampling accessories and FTIR as an analytical tool to rapidly characterize lipid–protein ratios and detergent levels in solubilized membrane protein samples for protein crystallization experiments (see Section 11.4.6) [7].

11.2.2 Detecting Changes in Side Chain Structure/Environment During Protein Conformational Change

The changes in vibrational frequency and/or intensity that occur upon the conversion of a protein from one conformation (state A) to another (state B) usually represent less than 0.1% of the total protein spectral intensity at a given frequency. Therefore, a direct comparison of spectra recorded from a protein in states A and B rarely reveals the vibrational changes that are associated with the conformational transition (Fig. 11.2a). To detect these subtle differences, the spectrum of state A must be digitally subtracted from the spectrum of state B (or vice versa) to eliminate the vibrational bands from those residues whose structures are unaffected by

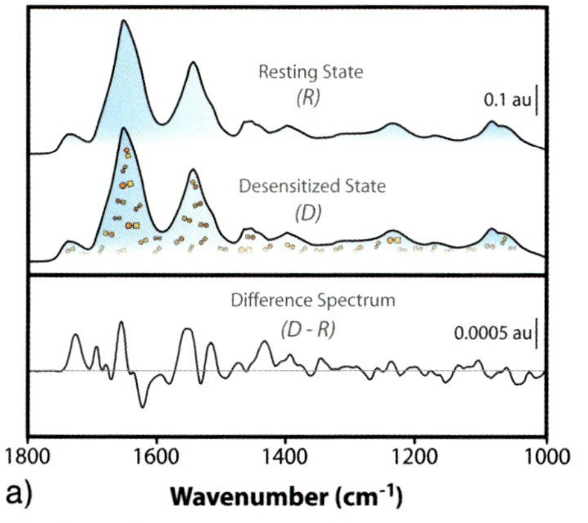

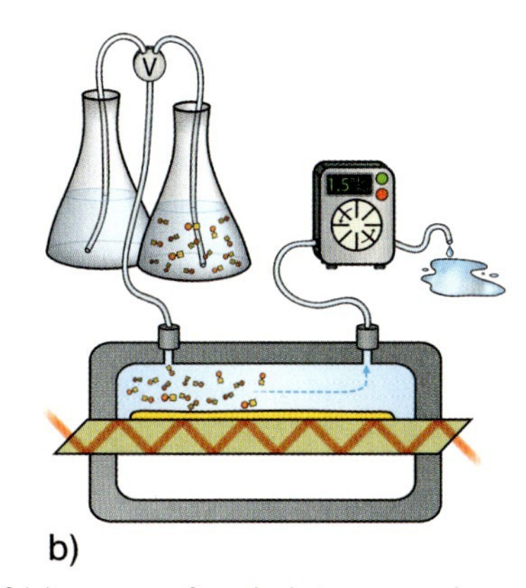

Fig. 11.2 FTIR monitors membrane protein conformational change. (a) FTIR spectra of the nAChR recorded in the resting (*R*, top trace) and ligand-bound desensitized (*D*, middle trace) states are essentially identical because most residues in the protein are unaffected by the conformational change. Subtle differences between the spectra can be detected by digitally subtracting the two spectra (lower trace). Note the difference in absorption scales. (b) To obtain sufficiently high fidelity spectra for calculating spectral differences, films of the nAChR are deposited on a germanium IRE. Spectra in the ligand-free (resting) and ligand bound (desensitized) states are recorded while alternately flowing buffer either with or without the ligand of interest (in this case the acetylcholine analogue, carbamylcholine), past the nAChR film. Buffer flow is controlled using a peristaltic pump (flow rate = 1.5 mL min^{-1}) and an electronic valve (V).

the change in conformational state. The resulting difference spectrum exhibits the vibrational bands from only those residues whose structures and/or environments differ between the two states A and B, and thus provides a spectral map of the conformational change (Fig. 11.2a).

Accurate spectral differences can only be measured under conditions where intensity variations from one spectrum to the next (due to thermal fluctuations, changes in sample concentration, etc.) are much less intense than the changes in vibrational intensity that result from the protein conformational change itself. In most cases, this high degree of spectral reproducibility can only be achieved if the conformational change is triggered while the protein of interest remains inside an infrared sampling device within the FTIR spectrometer. The ability to repetitively cycle a protein between two conformational states inside the FTIR spectrometer is often essential, as successive difference spectra measured between states A and B must be averaged to achieve a sufficient signal-to-noise ratio.

FTIR difference spectroscopy was originally developed to monitor the structural changes that occur in light-activated proteins, such as bacteriorhodopsin and the photosynthetic reaction center [8–11]. These proteins are particularly well suited for difference spectroscopy because conformational change can be repetitively triggered inside the infrared sampling device with a flash of ultraviolet (UV) or visible light. Difference spectroscopy has detected changes in both the protonation state and the strength of hydrogen bonding of a number of amino acid side chains upon light activation of both proteins. In many cases, these subtle changes in structure and environment have been monitored in real time using time-resolved FTIR techniques. FTIR studies of the light-induced protein conformational change remains an active and important area of research.

In order to apply the difference technique to proteins that lack intrinsic activatable chromophores, a variety of technically innovative methods have since been developed. These methods include: (i) the light-induced release of an effecter ligand from a caged precursor [12]; (ii) stopped- and continuous-flow measurements [13]; (iii) temperature and pressure jump experiments [14, 15]; (iv) equilibrium electrochemistry [16]; (v) light-induced photo-reduction [17]; and (vi) attenuated total reflection (ATR) with buffer exchange [5, 18, 19]. For a recent review of the different methodologies, see Ref. [20].

The ATR approach with buffer flow is the most versatile method for probing membrane protein conformational change, and can be used to probe the molecular details of membrane receptor–ligand interactions. Because of the limited penetration of infrared light into a biological membrane film deposited on an IRE surface, spectra can be recorded in the presence of bulk aqueous solution. Ligands added to the bulk solution percolate through hydrated membrane films and bind to proteins located within the lipid multi-bilayers (Fig. 11.2b). In addition, biological membranes adhere to hydrophilic IREs, thus allowing the acquisition of reproducible spectra in the presence of *flowing* bulk solution. The ability to record reproducible spectra in the presence of flowing buffer provides a convenient method for triggering ligand-induced conformational change. By alternately flowing buffer either with or without a ligand of interest past the membrane film, the difference between spectra of the ligand-bound and ligand-free states can be measured, allowing ligand-induced structural change to be probed.

The structural changes associated with the binding of almost any water-soluble small ligand can be studied using the ATR approach. Ligand binding can be probed at varying pH and ionic strength. Ligand-induced changes in functional group orientation can be determined by recording difference spectra with linearly polarized infrared light. A number of different IRE materials and dimensions are available, which allows the experimental approach to be optimized for samples with varying quantities of protein. The difference technique, however, requires the formation of a stable membrane film on an IRE surface. Rigorous care must also be taken to avoid variations in flowing buffer temperature, which lead to baseline distortions. A detailed discussion of the various factors that influence the acquisition of difference spectra using the ATR technique is available in Ref. [19].

An example of the use of the ATR method of recording difference spectra in the presence of flowing buffer to monitor the vibrational changes that occur upon ligand binding to the nAChR is presented in Fig. 11.2a. The difference between spectra of the nAChR recorded in the presence and absence of the agonist analogue carbamylcholine (referred to as a *Carb difference spectrum*) exhibits a complex pattern of positive and negative difference bands (Fig. 11.3a) [21]. These bands reflect:

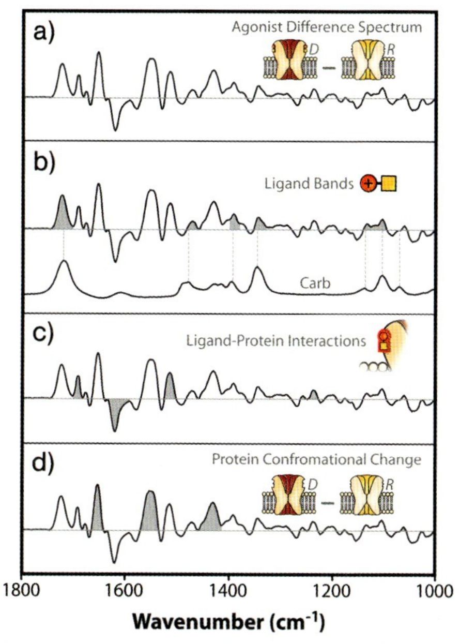

Fig. 11.3 Agonist-induced changes in nAChR structure revealed by FTIR. (a) The difference between spectra of the *Torpedo* nAChR recorded in the absence and presence of the agonist, Carb exhibit bands due to (b) the vibrations of nAChR-bound Carb, (c) the vibrational shifts that result from the formation of physical interactions between Carb and the nAChR, and (c) the vibrational shifts that result from the conformational change from the resting (*R*) to the desensitized (*D*) states.

- vibrations of nAChR-bound carbamylcholine (Carb) (Fig. 11.3b)
- vibrational changes associated with the formation of physical interactions, such as hydrogen bonds, etc., between Carb and neurotransmitter binding site residues (Fig. 11.3c), and
- vibrational changes associated with the resting-to-desensitized conformational transition (Fig. 11.3d).

Positive difference bands centered near 1663, 1655, 1547, 1430, and 1059 cm^{-1} serve as markers of the Carb-induced transition from the resting to the desensitized state [22–27]. The latter marker bands have been used to investigate how lipids and various drugs influence the ability of the nAChR to undergo the ligand-induced resting to desensitized state conformational change, as discussed below.

11.2.3 Probing the Orientation of Functional Groups

The orientation of a functional group in a membrane protein relative to the bilayer surface can be determined from oriented planar lipid bilayers using linear dichroism ATR measurements [4, 28]. Linear dichroism refers to the differential absorption of infrared light that is linearly polarized either parallel or perpendicular to the plane of incidence.

The dichroic ratio, R, of a given absorption band is defined as the ratio of the integrated absorbance intensity obtained with infrared radiation polarized either parallel or perpendicular to the plane of incidence (Fig. 11.4). The plane of incidence is defined by the incoming and reflected infrared beam (xz-plane), where the x-axis denotes the direction of propagation through the IRE and the z-axis is perpendicular to the IRE surface. The angle of incidence, θ, is defined as the angle between the incoming infrared beam and the z-axis. For *thick membrane*

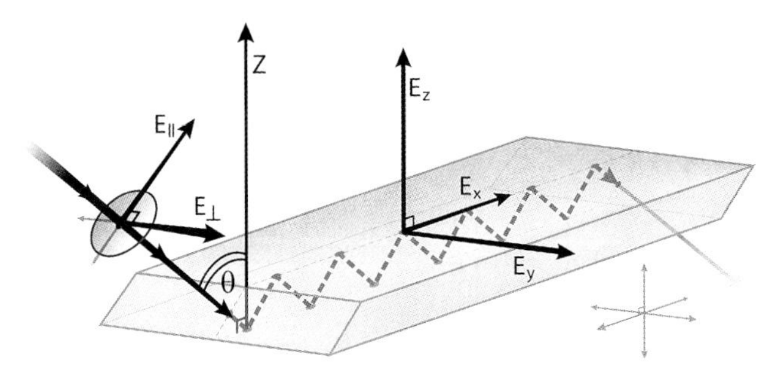

Fig. 11.4 FTIR can be used to monitor the orientation of protein functional groups relative to the bilayer normal. Linear dichroism ATR measurements utilize infrared light that is polarized either parallel ($E_{//}$) or perpendicular ($E_\perp$) to the plane of incidence (defined by the incoming infrared beam and the normal to the IRE surface, Z). θ is the angle of incidence.

films (films that extend well beyond d_p) formed on a germanium IRE ($n_1 = 4$) with an angle of incidence $\theta = 45°$ and a sample refractive index $n_2 = 1.44$, R can be interpreted in terms of an order parameter f as follows:

$$f(\beta) = \frac{R - 2.00}{R + 1.45}$$

where:

$$f(\beta) = \frac{1}{2} \cdot (3\cos^2 \beta - 1)$$

and β is the angle between the vibrational transition dipole and the bilayer normal. A molecule that rotates isotropically (e.g., bulk H_2O) within the penetration depth of the IRE will exhibit an R-value, $R_{isotropic}$, of 2.00. An R-value of 2.00 is also observed if the vibrational transition dipole of the molecule is oriented at the "magic angle" angle of $\beta = 54.7°$ relative to the IRE surface/z-axis. An R-value greater than 2.00 corresponds to an average tilt angle β less than 54.7° relative to the IRE surface/z-axis, and thus suggests a preferred orientation normal to the bilayer surface. An R-value less than 2.00 corresponds to an average tilt angle β greater than 54.7°, and corresponds to a preferred orientation parallel to the bilayer surface.

A mathematical interpretation of linear dichroism data in terms of the structural orientation of functional groups relative to the bilayer surface is also possible, but requires knowledge of the angle between the vibrational transition dipole and the corresponding functional group on the protein. For example, the angle between the peptide amide I transition dipole and the long axis of an α-helix has been measured to be 39°, which allows an estimation to be made of the overall net tilt of α-helices in a protein [29]. A rigorous interpretation of linear dichroism data, however, requires an estimate of the orientational uniformity of the lipid multi-bilayers on the IRE surface (the mosaic spread). Anisotropic motion will average the measured dichroic ratio, R, closer to $R_{isotropic}$ and thus the calculated angle β closer to the magic angle of 54.7°. The motions of membrane lipids, for example, must be taken into account in the calculation of tilt angles. Finally, the value of $R_{isotropic}$ depends on the refractive index of the IRE, the angle of incidence of the infrared light, and the thickness of the membrane film. The use of membrane films that extend well beyond d_p (i.e., thick films) avoids complications in the interpretation of linear dichroism data. A more in-depth discussion of linear dichroism data recorded from thick films using the ATR technique may be found in Ref. [30].

11.3 Vibrational Spectra of Membrane Proteins

Representative infrared spectra recorded from the nAChR dried from ^1H_2O and in ^2H_2O buffer (the latter after buffer subtraction) are presented in Fig. 11.5. These

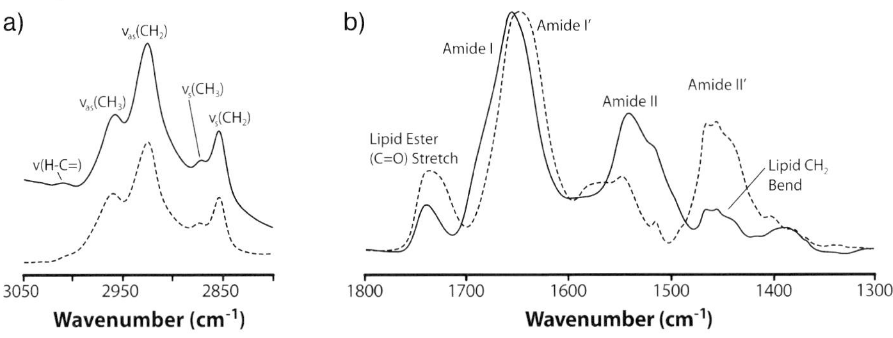

Fig. 11.5 Representative FTIR spectra of a membrane protein. Infrared spectra of the nAChR recorded in ^1H_2O (solid line) and ^2H_2O (dashed line). (a) Lipid acyl chain stretching region (2800–3050 cm^{-1}). (b) Lipid ester carbonyl stretching region and protein amide I/I′ and II/II′ regions. Note that the protein amide II′ vibration overlaps with a Lipid CH_2 vibration (scissoring/bending). ν_s = symmetric stretching; ν_{as} = asymmetric stretching.

two regions of the infrared spectrum exhibit four main features corresponding to the lipid acyl chain C—H (both the methyl and the methylene symmetric and asymmetric) stretching vibrations between 3000 and 2800 cm^{-1}, the lipid C=O stretching vibrations between 1740 and 1720 cm^{-1}, the protein amide I vibrations between 1700 and 1600 cm^{-1} (referred to as amide I′ in ^2H_2O), and the protein amide II vibrations between 1580 and 1520 cm^{-1} (these vibrations shift to near 1450 in ^2H_2O, referred to as amide II′) [3, 31, 32]. The noted vibrations are relatively intense because the functional groups that give rise to them are relatively abundant in membrane protein samples. The noted bands are the most commonly studied in infrared spectra recorded from membranes, and each can provide some insight into either membrane or membrane protein structure–function. There are also many vibrations due to protein side chains that absorb infrared light in the 1800 to 1000 cm^{-1} region of the infrared spectrum – the most visible being the tyrosine ring-stretching vibrations near 1515 cm^{-1}. Most side-chain vibrations are exquisitely sensitive to local chemistry (e.g., protonated versus deprotonated forms) and local environment (hydrogen bonding, polarity, etc.), and can be monitored to gain detailed insight into the nature of protein conformational change.

In the following sections, details are provided of how the four main groups of infrared vibrations noted above can be analyzed to probe membrane structure–function. A brief survey is also provided of the main side chain vibrations that can be studied to shed light on changes in side-chain chemistry and environment during protein conformational change.

11.3.1 Lipid Vibrations

11.3.1.1 Lipid Ester C=O

The lipid ester carbonyl stretching vibration typically gives rise to two overlapping bands centered near 1740 and 1730 cm^{-1} that reflect non-hydrogen-bonded and hydrogen-bonded lipid ester carbonyls, respectively. These two vibrations can be visualized using resolution enhancement techniques (as shown in Fig. 11.10a). Membranes with tightly packed lipids have less water penetration into the interfacial region between the polar headgroups and the apolar fatty acyl chains. Such membranes have a lower proportion of hydrogen-bonded ester carbonyls vibrating near $1730\,cm^{-1}$. For example, tightly packed gel-phase membranes composed of phosphatidylcholine (PC)/phosphatidic acid (PA)/cholesterol (Chol) (3:1:1, mol:mol:mol) membranes exhibit a lower proportion of hydrogen-bonded ester carbonyls near $1730\,cm^{-1}$ than is observed in the same membranes in the liquid crystalline phase [30]. The lipid ester carbonyl stretching vibrations provide a simple qualitative probe of lipid bilayer packing. It has been found that incorporation of the nAChR into some membranes leads to a change in the physical packing of the bilayers and to an increase in the proportion of non-hydrogen-bonded lipid ester carbonyls, which suggests a lateral tightening of the lipid bilayer (see Fig. 11.10a). Surprisingly, these changes occur in a lipid-specific manner [33, 34].

11.3.1.2 Lipid Methylene C—H

The lipid methylene symmetric stretching vibrations undergo a shift up in frequency with increasing acyl chain disorder (i.e., increasing methylene *trans–gauche* isomerizations), and thus provide a sensitive probe of the transition from the ordered gel phase to the relatively disordered liquid crystalline phase (see Fig. 11.10b).

In addition, the linear dichroism of these lipid vibrations can be used to probe the orientational uniformity (mosaic spread) of membrane films deposited on planar IRE surfaces. The lipid methylene C—H stretching transition dipole is oriented in the H—C—H plane, which is perpendicular to the long axis of an all-*trans* fatty acyl chain. With a perfectly ordered saturated acyl chain oriented normal to the bilayer surface in membranes that are parallel to the IRE surface, the expected dichroism, R, of the methylene stretching vibrations should approach a value of $R = 0.85$. A membrane film with a methylene stretching dichroism that approaches 0.85 thus indicates a uniformly oriented membrane film parallel to the IRE surface. Increasing dichroism (i.e., $R > 0.85$) of this vibration could reflect either a poor orientational uniformity or an increasing disorder of the lipid acyl chains within the lipid bilayer due to anisotropic motions [30].

11.3.2 Protein Backbone Vibrations

11.3.2.1 Amide I

The protein amide I band is due primarily to the peptide backbone C=O stretching vibration, which is sensitive to hydrogen bonding and thus to protein secondary structure [35–38]. Both, empirical and theoretical studies have shown that α-helical amide I vibrations absorb in the 1645 to $1660\,cm^{-1}$ region, while β-sheet vibrations absorb in the 1625 to $1640\,cm^{-1}$ region, with a second weaker vibration near

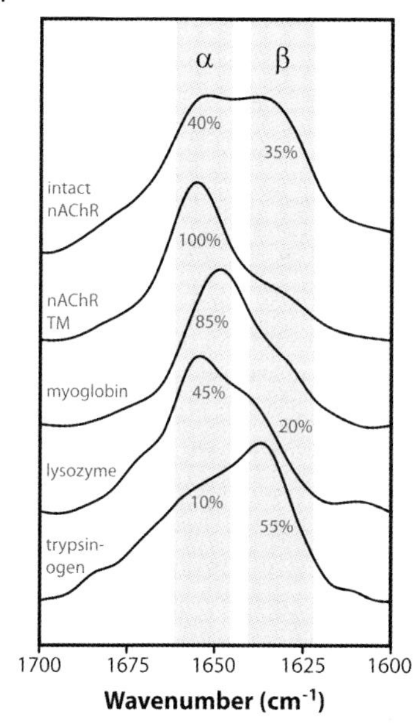

Fig. 11.6 FTIR used to elucidate protein secondary structure. Resolution-enhanced spectra of intact *Torpedo* nAChR (top trace), and Proteinase K-treated nAChR (removes extramembranous regions of the receptor, second trace from top). For comparison, also shown are resolution-enhanced spectra from myoglobin, lysozyme, and trypsinogen. The main frequency regions of the amide I bands typically attributed to α-helix (α) and β-sheet (β) are shaded. The relative proportion of α-helical and β-sheet secondary structure (as determined by other methods) for each of the proteins is also indicated. The resolution-enhanced spectra of the nAChR presented in Figs. 11.6, 11.7, and 11.9b differ slightly because they were acquired under different experimental conditions and thus were not processed using the same deconvolution parameters. (Adapted from Ref. [41].)

1680 cm^{-1} (Fig. 11.6). Predominantly α-helical proteins give rise to a symmetric amide I band shape centered near 1650 cm^{-1}, whereas predominantly β-sheet proteins give rise to an asymmetric band shape with an intense maximum centered near 1630 cm^{-1} and a shoulder located near 1680 cm^{-1}. Proteins with a mixture of both types of secondary structure predictably give rise to spectra that are a combination of the two extremes. The band shapes recorded from predominantly α-helical or β-sheet proteins tend to be relatively insensitive to solvent (^1H_2O and ^2H_2O), although the peak maxima shift down in frequency by 5 to 10 cm^{-1} upon exchange of peptide N—^{1}H for N—^{2}H in ^2H_2O. The band shapes of membrane proteins with more unordered structures, such as the nAChR, tend to undergo more substantial changes in band shape upon transfer from ^1H_2O to ^2H_2O buffer.

A simple visual inspection of amide I band shapes can provide rapid, qualitative insight into the secondary structural content of a protein. Careful analysis of amide I band profiles can also lead to numerical estimates of the relative content of different secondary structures. The typical approach is to use band narrowing/resolution enhancement techniques to identify the number and frequencies of the component bands (α-helix, β-sheet, turn, etc.) contributing to the broad amide I contour. This information is then used to curve fit the experimental data and thus to estimate the relative contribution of each band (each secondary structure element) to the amide I profile. While curve fitting is conceptually simple, such an analysis remains highly subjective as there are many curve fit solutions that will reproduce the experimental data. In order to test the curve fit results, the resolution-enhanced curve fit spectrum should be compared with the resolution-enhanced experimental data [39]. This is essential to ensure that reasonable values for the component band line widths and heights have been chosen. It is also important to note that resolution-enhancement techniques are very sensitive to the presence of noise and minute absorptions due to water vapor in the infrared spectra. Spectra should be analyzed for the presence of water vapor (see Ref. [40]) before applying resolution-enhancement techniques. Finally, it should be noted that the assignment of amide I component bands to different secondary structures is based primarily on empirical correlations. Even exclusively α-helical proteins exhibit weak bands in regions of the spectrum typically attributed to β-sheet [41] (Fig. 11.6). Hence, care must be taken not to over-interpret the spectral data.

The amide I vibration is sensitive to protein denaturation. Denaturation of both soluble and integral membrane proteins leads to a decrease in the intensities of the amide I component bands due to α-helix and β-sheet, and to a corresponding increase in the intensities of component bands near 1620 and 1680 cm^{-1} that have been attributed to unordered strands involved in intermolecular hydrogen bonds. The thermal denaturation temperature of the nAChR obtained by monitoring changes in amide I band shape (roughly 55 °C) is consistent with the thermal denaturation temperature that has been observed for the nAChR using other physical methods [42] (Fig. 11.7).

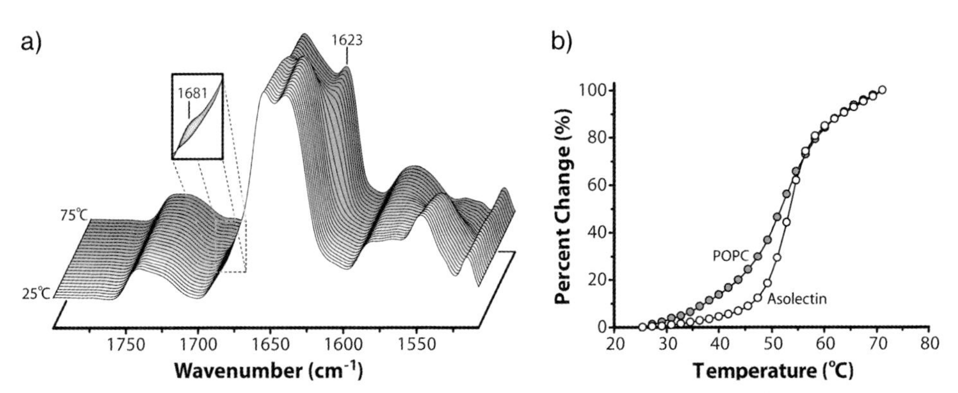

Fig. 11.7 FTIR used to monitor protein thermal denaturation. (a) Deconvolved spectra of the nAChR recorded with increasing temperature (25 °C to 75 °C) exhibit spectral changes indicative of thermal denaturation. (b) Percentage change in intensity at a given frequency (1681 cm^{-1}) as a function of temperature gives a cooperative thermal denaturation curve with a mid point close to 55 °C (nAChR in asolectin membranes). The thermal denaturation upon reconstitution into 1-palmitoyl-2-oleoyl-phosphatidylcholine) (POPC) is less cooperative and occurs at a slightly lower temperature. (Adapted from Ref. [42].)

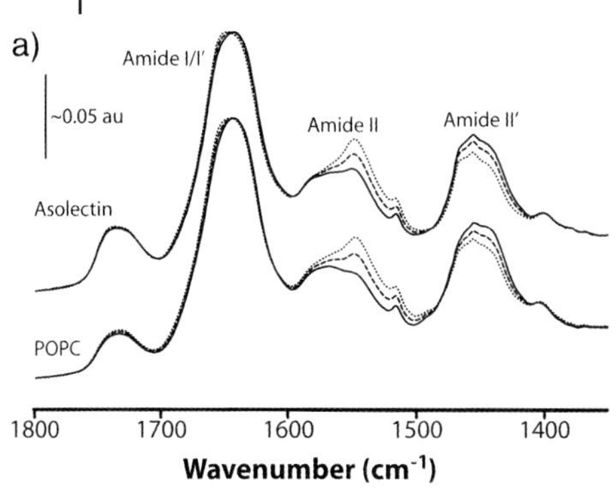

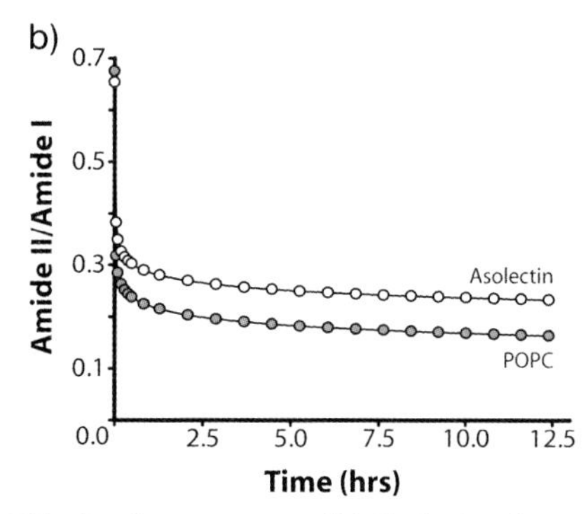

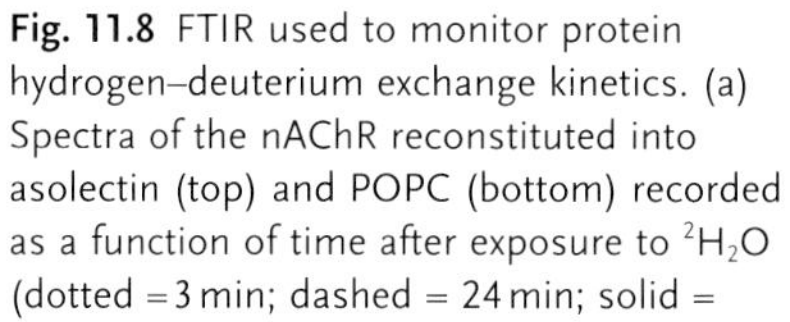

Fig. 11.8 FTIR used to monitor protein hydrogen–deuterium exchange kinetics. (a) Spectra of the nAChR reconstituted into asolectin (top) and POPC (bottom) recorded as a function of time after exposure to ^2H_2O (dotted = 3 min; dashed = 24 min; solid = 750 min after exposure to ^2H_2O). A plot of amide II/amide I ratio as a function of time shows increased hydrogen exchange when the nAChR is reconstituted into POPC membranes. (Adapted from Ref. [42].)

11.3.2.2 **Amide II**

The protein amide II band corresponds mainly to the peptide backbone N–H bending vibration. Although the amide II band shape is sensitive to protein secondary structure, the amide II band is more commonly used to monitor peptide hydrogen exchange kinetics. The frequency of this vibration downshifts from near 1545 cm^{-1} to 1450 cm^{-1} upon exchange of the peptide N–^{1}H for N–^{2}H. In a typical experiment, the intensity of the amide II vibration near 1545 cm^{-1} is measured relative to the intensity of the amide I band as a function of time [43] (Fig. 11.8). It should be noted, however, that in order to compare hydrogen exchange kinetics from one protein to another, each amide II/amide I band intensity ratio must be converted into the percentage of hydrogens that have exchanged for deuterium. This conversion requires accurate measurement of the amide II/amide I band intensity ratios for a fully protiated and a fully deuterated samples. Note also that the amide II band overlaps with a weak vibration of ^2H_2O. Accurate and consistent subtraction of the absorption bands of ^2H_2O buffer from the spectra is required for an accurate measurement of amide II/amide I band intensity ratios.

11.3.3
Protein Side-Chain Vibrations

There are many protein side-chain vibrations which absorb in the 1800 to 1000 cm^{-1} region that can provide insight into the chemistry and local environment of individual residues. For example, the C=O and C–O stretching vibrations of protonated aspartic and glutamic acid residues vibrate in the 1710 to 1790 cm^{-1} and 1120 to 1250 cm^{-1} regions, respectively. The frequency of the C=O stretching vibration shifts down in frequency by as much as 25 cm^{-1} with hydrogen bonding, and can thus be used to monitor changes in hydrogen bond strength during conformational change. In addition, the asymmetric and symmetric stretching vibrations of deprotonated aspartate and glutamate side chains vibrate near $1580\,cm^{-1}$ and $1400\,cm^{-1}$, respectively. Changes in the protonation states of aspartic acid and glutamic acid side chains are easily detected in FTIR spectra using difference techniques (see Section 11.2.2).

A list of the most intense protein side chain vibrations and their frequencies in both $^{1}H_2O$ and $^{2}H_2O$ (where available) is provided in Table 11.1. Side chains with distinct chemical properties give rise to characteristic vibrations; for example, the vibrations of aliphatic amino acid side chains are generally weaker and of lesser utility. It should be noted that the vibrational frequencies of most side-chain vibrations will change depending on the local environment; hence, the frequency–functional group correlations listed in Table 11.1 are useful only as a guide. Either site-directed mutagenesis or site-directed isotope labeling is usually required to assign vibrational shifts detected in difference spectra to individual amino acid side chains. FTIR difference techniques combined with molecular biology can provide residue-specific information regarding protein conformational change. A more comprehensive discussion of side-chain vibrations is presented by Barth [44].

11.4 Applications of FTIR To Membrane Proteins

11.4.1 Testing Protein Structural Models and Validating the Structures of Mutant Proteins

FTIR spectroscopy has been used extensively to probe membrane protein secondary structure, in order to test structural models. A poignant example comes from studies on the nAChR, the prototypic member of a super-family of ligand-gated ion channels that are central to inter-neuronal communication. Low-resolution electron microscopy images and biophysical data originally suggested that the nAChR exhibits a novel transmembrane fold composed of an inner ring of five α-helical segments encircled by a lipid-facing ring of transmembrane β-strands [45, 46]. To test this model, the structure of the nAChR was examined before and after proteolysis, the latter to remove the extramembranous regions of the receptor.

The *intact* nAChR exhibits a broad amide I band shape with major component bands centered at 1655 and $1630\,cm^{-1}$ indicative of α-helix and β-sheet structures, respectively. The relative intensities of the two main component bands suggest that the nAChR has a mixed α-helix/β-sheet structure (see Fig. 11.6). Curve fitting supports this contention, showing that the nAChR contains ~40% α-helix, which is sufficient to account for the entire transmembrane domain of the nAChR (~25% of the total protein mass) [39], as well as a significant proportion of the extramembranous domains. The residual amide II band intensity suggests that between 20 and 30% of the nAChR peptide hydrogens are resistant to peptide hydrogen/

Table 11.1 Absorption frequencies and molar extinction coefficients of prominent side-chain vibrations in both ^1H_2O and ^2H_2O.[(a),(b)]

Amino acid	^1H_2O		^2H_2O	
	Position (cm^{-1})	*ξ in $M^{-1}cm^{-1}$*	*Position (cm^{-1})*	*ξ in $M^{-1}cm^{-1}$*
1. Aspartic acid				
Asp protonated				
$\nu(C{=}O)$	1716–1788	(280)	1713–1775	(290)
$\nu_s(C{-}O)$	1120–1253	(100–200)	1270–1322	
$\delta(COH)$	1264–1450		955–1058	
Asp deprotonated				
$\nu_{as}(COO^-)$	1574–1579	(290–380)	1584	(820)
$\nu_s(COO^-)$	1402	(256)		1404
2. Glutamic acid				
Glu protonated				
$\nu\ (C{=}O)$	1712–1788	(220)	1706–1775	(280)
$\nu_s(C{-}O)$	1120–1253	(100–200)	1270–1322	
(COH)	1264–1450		955–1058	
Glu deprotonated				
$\nu_{as}(COO^-)$	1556–1660	(450–470)	1567	(830)
$\nu_s(COO^-)$	1404	(316)	1407	
3. Asparagine				
Asn, $\nu(C{=}O)$	1677–1678	(310–330)	1648	(570)
Asn, $\delta(NH_2)$	1612–1622	(140–160)		
4. Glutamine				
$\nu(C{=}O)$	1668–1687	(360–380)	1635–1654	(550)
$\delta(NH_2)$	1586–1610	(220–240)	1163	
$\nu(CN)$	1410		1409	
5. Lysine				
$\delta_{as}(NH_3^+)$	1626–1629	(60–130)	1201	
$\delta_s(NH_3^+)$	1526–1527	(70–100)	1170	
6. Arginine				
$\nu_{as}(CN_3H_5^+)$	1672–1673	(420–490)	1608	(460)
$\nu_s(CN_3H_5^+)$	1633–1636	(300–340)	1586	(500)
7. Histidine				
$HisH_2^+$				
$\nu(C\ C)$, $\nu(CC)$	1631	(250)	1600	
$\delta(CH)$, $\nu(CN)$, $\delta(NH)$	1199		1239	
$\nu(CN)$, $\delta(CH)$	1094		1110	
HisH				
$\nu(C{=}C)$, $\nu(CC)$	1575–1594	(70)	1569–1575	
$\nu(C{=}N)$, $\delta(CH)$	1490		1485	
$\delta(CH)$, $\nu(CN)$, $\delta(NH)$	1229		1223	
$\nu(CN)$, $\delta(CH)$	1090–1106		1096–1107	

Table 11.1 *Continued*

Amino acid	*^1H_2O*		*^2H_2O*	
	Position (cm^{-1})	*ξ in $M^{-1}cm^{-1}$*	*Position (cm^{-1})*	*ξ in $M^{-1}cm^{-1}$*
8. Tyrosine				
Tyr-OH				
ν(C—C), δ(CH)	1614–1621	(85–150)	1612–1618	(~160)
ν(C—C)	1594–1602	(70–100)	1590–1591	(<50)
ν(C—C), δ(CH)	1516–1518	(340–430)	1513–1517	(500)
ν(C—O), ν(CC)	1235–1270	(200)	1248–1265	(150)
δ(COH)	1169–1260	(200)		
Tyr-O^-				
ν(C—C)	1599–1602	(160)	1603	(350)
ν(C—C), δ(CH)	1498–1500	(700)	1498–1500	(650)
ν(C—O) ν(CC)	1269–1273	(580)		
9. Tryptophan				
ν(CC), ν(C=C)	1622	1618		
ν(CC), δ(CH)	1496			
δ(CH), ν(CC), ν(CN)	1462	1455 (200)		
δ(NH), ν(CC), δ(CH)	1412–1435	1382		
ν(CC), ν(CN), δ(CH)	1352–1361			
ν(CC), ν(CN)	1334–1342	1334 (100)		
δ(NH), ν(CN), δ(CH)	1276			
δ(CH), ν(CC)	1245			
ν(CC)	1203			
δ(CH)ν(NC)	1092			
ν(NC), δ(CH), ν(CC)	1064			
ν(CC), δ(CH)	1012–1016	1012		
10. Phenylalanine				
ν(C—C)	1494 (80)			
11. Serine				
ν(C—O)	1030			
12. Threonine				
ν(C—O)	1075–1150			
13. Cysteine				
ν(S—H)	2551	1849		
14. Aliphatic				
$\delta_{as}(CH_3)$	1445–1480			
$\delta_s(CH_3)$	1375 or 1368 + 1385 for adjacent CH_3 in Val, Leu			
$\delta(CH_2)$	1425–1475			
δ(CH)	1315–1350			
$\gamma_w(CH_2)$	1170–1382			
$\gamma_t(CH_2)$	1063–1295			
15. Proline				
ν(CN)	1400–1465			

a ν = stretching vibration; ν_{as} = asymmetric stretching vibration; ν_s = symmetric stretching vibration; δ = in-plane bending; γ_w = wagging vibration; γ_t = twisting vibration; γ_r = rocking vibration.

b Data from Ref. [44].

deuterium exchange after three days in ^2H_2O. Significantly, substantial α-helical amide I component band intensity remains at 1655 cm^{-1} after three days' exposure to ^2H_2O, but undergoes a 5 to 10 cm^{-1} downshift in frequency under conditions that enhance further peptide $N{-}^1H/N{-}^2H$ exchange [47, 48]. These findings show that there exist a percentage of exchange resistant residues in α-helices within the nAChR. Linear dichroism also shows that these exchange-resistant α-helices have a predominant orientation parallel to the bilayer normal suggesting, in contrast to the proposed model, that the transmembrane domain is composed entirely of α-helices.

Treatment of the nAChR with Proteinase K to remove the extramembranous domains, provided further evidence for an α-helical transmembrane domain. Proteolysis leads to a dramatic increase in the α-helical content of the nAChR, further suggesting that the transmembrane domain is formed exclusively from α-helical structures and, consequently, that the extramembranous domain contains extensive β-sheet (see Fig. 11.6). Again, linear dichroism measurements showed that the remaining α-helical structures in the transmembrane domain are preferentially oriented parallel to the bilayer normal. No definitive evidence was found for the existence of transmembrane β-strands. These findings supported an exclusively α-helical transmembrane domain, which was later confirmed by a higher-resolution structural model [49].

Of note, FTIR can be used as a rapid tool for the validation of the structures of expressed proteins. As described in Section 11.4.6, microsampling ATR accessories can be used to record spectra from small volumes of a membrane protein solution (5–10 μL is sufficient) [7]. Within a few minutes, the amide I band of a protein can be analyzed qualitatively to assess the structural integrity of a newly purified membrane protein or to provide secondary structural estimates of a newly cloned protein with an unknown fold.

11.4.2 Lipid–Protein Interactions

One advantage of FTIR spectroscopy is that the structural properties of both membrane lipids and membrane proteins can be probed in a single sample using the same technique. Previously, FTIR has been used to characterize lipid–protein interactions at the nAChR, with surprising results.

FTIR difference spectroscopy was first used to examine how lipids influence the ability of the nAChR to undergo the resting to desensitized conformational change in response to the binding of the agonist, carbamylcholine (Carb). Figure 11.9a shows Carb difference spectra recorded from the nAChR reconstituted into phosphatidylcholine (PC) membranes with and without the anionic lipid, phosphatidic acid (PA). For comparison, the top trace was recorded from the nAChR in PC/PA/Cholesterol (Chol) membranes, a membrane that supports a functional nAChR. The bottom trace obtained from the nAChR in PC membranes is similar to Carb difference spectrum recorded from the nAChR in PC/PA/Chol, but while the receptor is exposed to the local anesthetic, procaine, which stabilizes a non-

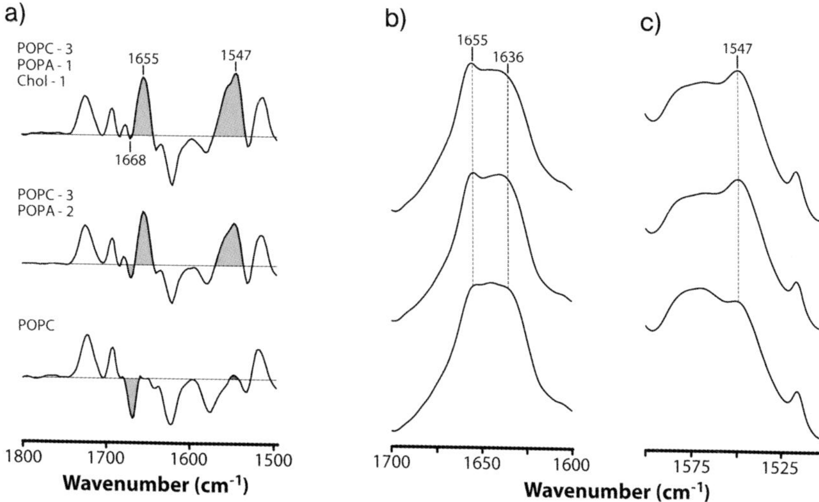

Fig. 11.9 FTIR used to probe the structure and functional ability of the nAChR in different membrane environments. (a) Carb difference spectra recorded from the nAChR reconstituted into membranes composed of POPC/POPA (1-palmitoyl-2-oleoyl-PA)/Chol 3:1:1, POPC/POPA 3:2, and POPC (top to bottom). The shaded areas denote band intensity in the amide I and II regions which reflects the resting to desensitized conformational change (see Fig. 11.3). This intensity is absent in the Carb difference spectra recorded from the nAChR in POPC membranes, indicating that the nAChR cannot undergo agonist-induced conformational change. (b) Resolution-enhanced amide I band of the nAChR in POPC/POPA/Chol 3:1:1, POPC/POPA 3:2, and POPC (top to bottom). The α-helical component band is weaker in the spectra recorded from the nAChR in POPC, most likely due to enhanced hydrogen–deuterium exchange of the transmembrane α-helices after three days' exposure to ^2H_2O and a concomitant 5 to 10 wavenumber downshift in frequency of the α-helical component band. (c) The increased hydrogen–deuterium exchange of the nAChR in POPC membranes is shown by the lower residual amide II band intensity in non-deconvolved spectra recorded from the nAChR in POPC (lowest trace). The residual intensity at 1547 cm^{-1} is due to vibrations of unexchanged peptide N ^1H. (Adapted from Ref. [33].)

functional desensitized state that does not undergo conformational change upon Carb binding (second from bottom trace in Fig. 11.11b). The similarity of the two difference spectra suggests that the nAChR in PC membranes is stabilized in a conformation that cannot respond to agonist binding. Increasing levels of PA in PC membranes shift an increasing proportion of nAChRs into a functional state that responds to Carb binding by undergoing the resting to desensitized conformational transition [23, 33].

Absorbance spectra of the nAChR in each lipid environment were recorded to validate the structure of the purified protein. Surprisingly, the spectra exhibit subtle changes in amide I band shape that are particularly evident if the spectra of the nAChR recorded in PC/PA/Chol are compared to PC-alone membranes (Fig. 11.9b). Some have attributed these changes in amide I band shape to lipid-induced changes in nAChR secondary [50], although subsequent analysis showed that the variations in amide I band shapes are due entirely to varying rates/extent of peptide hydrogen exchange in the different lipid environments (Fig. 11.9c) [33, 51]. It has also been shown that the thermal denaturation of the nAChR in PC membranes is less cooperative and occurs at a slightly lower temperature than the thermal denaturation of the nAChR in membranes that stabilize a functional nAChR (see Fig. 11.7). The hydrogen exchange kinetics are also faster when the nAChR is reconstituted into membranes composed of PC (Fig. 11.8). The latter two findings suggest that the non-functional nAChR in PC membranes has a less compact tertiary structure than the nAChR in functional PC/PA/Chol membranes. With a new atomic-resolution model of the nAChR, these biophysical observations may lead to an insight into the mechanisms by which lipids influence the structure and function of the nAChR.

An interesting finding of these preliminary FTIR analyses is that lipids not only affect nAChR structure and function, but also that the nAChR influences the physical properties of the lipid bilayer (Fig. 11.10). Incorporation of the nAChR into membranes containing the anionic lipid PA leads to an increase in the gel to liquid-crystal-phase transition of the lipid bilayer, as well as a lateral tightening of the bilayer. The latter effect is indicated by a decrease in the proportion of hydrogen-bonded ester carbonyls, which reflects a diminished water penetration into the bilayer interfacial region [33]. Incorporation of the nAChR into membranes containing the anionic lipid, phosphatidylserine, reveal different and unique changes in bilayer structure [34]. The surprising observation that the nAChR influences the packing of its membrane environment in a lipid-specific manner suggests that the nAChR may play a role in modulating the structure and/or composition of its lipid microenvironment. The biological implications of these findings for nAChR function *in vivo* remain to be defined.

11.4.3
Receptor–Drug Interactions

FTIR difference spectroscopy can also be used to monitor the changes in the structure of membrane receptors upon the binding of a variety of agonists and antagonists. The mechanisms by which local anesthetics interact with the nAChR have been studied using a combination of FTIR difference spectroscopy and a classical pharmacological approach. In this experiment, the nAChR is first preincubated with a given concentration of the local anesthetic. The effects of the local anesthetic on the ability of the agonist Carb to trigger nAChR conformational change are then monitored – in this case using FTIR difference spectroscopy [24].

The stacked plot in Fig. 11.11a shows Carb difference spectra recorded in the presence of increasing concentrations of the local anesthetic, procaine. Such increases diminish the intensities of difference bands in the amide I and II regions that are indicative of nAChR conformational change, suggesting that procaine stabilizes the nAChR in a non-functional desensitized state. The difference spectra also exhibit a number of negative vibrational bands, including one at 1605 cm^{-1},

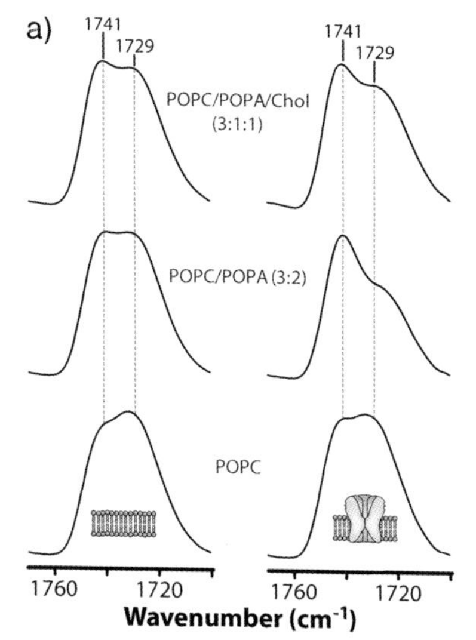

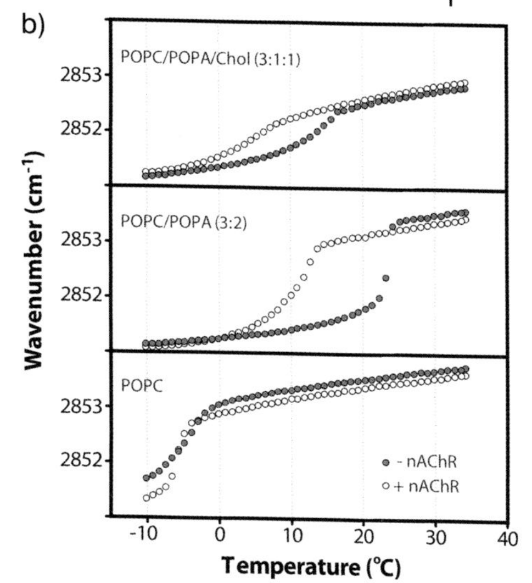

Fig. 11.10 FTIR used to probe the effects of the nAChR on lipid bilayer packing. Incorporation of the nAChR into POPC membranes containing POPA increases the lateral packing density of the lipid bilayer. (a) The resolution-enhanced lipid ester carbonyl vibration reveals two peaks due to hydrogen bonded (1729 cm^{-1}) and non-hydrogen-bonded (1741 cm^{-1}) ester carbonyls (left and right spectra are of membranes either without or with the nAChR incorporated into the lipid bilayer, respectively). Incorporation (right spectra) of the nAChR into either POPC/POPA/Chol 3 : 1 : 1 or POPC/POPA 3 : 2 membranes leads to an increase in the proportion of non-hydrogen-bonded carbonyls (less water penetration into the bilayer due to a tighter lateral packing), whereas incorporation of the nAChR into POPC membranes has little effect. (b) The lipid acyl chain methylene C H stretching vibration shifts up abruptly in frequency upon transition from the gel to the liquid crystal phase. Incorporation of the nAChR (solid circles) into POPC/POPA/Chol 3 : 1 : 1 (top) and POPC/POPA 3 : 2 (middle) membranes leads to an increase in the gel-to-liquid crystal phase-transition temperature compared to bilayers lacking the nAChR (open circles). Incorporation of the nAChR into POPC (bottom) membranes has little effect on the phase transition. (Adapted from Ref. [33].)

that correspond to the vibrations of procaine itself. As procaine binds to the neurotransmitter binding site over the studied concentration range, the negative procaine vibrations were attributed to the displacement of procaine from the neurotransmitter binding site upon the addition of Carb. Procaine also diminishes the intensity of two bands near 1620 and 1515 cm^{-1} that reflect vibrational changes associated with Carb-neurotransmitter binding site physical interactions. The 1515 cm^{-1} vibration likely reflects the formation of cation-π electron interactions between Carb and a tyrosine in the agonist binding site. The 1620 cm^{-1} vibration could reflect similar interactions with a binding site tryptophan. Procaine likely reduces the intensities of these vibrations by forming similar interactions with these neurotransmitter binding site residues prior to the addition of Carb. It should be noted that the changes in the difference spectrum upon the addition of procaine follow a dose-dependent relationship with EC_{50}s that match the expected

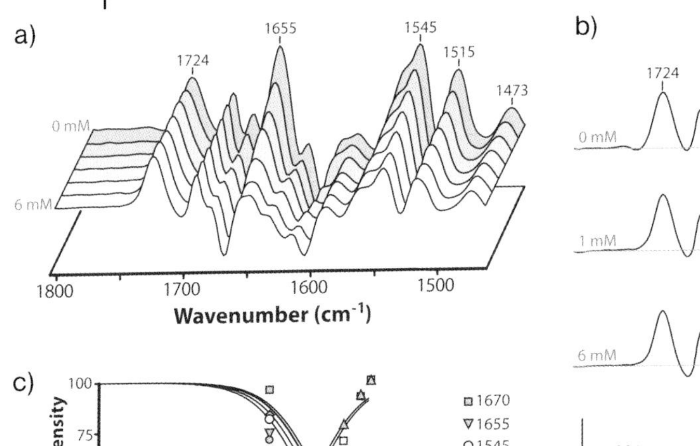

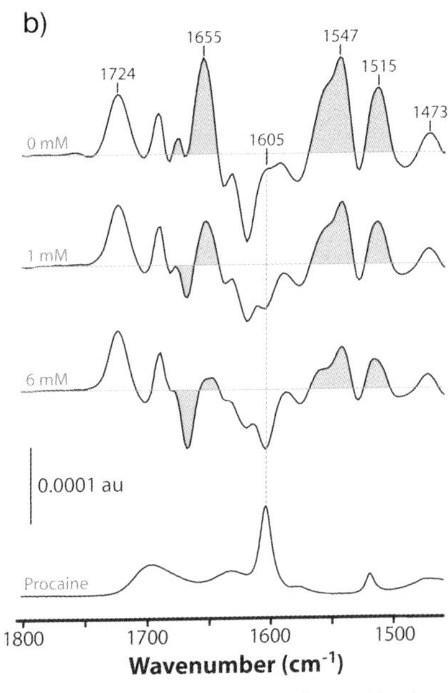

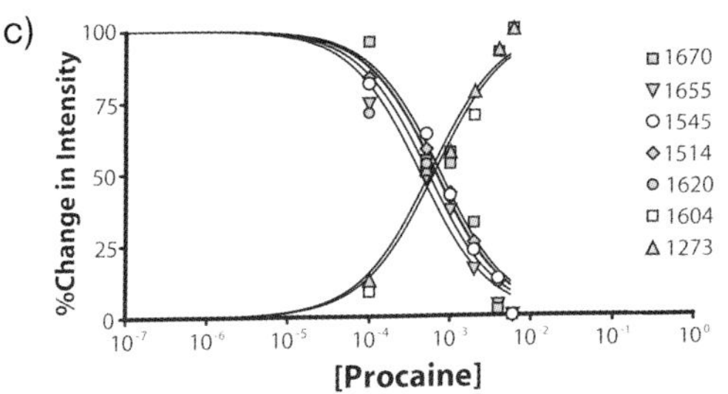

Fig. 11.11 FTIR used to elucidate the binding of drugs and structural changes induced by drug action at the nAChR. (a) Carb difference spectra recorded from the nAChR in egg PC/1,2-dioleoylPA/Chol membranes in the presence of increasing concentrations of the local anesthetic, procaine. For clarity, three representative difference spectra are shown in (b). The lowest spectrum in (b) is an absorption spectrum of procaine itself. Procaine reduces the intensities of conformationally sensitive amide I and amide II vibrations at 1655 and 1545 cm^{-1}, which reflect the resting to desensitized conformational change. The reduced intensities of these vibrations suggest that the drug stabilizes the nAChR in a desensitized state. The drug also reduces the intensities of vibrations near 1620 and 1515 cm^{-1} that reflect Carb–nAChR physical interactions. These vibrational intensities are reduced because procaine binds to the Carb binding site and mimics these interactions with the nAChR prior to Carb binding. Carb binding results in the displacement of procaine from the Carb binding site, and thus the appearance of a negative vibration (due to procaine) near 1610 cm^{-1}. (c) Spectral changes at a given frequency occur in a dose-dependent manner at concentrations of procaine consistent with its known K_D for binding.

EC_{50}s for procaine action at the nAChR. These results demonstrate that classical pharmacological studies can be carried out using FTIR spectroscopy. The advantage of FTIR, however, is that the technique monitors directly the drug-induced structural changes in the target receptor and can provide residue-specific information regarding the nature of receptor–drug interactions.

Studies of other local anesthetics have revealed more complex effects on nAChR conformation than have been detected using standard pharmacological approaches. Difference spectra suggest the existence of novel conformational states, some of which may play a significant role in how nAChR function is modulated *in vivo* [24, 25].

11.4.4 Chemistry of Receptor–Ligand Interactions

Some absorption bands in the Carb difference spectrum reflect vibrational changes that result from the formation of physical interactions between Carb and neurotransmitter binding site residues. These bands provide insight into the chemistry of Carb–nAChR interactions. In order to probe these interactions in more detail, Carb difference spectra were recorded from a desensitized nAChR film that was incubated continuously with the agonist analogue, tetramethylamine (TMA). In other words, the difference was measured between a desensitized nAChR to which either TMA or Carb was bound (Fig. 11.12) [52].

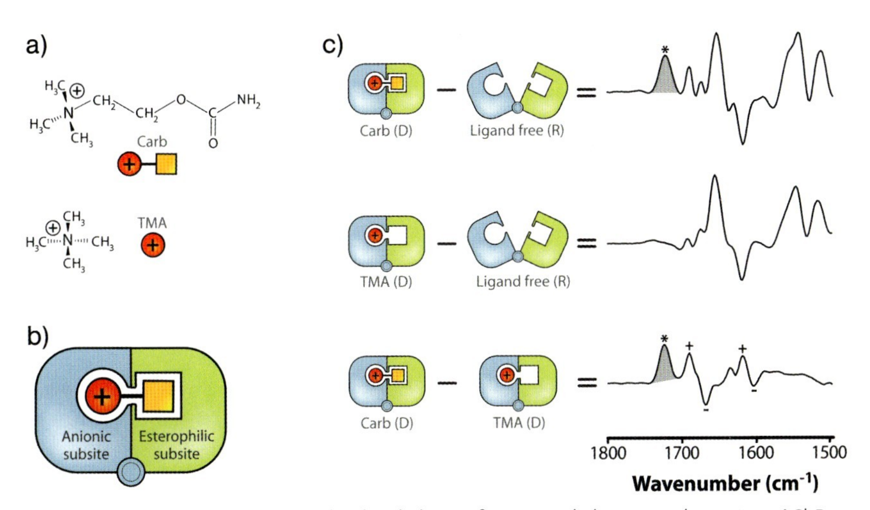

Fig. 11.12 FTIR used to elucidate the detailed chemistry of nAChR–ligand interactions. (a) Chemical structures of the two agonists carbamylcholine (Carb) and tetramethylammonium (TMA), which both contain a quaternary amine. (b) The agonist binding site on the nAChR for Carb is thought to have an anionic subsite (blue) for the charged amine (composed of aromatic π-electrons) and an esterophilic subsite (green) that binds the ester carbonyl. Carb difference spectra exhibit features indicative of (1) the agonist itself (asterisk), (2) the resting to desensitized conformational change (light shading), and (3) Carb–nAChR physical interactions (top trace in (c)). TMA difference spectra exhibit the features similar to those observed in the Carb difference spectrum due to the agonist, in this case TMA (not shown in this region), the resting to desensitized conformational change, and agonist–nAChR interactions (middle trace in (c)). A Carb-minus-TMA difference spectrum does not exhibit peaks due to the conformational change as both ligands stabilize the desensitized state (lower trace in (c)). Bands due to the physical interactions between the quaternary amine and the anionic subsite are also absent as both ligands form these interactions. The spectrum exhibits a peak due to the vibration of the ester carbonyl of Carb (asterisk) and two positive/negative couples (labeled +, –) which reflect vibrations of residues in the esterophilic subsite. The shift in frequencies of these two vibrations reflect the formation of physical interactions between the ester carbonyl of Carb and esterophilic subsite residues in the nAChR that are absent with bound TMA.

The resulting difference spectrum (referred to as a Carb-minus-TMA difference spectrum; see lowest trace/schematic in Fig. 11.12c) is simpler than a typical Carb difference spectrum. Bands due to the resting to desensitized conformational change are absent, as the nAChR is stabilized in a desensitized state in the presence of both Carb and TMA. Similarly, bands that reflect the physical interactions between the quaternary amide of Carb and the nAChR are absent, as these interactions form with both ligands. In the presented region, the Carb-minus-TMA difference spectrum exhibits a positive band due the ester carbonyl of Carb (asterisks). The difference spectrum also exhibits two negative/positive coupled peaks that reflect shifts in the vibrational frequencies of protein vibrations upon Carb, versus TMA binding. In other words, the peaks reflect the shifts in the vibrational spectrum of an amino acid side chain(s) that occur upon formation of a physical interaction between the side chain and the ester functional group on Carb. Specifically, the two negative peaks reflect the vibrations of the amino acid side chain(s) in the absence of a bound ester group, while the two positive peaks reflect the vibrations of the amino acid side chain(s) bound to the ester group. Carb-minus-TMA difference spectra recorded in ^2H_2O and at alternative pH values suggest preliminary assignment of the bands and provide insight into the nature of the physical interactions that occur between the ester carbonyl of Carb and the nAChR [52].

11.4.5
Changes in Orientation of Functional Groups During Conformational Change

Membrane films deposited on planar IREs exhibit a strong orientational preference parallel to the IRE surface [30]. Oriented planar films of the nAChR and linear dichroism FTIR spectroscopy was used to examine whether or not the transmembrane α-helices undergo a change in orientation upon desensitization of the nAChR. Carb-difference spectra recorded using either parallel or perpendicular polarized infrared light reveal subtle (but reproducible) variations in the intensities of several vibrations relative to Carb difference spectra recorded with unpolarized infrared light (Fig. 11.13). Of particular interest are the amide I and II difference bands near 1655 and 1545 cm^{-1}, respectively, that reflect changes in structure of the polypeptide backbone upon desensitization. In Carb difference spectra recorded with unpolarized infrared light, the amide I and II difference bands have similar vibrational intensities. In contrast, the amide I vibration near 1655 cm^{-1} is less intense than the amide II vibration centered near 1545 cm^{-1} in the Carb difference spectrum recorded using parallel polarized infrared light, whereas the relative intensities of the two vibrations are reversed in difference spectra recorded using perpendicular polarized infrared light [30].

Whilst a detailed discussion of the polarized difference spectra cannot be presented here, the data suggest several important features regarding the nature of the conformational change that occurs upon desensitization. First, the difference spectra indicate that there is a change in the conformation of the polypeptide backbone upon desensitization that involves a slight change in orientation relative

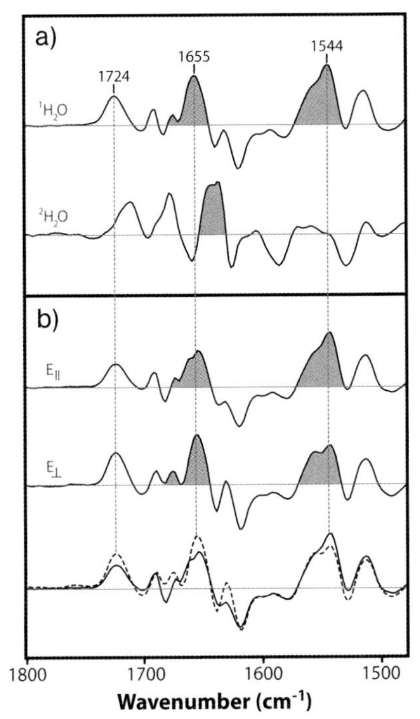

Fig. 11.13 FTIR used to probe the changes in transmembrane α-helix orientation upon nAChR desensitization. (a) Carb difference spectra recorded from the nAChR in ^1H_2O (top) and ^2H_2O (bottom) buffer. In the latter, the nAChR was exposed to ^2H_2O for 72 h prior to data acquisition. Similar spectra in ^2H_2O are observed after minutes exposure of the nAChR to ^2H_2O [52]. (b) Carb difference spectra recorded in ^1H_2O with infrared light polarized either parallel (top trace) or perpendicular (middle trace) to the plane of incidence (see Fig. 11.4). The two polarized Carb difference spectra are superimposed in the lower trace, with the parallel polarized Carb difference spectrum represented as a solid line and the perpendicular polarized spectrum as a dashed line. The varying relative intensities of the amide I (1655 cm^{-1}) and amide II (1545 cm^{-1}) difference bands suggests a slight tilt in the polypeptide backbone upon desensitization. The fact that the amide I and II difference bands undergo large downshifts in frequency within minutes of exposure of the nAChR to ^2H_2O (lower trace in (a) and Ref. [52]) suggests that this tilt does not involve transmembrane α-helices.

to the bilayer normal as well as a change in the actual structure (i.e., a change in ϕ, φ angles, possibly leading to different hydrogen bonding of the polypeptide backbone). Second, the differences in intensity of the amide I and II vibrations between the data acquired with parallel versus perpendicular polarized infrared light are very small, indicating that if the detected vibrational changes in amide I and II band intensity are due to transmembrane α-helices, the change in net tilt is at most a few degrees for one of the α-helices in each subunit. Finally, Carb difference spectra recorded in ^2H_2O show that the amide I and II difference bands centered near 1655 and 1545 cm^{-1} undergo a large downshift in frequency in difference spectra recorded in ^2H_2O buffer. These downshifts in frequency occur within minutes of exposure of the nAChR to ^2H_2O [53], and thus must reflect vibrational changes in highly solvent accessible regions of the polypeptide backbone. Although the possibility that pore-lining transmembrane α-helices exchange their peptide hydrogens rapidly with solvent cannot be ruled out, the most likely interpretation of these data is that changes in conformation and orientation of loop/random structures are being detected, and not a change in the orientation of exchange-resistant transmembrane α-helices (see Section 11.4.2). It is likely that desensitization involves mainly a change in conformation/orientation of a solvent-accessible loop that lines one surface of the ligand binding site [30].

11.4.6
A Tool in the Crystallization of Integral Membrane Proteins

The structural characterization of integral membrane proteins by both X-ray crystallography usually requires the solubilization of the proteins from the lipid bilayer using detergent. Finding suitable conditions that lead to the formation of stable protein–detergent or protein–lipid–detergent complexes, however, can be a daunting task.

Both, the level of endogenous lipid and the concentration/physical properties of the detergent are important factors that affect the solubilization of membrane proteins and thus ultimately their crystallization [54]. Some proteins crystallize in the presence of a number of specifically bound lipids, whereas others only crystallize in minimal lipid [55, 56]. Sufficient detergent must be present first to remove the protein from its membrane environment, and then to prevent the formation of non-specific protein–protein aggregates or precipitates. Excessive detergent, however, can potentially lead to both protein denaturation and the formation of protein-free micelles, both of which may interfere with crystal formation. Although membrane proteins are ideally crystallized in the presence of a minimal amounts of detergent and lipid, each protein likely has a select detergent concentration and lipid–protein ratio under which structural integrity and monodispersity are both maintained, and thus under which crystallization is favorable.

An initial step in the structural characterization of a membrane protein requires an understanding of how lipid–protein and detergent–protein ratios influence both the structural stability and solubility of the protein. FTIR spectroscopy has been used to monitor lipid–protein ratios and detergent concentrations in membrane protein solutions [57, 58]. An important advance was the commercial availability of a single-bounce diamond ATR accessory, which allows spectra to be recorded from detergent-solubilized membrane protein solutions using sample volumes of only 5 to 10 μL. This quick and simple method allows for rapid spectral acquisition from small aliquots, and has allowed the monitoring of both lipid–protein ratios and detergent concentrations of samples at all stages of the purification protocol.

For monitoring lipid–protein ratios, spectra can be recorded from dried solutions containing a constant known concentration of the protein lysozyme and increasing levels of the phospholipid POPC, thereby generating a standard curve.

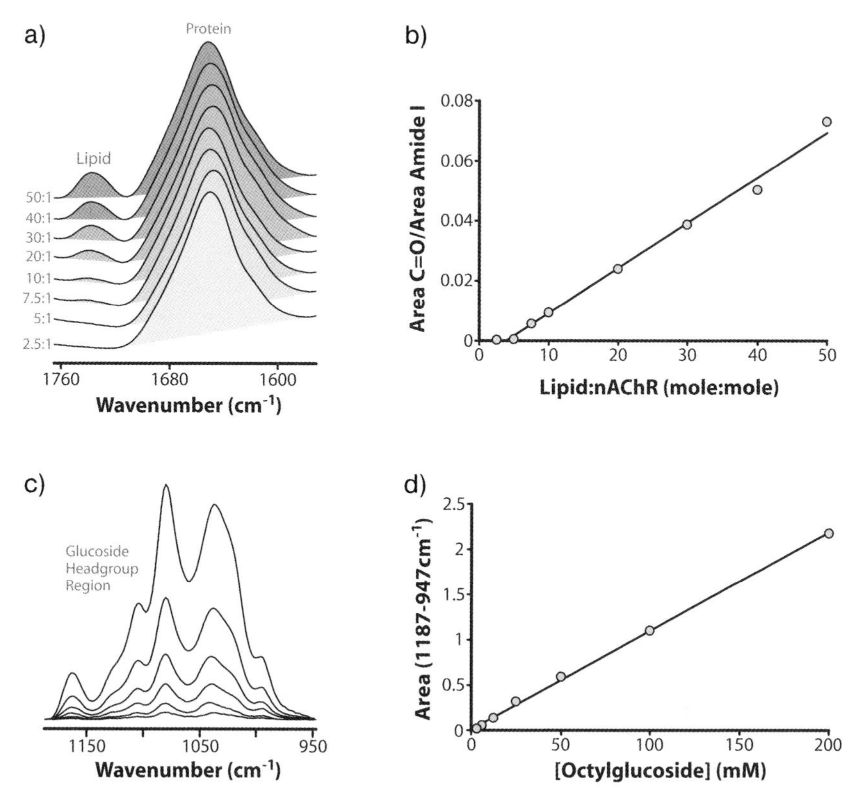

Fig. 11.14 FTIR used as an analytical tool for assessing lipid–protein ratios and detergent concentrations in solutions used for membrane protein crystallization. (a) Spectra recorded using a microsampling device (see Ref. [7]) from 10-μL aliquots of solutions of the protein lysozyme with increasing concentrations of the lipid POPC. The molar lipid to protein ratio at the left of each spectrum is calculated for a 300-kDa protein. (b) Standard curve comparing the relative areas of the lipid/protein vibrations with the lipid to protein molar ratio, calculated for a protein of 300 kDa (i.e. the nAChR). Note that the technique detects as little as ~5 molecules of lipid bound to a 300-kDa protein. (c) Spectra recorded from 10-μL aliquots from solutions containing increasing concentrations of the detergent, *n*-octyl-β-D-glucopyranoside. (d) Standard curve comparing the area of the detergent headgroup vibrations with the detergent concentration.

As shown in Fig. 11.14a, the relative area of the lipid C=O and the protein amide I band is directly related to the lipid–protein ratio (note that the molar lipid–protein ratio has been scaled for the nAChR, a 300 kDa protein). The technique is able to detect, accurately, lipid–protein ratios down to approximately five molecules of lipid per molecule of a 300-kDa protein.

In order to measure detergent concentrations, spectra can be recorded from solutions with increasing concentrations of a detergent such as *n*-octyl-β-D-glucoside. The area of the detergent vibration in the 1200 to 950 cm^{-1} region correlates in a linear fashion with the concentration of the detergent (Fig. 11.14b), showing that the infrared technique is a viable approach for measuring detergent concentrations in aqueous solutions. It should be noted that accurate spectra of this and other detergents have been recorded to well below their critical micellar concentrations (CMC) (for octylglucoside this is ~19–25 mM, but for dodecylmaltoside the CMC is only 0.18 mM [54]).

11.5 Conclusions and Future Directions

The information provided in this chapter illustrates the potential of ATR FTIR spectroscopy for the structural characterization of integral membrane proteins. The ability to monitor the structure of the nAChR in a membrane environment has provided important insights into the receptor's structure in the context of different lipids. The presented spectra also highlight both the extremely high signal-to-noise ratio and reproducibility that can be obtained in ligand-induced difference spectra recorded using the ATR approach. Difference spectra have shed light on the nature of nAChR conformational changes, and also on the detailed chemistry of nAChR–ligand interactions. These data have provided new insights into the mechanisms of both lipid and drug action at the nAChR. In addition, micro-sampling ATR accessories have allowed FTIR to be used as an analytical tool for analyzing membrane protein samples prior to crystallization.

The studies described here of ligand-induced nAChR conformational change, however, are only in their infancy. The inability to express mutant nAChRs in sufficient quantities for FTIR has, to date, prevented the assignment of bands observed in the Carb difference spectra to structural changes in specific residues. Today, the use of the micro-sampling methods are bringing the spectroscopic requirements in line with current expression capabilities (e.g., see Ref. [59]). The ability to combine modern molecular biological approaches with FTIR difference techniques will open up a whole new avenue of investigation, and should lead to detailed insight into the nature of localized structural change in the nAChR that occur upon ligand binding. In the near future, it will be possible to study receptors from neuronal sources, which are of important clinical and pharmacological interest, as well as a host of other integral membrane proteins.

Finally, one of the most exciting applications of FTIR difference spectroscopy involves the monitoring of protein conformational change in real time. Rapid-scan spectrometers are able to record spectra in the millisecond to second time scale [60], while step-scan methods can extend the measurement of spectral differences into the micro-second time regime [61]. Both of these time-resolved methods have been coupled to difference spectroscopy to follow protein conformational change for light-activated proteins, such as bacteriorhodopsin, in the milli- to micro-second time regimes. The light-sensitive chromophore in bacteriorhodopsin provides an intrinsic tool for the simultaneous and repetitive activation of an entire population of receptors, thereby allowing conformational changes to be monitored simultaneously using either approach. The ATR method of data acquisition is compatible with time-resolved FTIR measurements using a caged ligand approach (see Ref. [62]). The uniform release of a ligand from its caged precursor with a flash of visible light will allow the kinetics of ligand-induced conformational change to be studied with a fast time-resolution, whilst ATR with buffer flow will allow signal averaging of flash-induced kinetic experiments. Continued technical advances should extend the ATR difference approach to other membrane proteins, leading to more sophisticated studies examining ligand-induced changes in membrane protein conformation.

Today, the number of membrane protein structures which have been solved at or near atomic resolution is increasing at a rapid pace. FTIR, which can be used to probe dynamic changes in membrane protein structure at the single amino acid residue level, is poised to fill a unique niche that will help lead to a detailed understanding of membrane protein function.

Acknowledgments

These studies were supported by a grant from the Canadian Institutes of Health Research (to J. E. B.)

References

1 N. B. Colthup, L. H. Daly, S. E. Wiberley. *Introduction to Infrared and Raman Spectroscopy*, 3rd edn. Academic Press, San Diego, **1990**.

2 N. J. Harrick. *Internal Reflection Spectroscopy*. Wiley, New York, **1967**.

3 E. Goormaghtigh, V. Raussens, J. M. Ruysschaert. *Biochim. Biophys. Acta.* **1999**, *1422*, 105–185.

4 U. P. Fringeli, Hs. H. Günthard, in: *Membrane Spectroscopy*, Grell, E. (eds.), Springer-Verlag, Berlin, **1981**, pp. 270–332.

5 J. E. Baenziger, K. W. Miller, K. J. Rothschild. *Biophys. J.* **1992**, *61*, 983–992.

6 J. E. Baenziger, K. W. Miller, M. P. McCarthy, K. J. Rothschild. *Biophys. J.* **1992**, *62*, 64–66.

7 C. J. B. daCosta, J. E. Baenziger. *Acta Crystallogr. D Biol. Crystallogr.* **2003**, *59*, 77–83.

8 M. S. Braiman, K. J. Rothschild. *Annu. Rev. Biophys. Chem.* **1988**, *17*, 541–570.

9 K. J. Rothschild. *J. Bioenerg. Biomembr.* **1992**, *24*, 147–67.

10 W. Mäntele. *Trends Biochem Sci.* **1993**, *18*, 197–202.

11 F. Siebert. *Methods Enzymol.* **1995**, *246*, 501–526.

12 A. Barth, C. Zscherp. *FEBS Lett.* **2000**, *447*, 151–156.

13 A. J. White, K. Drabble, C. W. Wharton. *Biochem J.* **1995**, *306*, 843–849.

14 J. Backmann, H. Fabian, D. Naumann. *FEBS Lett.* **1995**, *364*, 175–178.

15 D. Reinstadler, H. Fabian, J. Backmann, D. Naumann. *Biochemistry* **1996**, *35*, 15822–15830.

16 D. Moss, E. Nabedryk, J. Breton, W. Mäntele. *Eur. J. Biochem.* **1990**, *187*, 565–572.

17 M. Lubben, K. Gerwert. *FEBS Lett.* **1996**, *397*, 303–307.

18 K. Fahmy. *Biophys. J.* **1998**, *75*, 1306–1318.

19 V. G. Gregoriou, M. S. Braiman. *Vibrational Spectroscopy of Biological and Polymeric Materials*. CRC Press, Florida, **2006**, pp. 325–351.

20 C. Zscherp, A. Barth. *Biochemistry* **2001**, *40*, 1875–1883.

21 J. E. Baenziger, K. W. Miller, K. J. Rothschild. *Biochemistry* **1993**, *32*, 5448–5454.
22 S. E. Ryan, C. N. Demers, J. P. Chew, J. E. Baenziger. *J. Biol. Chem.* **1996**, *271*, 24590–24597.
23 J. E. Baenziger, M. L. Morris, T. E. Darsaut, S. E. Ryan. *J. Biol. Chem.* **2000**, *275*, 777–784.
24 S. E. Ryan, J. E. Baenziger. *Mol. Pharmacol.* **1999**, *55*, 348–55.
25 S. E. Ryan, M. P. Blanton, J. E. Baenziger. *J. Biol. Chem.* **2001**, *276*, 4796–4803.
26 S. E. Ryan, H. P. Nguyen, J. E. Baenziger. *Toxicol. Lett.* **1998**, *100–101*, 179–183.
27 J. E. Baenziger, J. P. Chew. *Biochemistry* **1997**, *36*, 3617–3624.
28 W. Hübner, H. H. Mantsch. *Biophys. J.* **1991**, *59*, 1261–1272.
29 D. Marsh, M. Muller, F. J. Schmitt. *Biophys. J.* **2000**, *78*, 2499–2510.
30 D. G. Hill, J. E. Baenziger. *Biophys. J.* **2006**, *91*, 705–714.
31 L. K. Tamm, S. A. Tatulian. *Q. Rev. Biophys.* **1997**, *30*, 365–429.
32 H. H. Mantsch, R. Mendelsohn. *Progress in Protein-Lipid Interactions 2*. Elsevier Science Publishers BV, **1986**, pp. 103–146.
33 C. J. B. daCosta, A. A. Ogrel, E. A. McCardy, M. P. Blanton, J. E. Baenziger. *J. Biol. Chem.* **2002**, *277*, 201–208.
34 C. J. B. daCosta, I. D. Wagg, M. E. McKay, J. E. Baenziger. *J. Biol. Chem.* **2004**, *279*, 14967–14974.
35 W. K. Surewicz, H. H. Mantsch. *Biochim. Biophys. Acta* **1988**, *952*, 115–130.
36 W. K. Surewicz, H. H. Mantsch, D. Chapman. *Biochemistry* **1993**, *32*, 389–394.
37 M. Jackson, H. H. Mantsch. *Crit. Rev. Biochem. Mol. Biol.* **1995**, *30*, 95–120.
38 F. Dousseau, M. Pézolet. *Biochemistry* **1990**, *29*, 8771–8779.
39 N. Methot, M. P. McCarthy, J. E. Baenziger. *Biochemistry* **1994**, *33*, 7709–7717.
40 S. E. Reid, D. J. Moffatt, J. E. Baenziger. *Spectrochim. Acta* **1996**, *52*, 1347–1356.
41 N. Methot, B. D. Ritchie, M. P. Blanton, J. E. Baenziger. *J. Biol. Chem.* **2001**, *276*, 23726–23732.
42 C. J. B. daCosta, D. E. Kaiser, J. E. Baenziger. *Biophys. J.* **2005**, *88*, 1755–1764.
43 J. E. Baenziger, T. E. Darsaut, M. L. Morris. *Biochemistry* **1999**, *38*, 4905–4911.
44 A. Barth. *Prog. Biophys. Mol. Biol.* **2000**, *74*, 141–173.
45 N. Unwin. *J. Mol. Biol.* **1993**, *229*, 1101–1124.
46 U. Gorne-Tschelnokow, A. Strecker, C. Kaduk, D. Naumann, F. Hucho. *EMBO J.* **1994**, *13*, 338–341.
47 J. E. Baenziger, N. Methot. *J. Biol. Chem.* **1995**, *49*, 29129–29137.
48 N. Methot, J. E. Baenziger. *Biochemistry* **1998**, *37*, 14815–14822.
49 A. Miyazawa, Y. Fujiyoshi, N. Unwin. *Nature* **2003**, *423*, 949–955.
50 G. Fernandez-Ballester, J. Castresana, A. M. Fernandez, J. L. Arrondo, J. A. Ferragut, J. M. Gonzalez-Ros. *Biochemistry* **1994**, *33*, 4065–4071.
51 N. Methot, C. N. Demers, J. E. Baenziger. *Biochemistry* **1995**, *34*, 15142–15149.
52 S. E. Ryan, D. G. Hill, J. E. Baenziger. *J. Biol. Chem.* **2002**, *277*, 10420–10426.
53 J. E. Baenziger, J. P. Chew. *Biochemistry* **1997**, *36*, 3617–3624.
54 M. le Maire, P. Champeil, J. V. Moller. *Biochim. Biophys. Acta* **2000**, *1508*, 86–111.
55 R. M. Garavito, J. P. Rosenbusch. *Methods Enzymol.* **1986**, *125*, 309–328.
56 W. Kuhlbrandt. *Q. Rev. Biophys.* **1988**, *21*, 429–477.
57 E. Goormaghtigh, V. Cabiaux, J. M. Ruysschaert. *Eur. J. Biochem.* **1990**, *193*, 409–420.
58 A. M. Pistorius, F. M. Stekhoven, P. H. Bovee-Geurts, W. J. de Grip. *Anal. Biochem.* **1994**, *221*, 48–52.
59 S. E. Plunkett, R. E. Jonas, M. S. Braiman. *Biophys. J.* **1997**, *73*, 2235–2240.
60 E. T. Nibbering, H. Fidder, E. Pines. *Annu. Rev. Phys. Chem.* **2005**, *56*, 337–67.
61 C. Kotting, K. Gerwert. *Chemphyschem* **2005**, *6*, 881–888.
62 J. Heberle, C. Zscherp. *Appl. Spectrosc.* **1996**, *50*, 588–596.

12
Resonance Raman Spectroscopy of a Light-Harvesting Protein

Andrew Aaron Pascal and Bruno Robert

12.1
Introduction

The aim of this chapter is to show how vibrational techniques, such as resonance Raman spectroscopy, may be used as complementary methods to characterize crystallized proteins, in particular those which contain chromophores. Details will be addressed in the case of the major light-harvesting protein from higher plants, for which resonance Raman could be used to show that the conformation of this protein was different in the crystal from that it adopts in solution. It has been shown that the crystallization of LHCII induces a dramatic change in its function, most likely associated with this conformational modification. It seems that this protein, the role of which is to harvest and transfer light energy with high efficiency, possesses an ability to dissipate this energy within the crystal. Based on these findings, it was concluded that this protein also plays a role in higher plant photoprotection. Following a brief introduction on resonance Raman spectroscopy, the vital importance of photoprotection mechanisms for higher plants will be described in some detail, and an explanation provided of how a study on crystallized LHCII, combining resonance Raman and fluorescence, led to a completely new concept for the regulation of energy flux in photosynthetic organisms.

12.2
Principles of Resonance Raman Spectroscopy

The Raman effect is the change of frequency observed when monochromatic light is scattered by polyatomic molecules. During Raman scattering, energy is exchanged between the incoming photon and the scattering molecule. As the energy levels of the latter are discrete, and if the frequency of the incident light is ν_o and that of the scattered light is ν_r, the energy $h^*\Delta\nu = h^*(\nu_o - \nu_r)$ must correspond to that of a transition between energy levels of the scattering molecules (Fig. 12.1). Raman spectroscopy thus yields information on the energy of the vibrational levels of a given electronic state usually the ground state, although this can be any

Biophysical Analysis of Membrane Proteins. Investigating Structure and Function. Edited by Eva Pebay-Peyroula

ISBN: 978-3-527-31677-9

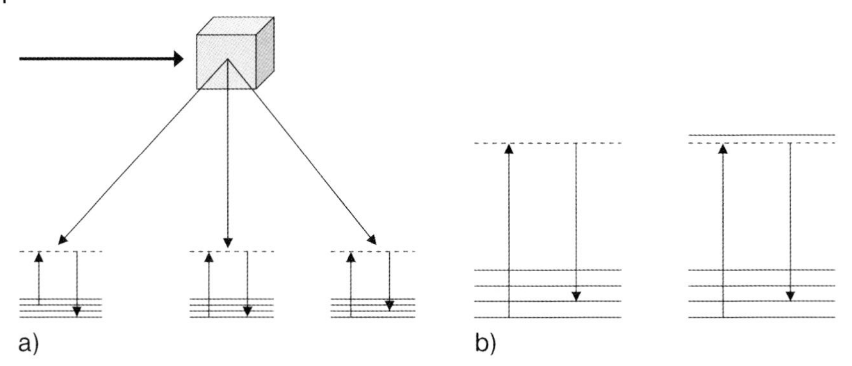

Fig. 12.1 Principle of resonance Raman spectroscopy: energy levels involved in the Raman scattering phenomenon. (A) From left to right: anti-Stokes Raman scattering, Rayleigh scattering, Stokes Raman scattering. (B) Raman (left) and resonance Raman (right).

excited electronic state if the latter has been populated before Raman scattering (by a first laser pulse, for instance). The vibrational levels of a particular molecule depend intimately on its structure – that is, on the nature of its constituent atoms, on the bonds between these atoms, and on its molecular symmetry. Raman spectroscopy can thus be used as an analytical method for determining the chemical structure of molecules. As a molecular spectroscopy, Raman may also provide indications on the conformation of the scattering molecules, and/or on the intermolecular interactions that they form with their immediate surrounding, such as H-bonds. The Raman effect is, however, a very low-probability process, and as a consequence a major drawback of Raman spectroscopy is that the signal measured is usually very weak, and can be blurred by traces of fluorescing molecules present in the sample. On the other hand, as Raman-active molecular vibrations are those which involve changes in molecular polarizability, the Raman signal of water seldom interferes with that of the biological molecules being studied. This constitutes an important advantage for the Raman technique over, for example, infrared absorption spectroscopy.

In classical Raman spectroscopy, the signal only depends on the frequency ν_0 of the light used for inducing the Raman effect as scattering – that is, its intensity varies according to the fourth power of this frequency ν_0^4. However, when this frequency matches an electronic transition of the irradiated molecule, an enhancement of a subset of Raman-active modes is observed which may reach six orders of magnitude: this is the *resonance effect*. In resonance Raman spectroscopy, it is thus possible to selectively observe a molecule in a complex medium, by matching the energy of the exciting photons with the absorption transition of this molecule. This unique property explains why resonance Raman is widely used for studying the interactions assumed by, or the conformation of, chromophores such as hemes, iron-sulfur clusters, chlorophylls and carotenoid molecules within proteins, as it selectively yields the contributions of these chromophores, without interference with the signal of the protein (for a review, see Ref. [1]). In resonance conditions, only a fraction of the vibrational modes of the scattering molecule are enhanced. In the simplest case, when only one electronic state is involved in the resonance phenomenon, the resonance Raman signal arises from those vibrational modes involving nuclei motions which correspond to distortions experienced by the molecule during transition between the ground- and the excited state used for inducing the resonance [2]. This intra-mode selection may be considered as a limitation; for example, if a domain of the molecule is not involved in the electronic transition, resonance Raman will not yield any information about this domain. However, for most biological chromophores the functional part of the molecule consists of those atoms which are conjugated with the electronic transition. Resonance Raman will therefore yield selective information on these biologically active submolecular structures and, from that point of view, the intra-mode selection may be considered as a clear advantage of this technique. Moreover, the analysis of the resonance Raman-active modes observed upon excitation with a given electronic transition will thus yield information about this transition. They may be used, for example, to attribute the molecule to which this transition belongs (in a complex, biological medium), or they may provide information on the nature of the excited state. This was used in particular, for attributing the electronic absorption transition in light-harvesting complexes from higher plants [3]. In short, in resonance Raman, the position of bands in the spectra will yield information about the vibrational structure of the ground electronic state involved in the transition used for inducing the resonance, whilst the intensity of these peaks will yield information about the excited electronic state involved in this transition.

12.3 Primary Processes in Photosynthesis

Photosynthesis comprises the molecular events which allow a large class of organisms (plants, but also algae and bacteria) to live from the use of solar energy. It is traditionally divided into two phases: (i) the primary processes, during which the energy of the incoming solar photons is transduced into chemical potential energy, in the form of both ATP and reducing power; and (ii) the secondary processes, during which this stored energy is used to produce metabolites (such as sugars) from CO_2. The primary processes start with the absorption of an incoming photon by a specialized chromophore, which consequently reaches an excited state. This first energy conversion – from light energy to excitation energy – is the only endergonic step of photosynthesis. This light-harvesting molecule is generally located within a protein, which binds many chromophores, and maintains them all at fixed relative distances and geometries. Consequently, immediately upon absorption of a photon there occurs a rapid (sub-picosecond) equilibrium between the different chromophores in this so-called antenna (or light-harvesting) protein (e.g., see Ref. [4]). The antenna system of photosynthetic organisms generally involves a complex network of such chromophore-bearing proteins, within the photosynthetic membrane and (in some cases) at the perimembrane interface. The antenna proteins are interconnected in such a way that the excitation energy may be rapidly transferred between them – within a few picoseconds – and quite often in a vectorized way, until it reaches another class of specialized protein, the *reaction center*.

Reaction centers are also membrane proteins, containing a chain of electron transporters, closely connected to a chromophore structure called the *primary electron donor*. Upon excitation, the primary donor becomes oxidized, providing an electron which is quickly transferred along the transport chain from one side of the membrane to the other. During this stabilizing step, the energy is thus transduced from excitation energy into chemical potential energy. The electron donor is subsequently reduced by secondary donors located on the other side of the photosynthetic membrane. The terminal acceptor in the reaction center transfers the electron either to stored reducing power, in the form of NADH or NADPH, or to establish, at the level of cytochrome bc_1 or b_6f, a gradient of protons through this membrane, which will be used by the ATPase to produce ATP. This general description of the primary processes is valid for all photosynthetic organisms on Earth. However, depending on the organism considered, quite large variations may be observed in the composition of the macromolecules involved. Although all photosystems are built according to the same principles, they exhibit large differences in the redox potentials generated, and thus in the molecules they may reduce/oxidize. Photosynthetic organisms have adapted to the different ecotopes that they have colonized, mainly by adapting their light-harvesting system, the architecture and composition of which may thus exhibit dramatic variations from one organism to another. The structure and function of a light-harvesting protein from higher plants is outlined in the following section, after which the organization of the photosynthetic apparatus of higher plants will be described in more detail.

12.4
Photosynthesis in Plants

Photosynthesis in plants involves two photosystems working most of the time in series, namely photosystems I and II. In the thylakoid membrane, photosystem II is found in the granal stacks, formed from appressed regions of the membrane, whilst photosystem I is mainly found in the lamellae (non-appressed regions).

Photosystem I generates reducing power by transferring electrons from plastocyanin to ferredoxin. The reducing power created is strong enough so that ferredoxin is able to reduce $NADP^+$ into NADPH. The resulting oxidized plastocyanin is reduced by cytochrome b_6f, itself receiving the electron from photosystem II (for a review on oxygenic photosynthesis, see Ref. [5]).

Photosystem II is among the most fascinating biological macromolecules, as it is able to extract electrons from water, to reduce a plastoquinone molecule. Once four electrons have been extracted from two water molecules, molecular oxygen is created at the level of the so-called oxygen-evolving complex of photosystem II, and this is the origin of all the molecular oxygen on Earth. The oxygen-evolving complex uses a cluster of four manganese atoms to store the oxidizing power. However, in order to oxidize this cluster, very highly oxidizing species must be created upon photosystem II excitation. The primary electron donor in photosystem II is a chlorophyll species called P_{680}, and the redox potential of the couple P_{680}/P_{680}^{+} is greater than 1 V. The housing of such a species in a protein is clearly a challenge, as it is able to oxidize any amino acid side chain in its vicinity. This is actually what happens, as the electron transfer from P_{680}^{+} to the manganese cluster occurs via a tyrosine. Because of the danger inherent to the photogeneration of such an oxidant, many secondary electron donors exist in photosystem II to avoid damage in case the system were unable to quickly reduce P_{680}^{+}. Although no less than three different chemical species (a chlorophyll, a carotenoid and a cytochrome) are present in photosystem II to reduce P_{680}^{+} in case of malfunctioning [6, 7], the subunit which bears P_{680} has one of the fastest turnover rates observed in biology [8]. This fact alone illustrates, in a very straightforward manner, how difficult it is to maintain this system in a proper working state, and how difficult it is to avoid oxidative stress, at this level of the photosynthesis process.

12.5 The Light-Harvesting System of Plants

Plants synthesize complex light-harvesting systems for both photosystem I and II. Photosystem I is principally composed of two large polypeptides which bind, as well as the cofactors involved in electron transfer, about 100 antenna chlorophylls and 22 carotenoid molecules [9], which form an inner antenna system. In contrast, the electron transfer chain of photosystem II resides in a much smaller subunit which binds only a very limited number of chlorophylls (six, four of which are involved in primary electron transfer). However, this subunit is in very close contact with the antenna proteins CP43 and CP47 which bind a large number (40) of antenna chlorophylls. In cyanobacteria, for example, these subunits are the only membrane antenna system of photosystem II [10]. The ensemble photosystem II reaction center + CP47 and CP43 is also called the *photosystem II core*. In plants it has been shown, mainly by electron microscopy, that this photosystem II core is associated in a defined way in the thylakoid membrane with a number of antenna proteins, namely CP29, CP26 and CP24, and a number of LHCII proteins) [11, 12]. CP24, 26, and 29 (also called minor plant antenna proteins), and LHCII are members of the CAB protein family, which also comprises the LHCI proteins and a number of others, some synthesized during greening, some involved in photoprotection (see below). In photosystem I, the equivalent antenna system comprises four LHCI proteins, organized around the large PSI subunits. Finally, a peripheral antenna system exists in plants, constituted of LHCII proteins which are functionally connected to photosystem I or photosystem II, but do not form with these photosystems membrane protein architectures, or at least, these architectures have not yet been demonstrated. Depending on the quality of the light, a number of LHCII is able to functionally migrate from photosystem II to photosystem I, in order to keep the balance between the excitation energy delivered to these two photosystems. This phenomenon – which is also known as state-transitions – concerns a limited amount of LHCII antenna in plants (about 20%) [13], but may become very important in some organisms (e.g., in *Chlamydomonas reinhardtii*, a unicellular green algae, no less than 80% of the LHCII protein may functionally migrate from one photosystem to another [14]).

12.6 Protection against Oxidative Stress: Light-Harvesting Regulation in Plants

Plants, as terrestrial, non-motile organisms, must cope with a wide range of illumination conditions, which fluctuate with differing time scales (weather and shade conditions can vary seasonally, daily, hourly and even within minutes or seconds). They must be able to adapt to the light context, in order to keep their photosynthetic system in the best working conditions, and this is achieved by a very complex control of the light-harvesting process. This control may be achieved at the level of the organism (leaf motions, for instance, may reduce the amount of incoming photons by orders of magnitude), at a cellular level (depending on the illumination conditions, chloroplasts align parallel or perpendicular to the incoming light) or within the photosynthetic membrane. In the latter process, although the amount of antenna protein may vary in the long term, for seasonal adaptation, there is no way to adapt the antenna/photosystem stoichiometry instantaneously to account for rapid variation of illumination conditions. A small content of antenna proteins would fit high-illumination conditions, but this would result, in lower light, in an imbalance of the cell energetics. By contrast, a large antenna system would ensure the bioenergetics of the plant in dim light, but would induce stress in high-illumination conditions. If the plant absorbs more light that can be transformed into chemical potential energy, the photosynthetic system becomes saturated, and a number of harmful processes may occur, due to the presence of an excess of excitation energy. On one hand, saturating photosystem II generates highly oxidizing chemical species, which will react with the surrounding macromolecule, impairing its photosynthetic ability and resulting in subsequent oxidative damage for the photosynthetic organism. On the other, excitation energy in the photosynthetic membrane is mostly present in the form of lower excited singlet states of chlorophyll molecules. These singlet states usually relax in a few nanoseconds through electronic fluorescence, but they may also be converted, with a low but significant yield, into chlorophyll triplet states [15]. Direct relaxation of this triplet state to the ground electronic state is forbidden, and its intrinsic lifetime is thus much longer than that of the singlet state (micro to milliseconds). Triplet states may also be produced in photosystems upon charge recombination, when the electrons cannot be transferred along the transport chain. These chlorophyll

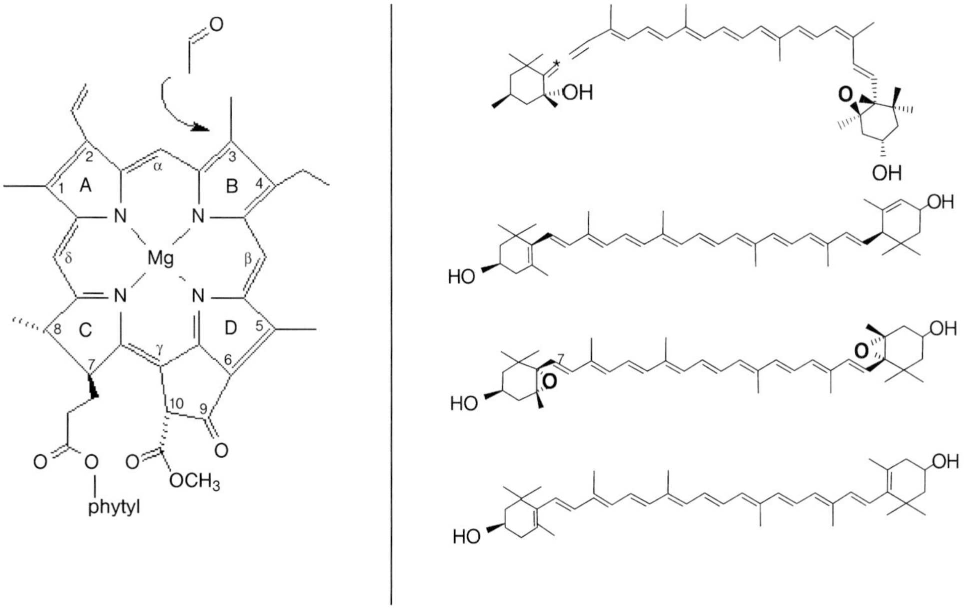

Fig. 12.2 Molecular structure of photosynthetic cofactors bound to LHCII. Left: Chlorophyll molecules: chlorophyll *b* possesses a formyl group in position 3, instead of a methyl group for chlorophyll *a*. Right: Carotenoid molecules. From top: 9-*cis* neoxanthin, lutein, violaxanthin, zeaxanthin.

triplet states represent a direct danger, as they may react with oxygen (and molecular oxygen is produced by photosystem II) to form oxygen singlet $^1O_2$, which is one of the most destructive reactive species of oxygen for living organisms. In photosynthetic plant proteins, the formation of singlet oxygen is limited by the presence of carotenoid molecules [16] (Fig. 12.2). These cofactors, in addition to acting as light-harvesters, play a photoprotective role as chlorophyll triplet state quenchers: the triplet state of carotenoid molecules is lower in energy than that of the chlorophylls or of singlet oxygen, and these molecules may thus quench both the chlorophyll triplet state and $^1O_2$. However, this mechanism alone does not seem sufficient to protect the photosynthetic membrane from damage induced by high illumination.

As a summary, an antenna system of a given size cannot function optimally in every light environment, particularly when illumination conditions vary rapidly. As a result, plant antenna proteins have evolved the ability to quench excess excitation energy into heat, and to regulate the amount of energy delivered to photosystem II by a subtle balance between energy harvesting and quenching. This phenomenon, termed non-photochemical quenching (NPQ) – as opposed to the photochemical quenching implicit to the normal function of the photosystem – was discovered more than two decades ago in whole plants (for reviews, see Refs.

[17, 18]). Indeed, when leaves are suddenly illuminated, a rapid fall is observed in their intrinsic fluorescence. This was attributed to the formation of energy traps in the photosynthetic membrane, which quench the excitation in high light conditions. A drop in the fluorescence yield is observed, accompanied by an alteration of the fluorescence spectrum, which broadens and shifts to the red. Although this phenomenon was observed many years ago, the molecular mechanisms involved are still not fully understood. This is due, in particular, to the fact that it disappears upon isolation of the photosynthetic membrane, and even simple chloroplast purification results in a significant reduction of the size of the NPQ. Molecular studies of the NPQ phenomenon thus often require the use of intact leaves as biological material. Chloroplast acidification studies showed that the transmembrane ΔpH triggers excitation energy quenching [17]. However, the NPQ mechanism, being essential for plant survival, comprises a number of molecular mechanisms. Almost 20 years ago it was shown that the NPQ build-up is accompanied by a change in the chemical composition of the plant antenna system [19]. A specific carotenoid present in the antenna, violaxanthin, is doubly de-epoxidized by a specialized enzyme, violaxanthin de-epoxidase, to form zeaxanthin. As zeaxanthin is a carotenoid molecule with a longer conjugated chain than violaxanthin, it was proposed to be the energy quencher, accepting the excitation energy residing on the lower singlet state of chlorophyll through its forbidden, optically silent S1 state [20]. This is often referred to as the "molecular gear-shift" mechanism.

However, detailed studies on the relationship between non-photochemical quenching and zeaxanthin formation revealed a more complex situation. First, although the rise of NPQ appears to correlate with violaxanthin de-epoxidation, there are significant deviations during the first seconds of strong illumination. Second – and more importantly – when illumination returns to normal after a short period of high light, NPQ relaxes in seconds while the re-epoxidation of zeaxanthin can take up to several hours. On the other hand, the presence of zeaxanthin favors the "quenched" state of the photosynthetic membrane, thus acting as a "memory" [21]. If the plant has been submitted to strong illumination a few minutes earlier, then its response to a second period of high light is faster and the quenching is more efficient, due to the presence of protein-bound zeaxanthin already in the photosynthetic membrane. This observation led to the hypothesis that NPQ involves a protein conformational change, induced by the transmembrane ΔpH, and which is favored by the presence of zeaxanthin. It is of note that the rapid recovery of the plant from the quenched state to the normal, light-harvesting state is only observed after short periods of strong illumination (a few minutes). When plants are exposed to strong illumination for longer times, zeaxanthin accumulates in the photosynthetic membrane, and NPQ does not relax as quickly. Under these conditions, NPQ relaxation becomes dependent on zeaxanthin epoxidation, and it appears that a different mechanism is implicated which may eventually involve the molecular gearshift mechanism. Finally, recent progress in plant molecular biology has allowed the production of random plant mutants, which were screened on the basis of their intrinsic leaf fluorescence during strong illumination (plants deficient in the NPQ mechanism should still fluoresce in strong illumination conditions, whilst the wild-type does not). Among the mutants impaired in the NPQ mechanisms, one is deficient in the synthesis of a small subunit of photosystem II, PsbS. In the absence of this subunit, the extent of NPQ is dramatically reduced [22]. PsbS is distantly related to the LHC proteins, and it was proposed that it binds zeaxanthin [23]. However, its role remains as yet unclear – it could be, for instance, the sensor of the transmembrane pH [24].

From the description above, it appears that there are still many open questions concerning the molecular mechanisms underlying NPQ. Among these, the chemical nature of the quencher is still a matter of controversy. To date, only a few experiments have provided direct information on this molecular species, and there is still no agreement on their validity. A number of cofactors could play the role of the quencher, including chlorophyll molecules, chlorophyll dimers, and carotenoids either through their cation state, through charge transfer states, or through their lower singlet excited state (e.g., see Ref. [25]). Besides the chemical nature of the quencher, until recently, there was no general agreement on the precise nature of the proteins involved in the NPQ mechanism, although it has become increasingly obvious that at least a sizeable part of the quenching occurs in the major LHCII antenna protein (see below).

During the late 1980s, it was noted that the fluorescence yield of the major antenna protein from photosystem II, LHCII, depends heavily on its level of association. When this protein is in its trimeric state, it possesses a fluorescence yield of almost 100%, and its fluorescence lifetime is approximately 4 ns. However, upon detergent removal, its fluorescence yield decreases progressively to 10% or even lower, corresponding to a fluorescence lifetime of about 200 ps. It was further noted that the difference between the fluorescence spectra of trimeric and aggregated LHCII exhibits many similarities with the difference between the *in-vivo* fluorescence of non-quenched and quenched leaves [26]. In particular, upon association-induced quenching, the fluorescence maximum shifts to the red, in precisely the same way as that observed for NPQ *in vivo* (see Fig. 12.6). These experiments, which mainly were conducted by the group of Peter Horton, showed that the association of LHCII proteins promotes the formation of an excitation energy trap. However, their relevance to understanding the NPQ mechanism has been questioned for many years. From these studies it was proposed that LHCII association induces a conformational change in the protein, which in turn promotes the appearance of a quenching species. Resonance Raman spectroscopy was applied to this *in-vitro* system as early as 1993 in order to test the conformational change hypothesis [27].

12.7
Raman studies of LHCII

As an antenna protein, LHCII binds two classes of chromophores – carotenoids and chlorophylls. Each monomer of LHCII binds as carotenoids two luteins, one neoxanthin, and (more weakly) one carotenoid from the xanthophyll cycle (zeaxanthin or violaxanthin); plus eight molecules of chlorophyll *a* and six chlorophylls *b*. On such a complex system it is obviously difficult, even with the most selective techniques, to obtain information on the individual pigments. Carotenoid and chlorophyll molecules do not exhibit electronic transitions in the same spectral range: chlorophyll molecules display two intense electronic transitions in the blue (410–450 nm) and the red (620–700 nm) regions of the spectrum, whilst carotenoids, which act as complementary light-harvesters, absorb mainly in the blue–green range (450–510 nm). Hence, it is possible to observe the signals of chlorophyll or carotenoid molecules selectively, by choosing the appropriate excitation wavelength. Similarly, as the Soret electronic transition of chlorophyll *b* is somewhat red-shifted relative to chlorophyll *a* (440 versus 420 nm), selectivity between these two classes of molecule is also possible, at least to a certain extent [28] (Fig. 12.3).

Resonance Raman spectra of chlorophyll molecules comprise more than 60 bands, most of them arising from combinations of the vibrational modes of the conjugated macrocycle. The spectral region which has up to now been used the most in biological studies corresponds to the highest frequency region (1620–1710 cm^{-1}). Bands contributing in this region arise from the stretching modes of the conjugated carbonyl groups of chlorophylls *a* (keto) and *b* (formyl and keto) [28]. Because these modes involve an oxygen atom, their frequency is higher than the other modes of the molecule, and they combine only poorly with them. The

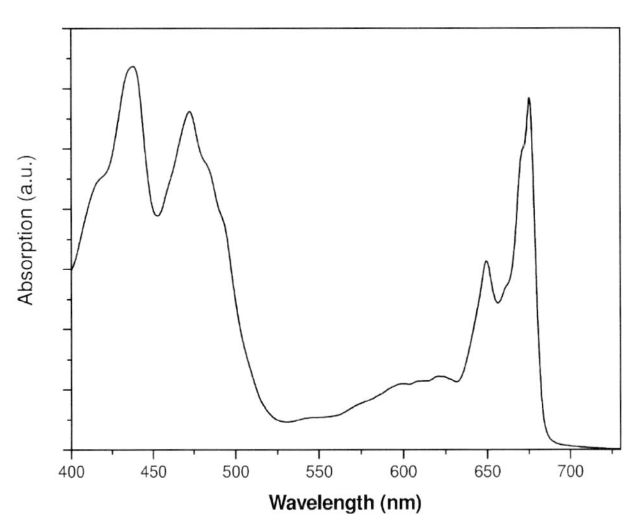

Fig. 12.3 Electronic absorption of LHCII (recorded at 77 K). In this spectrum, the main contributions to electronic transitions arise from (from left to right): the Soret electronic transitions from chlorophyll *a* and chlorophyll *b* molecules (420–450 nm); the $S_0 \rightarrow S_2$ transitions of the carotenoid molecules (460–510 nm); and the Q_y transitions of chlorophyll *b* molecules (650 nm) and of chlorophyll *a* molecules (668–680 nm).

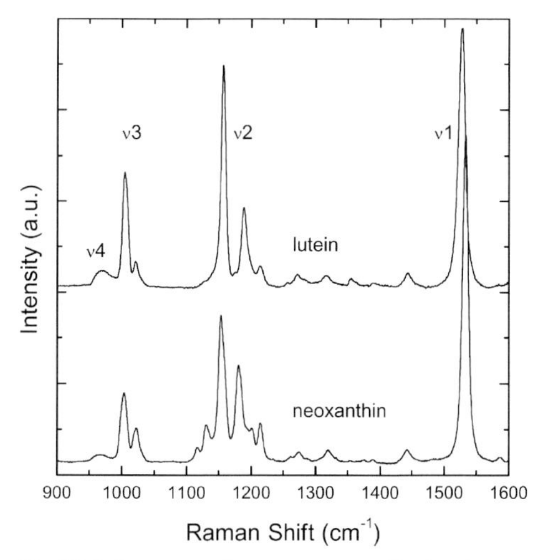

Fig. 12.4 Resonance Raman spectra of isolated carotenoid molecules, recorded at resonance (488 nm) excitation. Top: lutein spectrum, bottom: 9-*cis* neoxanthin. The different spectral regions are denoted ν_1 to ν_4. In the ν_2 the 9-*cis* conformation of neoxanthin induces the presence of new bands, which are symmetry-forbidden in the all-*trans* conformation.

frequency of the conjugated carbonyl depends tightly on the intermolecular interactions (H-bonds) in which these groups are involved, and the analysis of this spectral region thus allows a precise diagnostic of how the chlorophyll molecules interact with their environment. For instance, the formyl group of a chlorophyll *b* molecule vibrates at approximately 1660 cm^{-1} when free from intermolecular interactions, and it may shift down to 1620 cm^{-1} upon H-bond formation [28].

Resonance Raman spectra of carotenoid molecules display four main groups of intense bands, which provide information on the conformation and configuration of these molecules (Fig. 12.4) (for a review, see Ref. [29]). The ν_1 band around 1530 cm^{-1} arises from the stretching vibrations of the C=C bonds of the carotenoid. Its position is sensitive to the length of the conjugated system of the scattering carotenoid, and to its molecular conformation (*trans*/*cis*). The ν_2 bands mainly arise from the stretching modes of the C–C bonds. The structure of this complex cluster also depends on the molecular conformation of the scattering carotenoid. Another important region of these spectra is the so-called ν_4 band (around 960 cm^{-1}); this arises from out-of-plane wagging motions of the C–H groups of these molecules which, for reasons of symmetry, are not coupled with the electronic transition of perfectly planar carotenoid molecules. This band will thus tightly depend on the molecular configuration of the carotenoid, gaining intensity when the molecule is distorted out of the plane.

One of the major problems in applying resonance Raman to the LHCII is to determine which resonance conditions will lead to observing contributions of a given molecule. Among the carotenoid molecules, those of the xanthophyll cycle may be easily lost during purification, or indeed washed off afterwards. It is thus possible to study LHCII both with and without these carotenoids, in order to determine at which wavelength they contribute, as well as their Raman contribution [30]. However, as the protein environment may influence the electronic properties of carotenoid, it is impossible to predict where the lutein and neoxanthin absorb in this protein. When bound to LHCII, neoxanthin adopts a 9-*cis* conformation. This is expected to affect both its ν_1 and ν_2 resonance Raman bands (Fig. 12.4). Scanning the Raman excitation through the carotenoid electronic transition induces net frequency changes of the ν_1 band, from that expected for 9-*cis* neoxanthin to that expected for all-*trans* lutein, indicating that the contribution of these carotenoids dominates the resonance Raman spectra alternately [3]. From such experiments, the precise position of the absorption transitions of each carotenoid could thus be easily deduced, as they must correspond to the excitation wavelengths where that carotenoid dominates the resonance Raman spectra. For instance, it could be concluded that one lutein, called lutein 2, possesses a red-shifted electronic absorption, with a 0-0 transition peaking at 510 nm, while the other lutein contributes at 495 nm. Once it is known which bound carotenoid dominates the resonance Raman spectrum obtained at a given wavelength, precise molecular information can then be obtained selectively for each of these molecules. In such spectra, the intensity and structure of the ν_4 band tightly depends on the configuration of the carotenoids. Thus, it is possible to evaluate to what extent these molecules are distorted from the planar by their protein-binding pockets. It was shown that the red-shift, in the absorption transition of lutein 2 is caused by the LHCII trimer formation. It could be concluded, from the analysis of the ν_4 region, that this lutein was distorted upon trimerization. Resonance Raman thus allows the selective observation of the three main carotenoids of LHCII, the blue and red lutein molecules, and neoxanthin [3].

Once the conditions of selective observation, by resonance Raman, of each of these molecules was determined, the molecular processes involved in LHCII aggregation (and thus eventually underlying the quenching mechanism) were studied (see Fig. 12.7). It was shown that further association of trimeric LHCII into small oligomers induces a distortion of a carotenoid molecule (Fig. 12.7). This distortion is more easily seen at wavelengths where the contribution of neoxanthin is expected to dominate the spectra [3]. Thus, it was concluded that, upon LHCII association, neoxanthin becomes distorted, and that this distortion probably reflects changes in the conformation of the protein binding site of this molecule. The structure of the LHCII is thus highly plastic and dependent on its state of self-association. By using excitation wavelengths that enhance chlorophyll *a* or *b* contributions, clear differences in the resonance Raman spectra of LHCII trimers and oligomers were observed [27]. These changes showed that the H-bonding state of the formyl carbonyl of at least one, and most likely two, molecules of chlorophyll *b* change during the LHCII association, as well as that of at least one chlorophyll *a* keto carbonyl (see Fig. 12.8). The self-association of LHCII thus induces a change in the binding sites of at least three of its cofactors. However, at that stage, a major question remained unsolved: were these changes in pigment interactions an artifact of the association process itself, or were they caused by conformational changes within LHCII trimers upon association? If the cofactors involved were bound to the periphery of LHCII, then the changes observed could merely represent new (possibly artifactual) interactions with the neighboring trimer. Only the precise study of LHCII crystals could help to answer this question.

12.8 Crystallographic Structure of LHCII

The LHCII structure has been the object of intense studies conducted for more than 15 years. In 1994, an initial structure of pea LHCII was obtained from electron diffraction of two-dimensional (2-D) crystals [31]. This provided a general description of the LHCII fold, which contains three transmembrane α-helices (A, B, and C), as well as the positioning of a large number of chromophores. However, because of the limited resolution, it was impossible in this structure to distinguish chlorophyll *a* from chlorophyll *b*, as well as to position the binding site of the neoxanthin molecule. This incomplete structure was nevertheless the only one available for almost ten years, and it proved extremely useful for improving our understanding of the LHCII function and in directing ongoing research. In 2004, a three-dimensional (3-D) structure of spinach LHCII was obtained using X-ray crystallography by Chang's group in Beijing, at 2.7 Å resolution [32]. Since then, a second structure has been obtained by Kuhlbrandt's group, at a slightly better resolution, which largely confirmed the LHCII description deduced from Chang's crystals [33]. The LHCII crystal structure first allowed a precise determination of the stoichiometry of the chlorophyll bound to this protein, and no less than eight chlorophyll *a* and six 6 chlorophyll *b* could be determined in the atomic model (compared to 12 chlorophyll in the first, 2-D, LHCII structure) (Fig. 12.5). These chlorophylls are organized into two layers, one on the stromal side and one on the lumenal side of the protein. As predicted from the first structure, the two luteins bound to LHCII form a cross-brace in the very heart of the protein, in the groove of the supercoil formed by the two major helices (A and B). 9-*cis* neoxanthin is bound near helix C, close to a chlorophyll *b*-rich domain of LHCII, in a cleft formed by hydrophobic amino acids, and the rings and phytyl tails of chlorophyll *b* molecules. Approximately one-third of this molecule sticks out of the protein. The fourth carotenoid found in this structure, violaxanthin, is bound at the monomer/monomer interface, with a 1:1 violaxanthin:LHCII monomer stoichiometry. The overall structure of LHCII reveals an extremely complex network of cofactors, with a number of strongly-interacting chlorophyll molecules, forming chlorophyll *a*/*a*, *b*/*a*, and *b*/*b* pairs (Fig. 12.5). Extensive calculations performed on this structure have given an evaluation of the interaction energies between these pairs, and suggested which of these pigments participate in the different excitation energy levels of the LHCII protein [34].

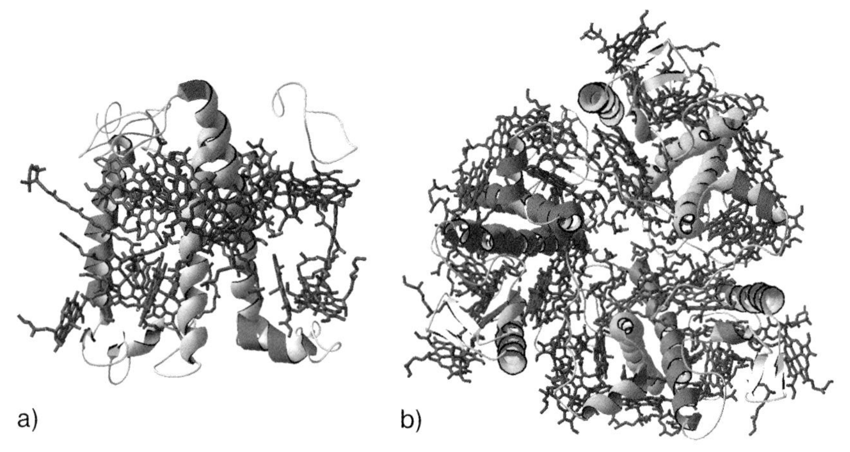

Fig. 12.5 Structure of LHCII. (a) LHCII monomer, viewed from the side (perpendicular to the membrane plane). (b) LHCII trimer, viewed in the plane of the membrane. For clarity, only the bound pigments are displayed in this structure, together with the protein backbone as a ribbon.

Besides its value to the study of LHCII, this investigation also allowed the description of a novel type of membrane protein crystal. In the crystals obtained by the Beijing team, LHCII is organized in icosahedral proteoliposomes, containing 20 LHCII trimers, oriented radially towards the center [32]. Contacts in the crystal between these icosahedral units are ensured by polar interactions involving the hydrophilic stromal domain of LHCII. Altogether, these crystals are thus constituted from the regular packing of LHCII proteoliposomes, with few protein–protein interactions between liposomes. As the inner diameter of these proteoliposomes is rather small (160 Å), the angle between adjacent trimers in this unit is large, and the molecular contacts between trimers is thus limited to a small region of their lumenal surfaces involving mainly digalactosyldiacylglycerol molecules. The only cofactors near this region are two weakly interacting chlorophyll *a*/*b* pairs. As described above, the properties of LHCII trimers depend heavily on their association state, being highly fluorescent when isolated, and highly quenched when forming large aggregates. It was thus interesting to study the properties of LHCII in these crystals, to determine whether they were fluorescent or quenched. Considering the limited nature of the interactions in the crystal, they were expected to be highly fluorescent rather than quenched.

12.9 Properties of LHCII in Crystal

When the fluorescence emission spectra of LHCII crystals were first obtained [35], they displayed – somewhat surprisingly – a main electronic transition dramatically shifted to the red as compared to that of LHCII trimers, peaking at 700 nm (680 nm

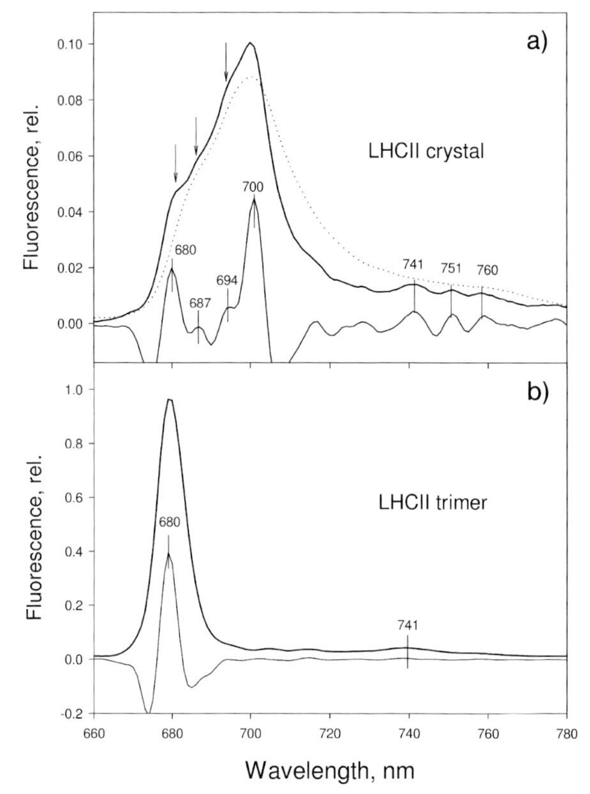

Fig. 12.6 Fluorescence properties of LHCII at 77 K. Lower panel: Fluorescence emission spectrum of LHCII trimers in solution (thin line = second derivative of spectrum). Upper panel: Fluorescence emission spectrum of LHCII in crystals (thin line = second derivative), compared to fluorescence emission spectrum of quenched LHCII oligomers (dotted line).

for LHCII trimers). Altogether, this fluorescent pattern is reminiscent of that of highly aggregated, quenched LHCII (Fig. 12.6). It is however, from this single result, impossible to draw conclusions with regards to the quenching state of LHCII in crystals. Indeed, because of the extremely high concentration of LHCII in these crystals, it is very difficult to determine its fluorescence yield.

The most direct way to determine the quenching state of a fluorescent protein is to measure its fluorescence lifetime, as this yields much more accurate information. This was achieved by performing fluorescence lifetime imaging (FLIM) of LHCII single crystals. FLIM is an imaging, confocal-based method which yields an estimation of the fluorescence lifetime with a resolution as high as 50 ps, with a spatial resolution of about 0.5 μm. The use of FLIM in a crystal is particularly important, as it allows discrimination between those proteins belonging to the crystal and those which are eventually left in the mother liquor, and also eventually between domains in the crystals, in case, for instance, some degradation is observed at the crystal edges. FLIM performed on single crystals of LHCII trimers led to an unambiguous determination of the quenching state of the proteins. The fluorescence lifetimes were found to be homogeneous over whole crystals, and were determined as 0.89 ns, compared to 4 ns for isolated LHCII trimers, and 0.65 ns for associated LHCII. From this result, it was thus concluded that, although the LHCII in the crystals are largely in their trimeric state, they were highly quenched – almost as much as in the LHCII aggregates [35].

When resonance Raman spectroscopy was performed on LHCII in its crystallized state, the neoxanthin conformation was seen to be even more distorted than in the associated LHCII (Fig. 12.7). At the same time, selective excitation of chlorophyll *b* (at 441.6 nm) showed that at least one (and probably two) chlorophyll *b* are involved in new H-bonds with their protein environment upon crystal formation (Fig. 12.8). As none of the contacts between LHCII trimers can account for these changes in conformation/intermolecular interactions, it was concluded that LHCII trimers possess the intrinsic ability to undergo conformational change, and that this conformational change was promoting the observed changes in cofactor properties – including the appearance of excitation quenchers. It was thus concluded that, more generally, aggregation-induced quenching is not a direct result of aggregation, but rather that both aggregation and crystallization select out or in some way induce the new protein conformation. That is, the LHCII protein behaves as a molecular switch, being able to shift from a state where it efficiently

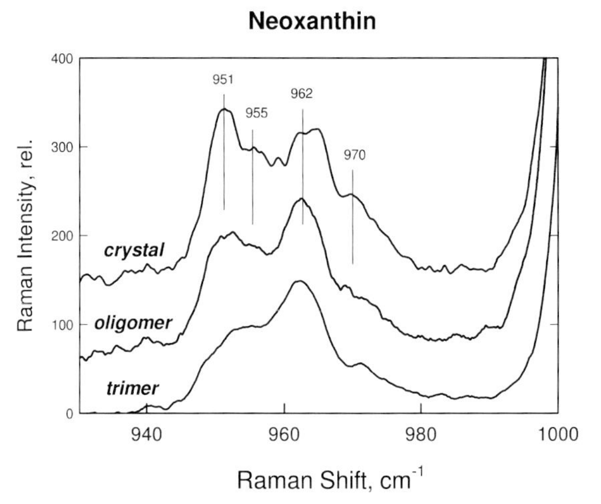

Fig. 12.7 Resonance Raman spectra (488.0 nm excitation, favoring neoxanthin contribution) in the ν_4 region, sensitive to out-of-plane deviation of the carotenoid molecule of (bottom to top): LHCII trimers, oligomers, and LHCII in crystals. The intensity of the 950 cm^{-1} Raman contribution increases upon LHCII oligomerization, indicating distortion of the neoxanthin molecule. In crystals, this band is even more intense than in self-associated LHCII, which is evidence of a net difference in structure between LHCII in solution and in crystals, at the level of the neoxanthin binding site.

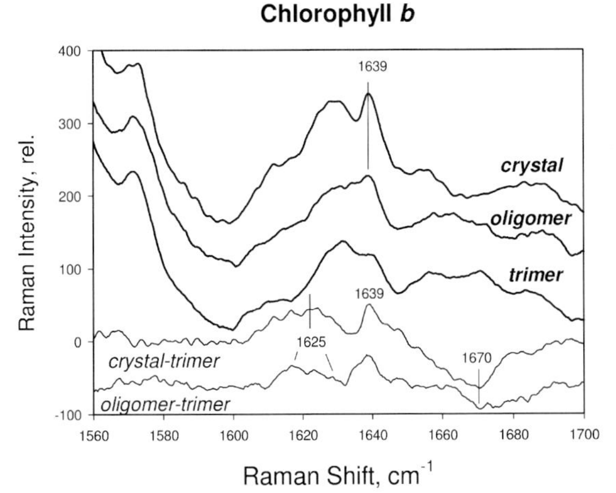

Fig. 12.8 Resonance Raman spectra (441.6 nm excitation, favoring chlorophyll *b* contributions in higher-frequency regions, where conjugated carbonyl stretching frequencies mainly contribute. From bottom to top (thick lines): spectra of LHCII trimers, oligomers, and LHCII in crystals. In oligomers, new contributions at 1639 cm^{-1} are observed, together with loss of intensity at 1670 cm^{-1}. This indicates a change of interaction of at least one chlorophyll *b*, upon LHCII self-association, where a free-from-interaction formyl carbonyl (vibrating at 1670 cm^{-1}) enters into H-bonding interactions (and downshifts to 1639 cm^{-1}). In LHCII crystals, the 1639 cm^{-1} band is even more prominent, and is associated with a second new contribution at 1625 cm^{-1}. This indicates that, during the crystallization process, the formyl group of one (and probably two) chlorophyll *b* enter into H-bonding interactions, indicating structural changes in the binding pockets of these molecules or reorganization of the chlorophyll *b* pairs.

harvests energy to a state where it is able to dissipate the excitation energy. The structure of LHCII in crystals thus reflects the molecular organization of this protein in its quenching form. It was further proposed that these conformational changes are responsible for the appearance of non-photochemical quenching in higher plant membranes *in vivo*.

12.10
Recent Developments and Perspectives

The LHCII conformational change mechanism for NPQ *in vivo* has been challenged by a number of alternative hypotheses. Among these, it was proposed that while interactions between neighboring LHCII in the crystal were weak, the quenching could be due to the large number of proteins connected in the crystalline state (R. van Grondelle, personal communication). Indeed, in this situation, only a few LHCII in a quenching state are necessary to provide efficient quenching to the whole crystal. During the past year, resonance Raman experiments have been conducted on chloroplasts and whole leaves, in order test for the same conformational changes in LHCII occurring *in vivo*. In parallel resonance Raman experiments were conducted on different-sized aggregates to relate the amount of quenching with the resonance Raman signal observed. The conclusion of these experiments is quite clear: *in vivo*, upon NPQ induction, a resonance Raman signal corresponding to the conformational change of LHCII is observed, with an intensity almost equal to that expected for the level of quenching observed (A.V. Ruban et al., unpublished results). This indicates that the conformational change observed in the crystal is most likely responsible for the appearance of NPQ, at least to a large extent. Ultrafast experiments, currently performed on isolated LHCII, should soon provide a clear indication of the chemical nature of the molecule which plays the role of the excitation quencher, *in vitro* in associated LHCII, as well as *in vivo*. In parallel, examination of the LHCII crystal fluorescence seems to indicate that their electronic properties evolve with time. This may indicate that the LHCII conformational change occurs secondary to the crystallization process, and that it might be possible, at mid-term, to elucidate the molecular structure of LHCII in its light-harvesting form [36].

References

1 P. R. Carey, *Biochemical applications of Raman and Resonance Raman spectroscopies*. Academic Press, New York, **1982**.
2 A. C. Albrecht, *J. Chem. Phys* **1961**, *34*, 1476–1484.
3 A. V. Ruban, A. A. Pascal, B. Robert, *FEBS Lett.* **2000**, *477*, 181–185.
4 R. van Grondelle, J. Dekker, T. Gillbro, V. Sundström, *Biochim. Biophys. Acta* **1994**, *1187*, 1–65.
5 B. A. Diner, G. T. Babcock, in: *Oxygenic Photosynthesis: the Light Reactions*, D. R. Ort, C. F. Yocum, (Eds.), Kluwer, Dordrecht, **1996**, pp. 213–247.
6 J. Hanley, Y. Deligiannakis, A. Pascal, P. Faller, A. W. Rutherford, *Biochemistry* **1999**, *38*, 8189–8195.
7 P. Faller, A. Pascal, A. W. Rutherford, *Biochemistry* **2000**, *40*, 6431–6440.
8 R. J. Ellis, *Annu. Rev. Plant Physiol.* **1981**, *32*, 111–137.
9 P. Fromme, P. Jordan, N. Krauss, *Biochim. Biophys. Acta* **2001**, *1507*, 5–31.
10 A. Zouni, T. H. Witt, J. Kern, P. Fromme, N. Krauss, W. Saenger, P. Orth, *Nature* **2001**, *409*, 739–743.
11 E. J. Boekema, H. van Roon, J. Dekker, *FEBS Lett.* **1998**, *424*, 95–99.
12 E. J. Boekema, J. F. L. van Breemer, H. Van Roon, J. Dekker, *J. Mol. Biol.* **2000**, *301*, 1123–1133.
13 J. F. Allen, *Biochim. Biophys. Acta* **1992**, *1098*, 275–335.
14 R. Delosme, J. Olive, F. A. Wollman, *Biochim. Biophys. Acta* **1996**, *1273*, 150–158.
15 D. Siefermann-Harms, *Biochim. Biophys. Acta* **1995**, *811*, 325–335.
16 D. Siefermann-Harms *Physiol. Plant.* **1987**, *69*, 561–568.
17 P. Horton, A. V. Ruban, R. G. Walters, *Annu. Rev. Plant Physiol. Plant Mol. Biol.* **1996**, *47*, 655–684.
18 K. K. Niyogi, *Annu. Rev. Plant Physiol. Plant Mol. Biol* **1999**, *50*, 333–360.
19 B. Demmig-Adams, *Biochim. Biophys. Acta* **1990**, *1020*, 1–24.
20 H. Frank, J. A. Bautista, J. S. Josne, A. J. Young, *Biochemistry* **2000**, *39*, 2831–2837.
21 A. V. Ruban, M. Wentworth, P. Horton, *Biochemistry* **2001**, *40*, 9896–9901.
22 X.-P. Li, O. Björkman, C. Shih, A. R. Grossman, M. Rosenquist, S. Jansson, K. K. Niyogi, *Nature* **2000**, *403*, 391–395.
23 M. Aspinall-O'Dea, M. Wentworth, A. A. Pascal, B. Robert, A. Ruban, P. Horton, *Proc. Natl. Acad. Sci. USA* **2002**, *99*, 16331–16335.

24 X.-P. Li, A. Gilmore, S. Caffari, R. Bassi, T. Golan, D. Kramer, K. K. Niyogi, *J. Biol. Chem.* **2004**, *279*, 22866–22874.

25 N. E. Holt, D. Zigmantas, L. Valkunas, X.-P. Lin, K. K. Niyogi, G. R. Fleming, *Science* **2005**, *307*, 433–436.

26 P. Horton, A. V. Ruban, D. Rees, A. A. Pascal, G. Noctor, A. J. Young, *FEBS Lett.* **1991**, *292*, 1–4.

27 A. V. Ruban, P. Horton, B. Robert, *Biochemistry* **1995**, *34*, 2333–2337.

28 M. Lutz, B. Robert B, in: *Biological Applications of Raman Spectroscopy*, T. G. Spiro (Ed.), John Wiley & Sons, New York, Vol. III, **1988**, pp. 347–411.

29 B. Robert, in: *Biophysical Techniques in Photosynthesis*, J. Amesz, A. J. Hoff (Eds.), Kluwer Academic Publisher, Amsterdam, **1998**, pp. 161–176.

30 A. V. Ruban, A. A. Pascal, P. J. Lee, B. Robert, P. Horton, *J. Biol. Chem.* **2002**, *277*, 42937–42942.

31 W. Kühlbrandt, D. N. Wang, Y. Fugiyoshi, *Nature* **1994**, *367*, 614–621

32 Z. Liu, H. Yan, T. Kuang, J. Zhang, L. Gui, X. An, W. Chang, *Nature* **2004**, *428*, 287–292.

33 J. Standfuss, A. C. Terwisscha van Scheltinga, M. Lamborghini, W. Kuhlbrandt, *EMBO J.* **2005**, *24*, 919–928.

34 V. I. Novoderezhkin, M. A. Palacios, H. van Amerongen, R. Van Grondelle, *J. Phys. Chem. B* **2005** *109*, 10493–504.

35 A. A. Pascal, Z. F. Liu, K. Broess, B. van Oort, H. Van Amerongen, C. Wang, P. Horton, B. Robert, W. Chang, A. Ruban, *Nature* **2005**, *436*, 134–137.

36 H. Yan, P. Zhang, C. Wang, Z. Liu, W. Chang, *Biochem. Biophys. Res. Commun.* **2007**, *355*, 457–463.

Part VI
Exploring Structure–Function Relationships in Whole Cells

13
Energy Transfer Technologies to Monitor the Dynamics and Signaling Properties of G-Protein-Coupled Receptors in Living Cells

Jean-Philippe Pin, Mohammed-Akli Ayoub, Damien Maurel, Julie Perroy and Eric Trinquet

13.1
Introduction

Membrane proteins at the cell surface play a critical role in cell physiology, being involved in solute transport, the control of membrane potential and, importantly, in the recognition of external stimuli and transduction of these signals inside the cell. For many years, the plasma membrane was considered as a fluidic mosaic in which proteins can freely diffuse. Signaling then occurs by chance when an activated protein encounters its partner in the plane of the membrane.

Among the signaling proteins at the cell surface are the G-protein-coupled receptors (GPCRs). These proteins, with seven transmembrane helices, are encoded by the largest gene family in the mammalian genomes, and are activated by a large variety of stimuli, from photons, to large proteins, through amino acids, small neurotransmitters and hormones, as well as by ions, lipids or steroids (Bockaert and Pin, 1999). These receptors are involved in most aspects of organism physiology, and as such constitute the main target for drug development, being already the target of about half of the drugs on the market (Drews, 2000; Schlyer and Horuk, 2006). They transmit external signals inside the cell by activating heterotrimeric G-proteins composed of an alpha subunit and a beta-gamma dimer. In the inactive state, the heterotrimeric G-protein is a GDP form, associated to the plasma membrane through lipid modifications of both the alpha and gamma subunits. According to the current model for GPCR-mediated signaling, agonist interaction in the receptor switch it to an active, G-protein high-affinity state, allowing its interaction with the G-protein, and the exchange of GDP for GTP. This in turn results in the dissociation of the protein complex, allowing both the alpha subunit and the beta-gamma dimer to activate their own effectors. The GTPase activity of the alpha subunit, further enhanced by regulators of G-protein signaling (RGS) proteins, ends the cycle, and results in the reconstitution of the inactive GDP-bound heterotrimer.

Biophysical Analysis of Membrane Proteins. Investigating Structure and Function. Edited by Eva Pebay-Peyroula

ISBN: 978-3-527-31677-9

Although supported by experimental data in reconstituted lipid bilayers, the fluidic mosaic model is difficult to reconcile with the requirement for rapid and highly specific signaling processes required for normal cell physiology, especially in the case of GPCR signaling (Neubig, 1994). The development of new techniques which would allow the direct recording of protein dynamics in the plasma membrane of living cells was therefore required to clarify this issue.

The Förster resonance energy transfer (FRET) (Förster, 1948) – also commonly known as fluorescence resonance energy transfer – appeared as a way to monitor protein proximity in living cells; however, such an approach requires the labeling of proteins with specific donor and acceptor fluorophores. FRET was therefore first used in reconstituted systems with purified proteins.

During the mid-1990s, the cloning of the green fluorescent protein (GFP) allowed the easy labeling of proteins with fluorophores, and their expression in mammalian cells (Prasher et al., 1992; Marshall et al., 1995). Then, the manipulation of GFP allowed the characterization of several mutants with different spectral properties, some being compatible with FRET (Miyawaki et al., 1997; Tsien, 1998). Such fluorescent proteins can easily be fused to the proteins of interest to examine their dynamic in living cells, and their interaction with other partners (Selvin, 2000; Vogel et al., 2006).

In this chapter, the principle of the resonance energy transfer will first be described, followed by details of the different technologies that have been developed to monitor protein dynamics in living cells. How these technologies have changed the concept of the functioning of GPCR signaling in a natural environment is then outlined. It will be seen that recent data are consistent with a precise organization of the various components of the signaling cascade, allowing a very rapid and highly specific signaling.

13.2 Fluorescence Resonance Energy Transfer (FRET)

FRET is a photophysical process in which energy is transferred from a fluorescent donor (D) to a suitable energy acceptor (A). FRET involves the resonant coupling of the donor emission and the acceptor absorption dipoles, and is therefore non-radiative (Stryer, 1978). In order to occur, FRET requests an energetic compatibility between the two components of the FRET pair. Such compatibility depends on the degree of spectral overlap between the absorption spectrum of the acceptor A and the emission spectrum of the donor D (Fig. 13.1). This can be calculated through the determination of J, the spectral overlap (also known as the "Förster overlap integral") using the following formula:

$$J = \frac{\int F_D(\lambda)\varepsilon_A(\lambda)\lambda^4 d\lambda}{\int F_D(\lambda)d(\lambda)}$$

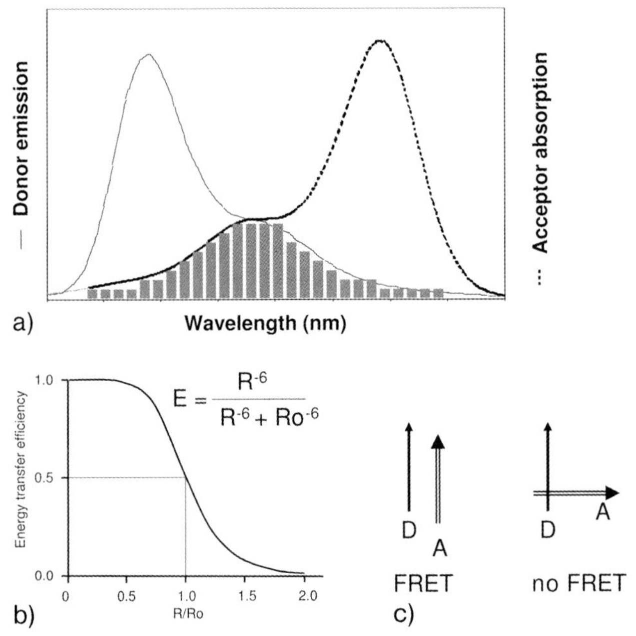

Fig. 13.1 Requirements for energy transfer between two fluorophores. (a) Spectral overlap between a fluorescent donor and an acceptor. The gray area represents the part of the acceptor absorption spectrum which overlaps the donor emission spectrum. (b) Distance between the two fluorophores. (c) Orientation between the two fluorophore dipoles; no transfer can occur if these are perpendicular.

where $F_D(\lambda)$ is the fluorescence intensity of the donor at wavelength λ, and $\varepsilon_A(\lambda)$ is the molar extinction coefficient of the acceptor.

According to the Förster theory, FRET is a distance-dependent process. Its efficiency (E) is dependent on the inverse sixth power of the distance separating the donor and the acceptor as shown in the following formula:

$$E = \frac{R^{-6}}{R^{-6} + R_0^{-6}}$$

where R is the separation distance between the donor and the acceptor and R_0 is the Förster radius. R_0 is defined as the distance at which the FRET process is 50% efficient, and can be calculated using the following formula:

$$\mathrm{R_0} = (\mathrm{J} \times 10^{-3} \times \mathrm{k}^2 \times \mathrm{n}^{-4} \times Q_D)^{1/6} \times 9730$$

where J is the integral overlap, n is the medium refractive index (n^{-4} is usually in the range of 1/3 to 1/5), Q_D is the fluorescence quantum yield of the donor in the absence of the acceptor, and k^2 is an orientation factor which is a function of the relative orientation of the donor's emission dipole and the acceptor's absorption dipole in space. Although k^2 can have a theoretical value between 0 and 4, a value of 2/3 is usually used in the determination of the R_0. k^2 becomes 2/3 when the donor or the acceptor are randomly oriented (Stryer, 1978). This condition is usually satisfied for fluorescent probes attached to biomolecules which can have a certain degree of rotational freedom (Dale et al., 1979).

For commonly used FRET pairs, a 5-nm R_0 value can be reached, giving an operative FRET distance in the range of 1 to 10 nm (Stryer and Haugland, 1967). Beyond this range, the FRET efficiency decreases very rapidly; hence, FRET could probe interactions over distances similar to the dimensions of biological molecules. To achieve this, each molecule involved in the interaction must be labeled by one component of the FRET pair.

The FRET signal resulting from the biological interaction can be discriminated by different ways from the other signals emitted by the free donor and/or the free acceptor (Fig. 13.1b) (Vogel et al., 2006). Detection of the variation in fluorescence intensities are the most commonly used FRET readouts: decrease of the donor fluorescence emission (Sokol et al., 1998); increase of the acceptor fluorescence emission (if the chosen acceptor is fluorescent) (Chan et al., 1979); or calculation of a ratio (e.g., acceptor fluorescence emission/donor fluorescence emission) (Miyawaki et al., 1997). Recent developments in confocal microscopy have allowed the measurement of a intracellular FRET process by determining the percentage increase in donor fluorescence after photobleaching of the acceptor moiety (He et al., 2003). Alternatively, as the donor fluorescence lifetime decreases proportionally to the increase of the FRET efficiency, this parameter can also be used to monitor FRET (Wallrabe and Periasamy, 2005). This could be achieved using specific instruments based on modulation-phase techniques or on photon-counting techniques. The competitive nature of FRET and photobleaching can also be used to enable an indirect measurement of FRET through its effect on donor photobleaching lifetimes using confocal microscopy (Patel et al., 2002b).

13.3 FRET Using GFP and its Various Mutants

The use of FRET technologies in biology requires the labeling of proteins with specific donor and acceptor fluorophores, and this has long been a main limitation to study protein dynamics in living cells. As such, the discovery and cloning of the GFP (a protein of 27 kDa) (Prasher et al., 1992) has been a major achievement, allowing the easy labeling of proteins in living cells (Marshall et al., 1995), and living animals. In addition, mutagenesis and systematic screening approaches have allowed the identification of a number of GFP mutants with different spectral properties and fluorescent intensities (Tsien, 1998). As illustrated in Fig. 13.2, several couples of GFP variants are compatible with efficient FRET measurements. Among these, the CFP (the cyan variant of GFP) and YFP (the yellow variant of GFP) variants identified in 1997 (Miyawaki et al., 1997) boosted the use of FRET in biological studies (Tsien, 1998; Selvin, 2000; Vogel et al., 2006). Due to the

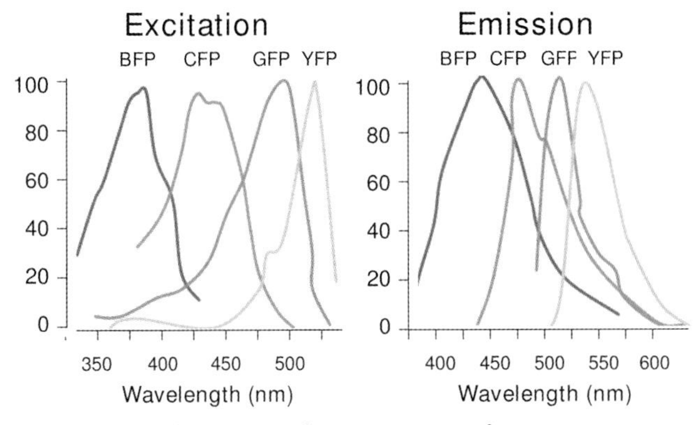

Fig. 13.2 Spectral properties of various variants of green fluorescent protein (GFP). For details of abbreviations, see the text.

overlap in their emission and excitation spectra, the excitation of CFP at 433 nm allows emission of the YFP at 535 nm if these two proteins are in close proximity (<100 Å) and in a compatible orientation. However, the overlap between the spectra of these two proteins is such that the signal-to-noise ratio measured at 535 nm (whether FRET occurs or not) is limited due to the still intense emission of CFP at this specific wavelength. As such, the ratio of fluorescent intensities measured at 480 nm (peak emission of CFP) and 535 nm (peak emission of YFP) is commonly used as an evidence for FRET. Recently, new variants of CFP and YFP (called Cerulean and Venus, respectively) with improved properties for FRET studies (improved maturation, folding and stability for Venus, improved brightness for Cerulean) have been described (Nagai et al., 2002; Rizzo et al., 2004).

Although easy to use because of the simplicity to generate fusion proteins, it must be borne in mind that the relatively large GFP proteins may affect the biological process under study. In the case of the GPCRs for example, fusion of these fluorescent proteins in a specific domain of the receptor may prevent their normal functioning, as observed in the case of the parathyroid hormone (PTH) or α_2 adrenergic receptor (Vilardaga et al., 2003) or the metabotropic glutamate receptor type 1 (Tateyama et al., 2004). As such, alternative labeling techniques should be used.

One very interesting feature of FRET using CFP and YFP, is that emission intensities are compatible with measurements under a confocal microscope. This allows the identification of the subcellular compartments where the phenomena generating the FRET signal occurs, in real time.

13.4
BRET as an Alternative to FRET

Bioluminescence resonance energy transfer (BRET) is a naturally occurring phenomenon, discovered in marine organisms (Shimomura et al., 1962). The energy transfer occurs between luminescent donor and fluorescent acceptor proteins. Oxidation of coelenterazine by *Renilla* luciferase (*R*luc) produces light at 480 nm (Prendergast, 2000). In the sea pansy *Renilla*, the close proximity of a GFP (*Renilla* GFP) allows a non-radiative energy transfer that results in light emission at 509 nm by the GFP (Morin and Hastings, 1971; Lorenz et al., 1991). BRET thus happens when part of the energy of the excited donor is transferred to an acceptor fluorophore, which in turn re-emits light at its specific wavelength. As with other non-radiative energy transfer, BRET only takes place if the emission spectrum of the donor molecule and the absorption spectrum of the acceptor molecule sufficiently overlap. Like FRET, BRET also depends on the distance between the donor and the acceptor (≤100 Å), and on their relative orientation. The crucial distance-dependence between donor and acceptor molecules for energy transfer makes BRET systems suitable for monitoring protein–protein interactions in living cells.

Depending on the type of *R*luc substrate, and the nature of the fluorescent protein acceptor, two generations of BRET have been developed. In the first generation (or BRET^1) methodology, the donor protein partner is fused to *R*luc, whereas the acceptor protein partner is fused to the YFP (Xu et al., 1999; Pfleger and Eidne, 2006; Prinz et al., 2006). Upon degradation of coelenterazine H, *R*luc emits light at 480 nm. If the two partners do not interact, only the signal emitted by the *R*luc can be detected after addition of its substrate at 530 nm. However, if the two partners are in molecular proximity (≤100 Å), then a resonance energy transfer occurs between the *R*luc and the YFP, and an increase in the emitted signal at 530 nm can be detected, due to the emission of YFP at this wavelength. The second-generation (or BRET^2) technology involves *R*luc and GFP as donor and acceptor molecules, respectively. BRET^2 takes advantage of the spectral properties of a *R*luc substrate known as Deep Blue coelenterazine (DeepBlueC), which allows a better separation between the *R*luc and GFP emission spectra (Bertrand et al., 2002; Mercier et al., 2002; Ramsay et al., 2002) (Fig. 13.3). Upon the catalytic degradation of DeepBlueC, the energy donor *R*luc emits light with a peak at 400 nm that allows the excitation of the energy acceptor, GFP. Once excited, GFP then re-emits fluorescence with a peak at 510 nm if the donor and acceptor molecules are within BRET-permissive distance (≤100 Å).

The better resolution between the emission peaks of *R*luc and GFP with BRET^2 is not the only difference between the two generations of BRET. In particular, the quantum yield of the *R*luc/DeepBlueC coelenterazine couple is lower than that of *R*luc/coelenterazine H, and the extent of overlap between the emission spectra of *R*luc and the excitation of the GFPs is better for BRET^2 than BRET^1. Accordingly, although the main advantage of BRET^2 is the better signal-to-noise ratio, the lower emission intensity limits its use to very sensitive equipment. Interestingly, BRET^1 and BRET^2 can also be combined to monitor two concomitant interactions (Fig. 13.4) (Perroy et al., 2004; Gales et al., 2006).

BRET is therefore a technique of choice to monitor protein–protein interaction as well as protein conformational changes. Compared to FRET, and because of the overlapping absorption and emission spectra of the existing mutants of GFP, the signal-to-noise ratio is significantly higher in BRET. In addition, the study of

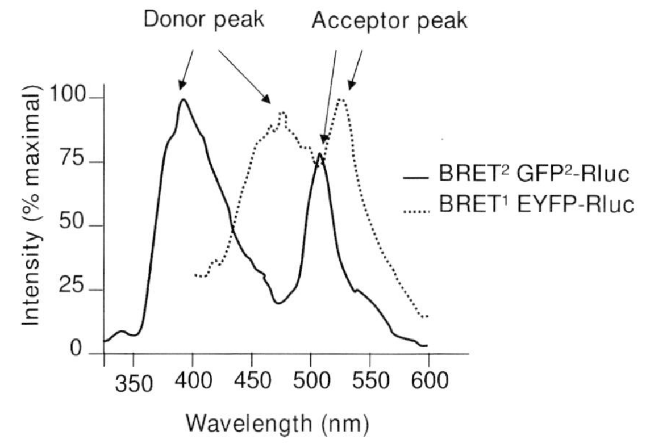

Fig. 13.3 Comparison of BRET[1] and BRET[2] emission spectra. The BRET[2] positive control GFP-*R*luc fusion or its BRET[1] counterpart (EYFP-*R*luc) was expressed in Chinese hamster ovary (CHO) cells. BRET[2] and BRET[1] emission spectra were recorded using a spectrofluorimeter with the light source turned off after addition of DeepBlueC and Coelenterazine H, respectively.

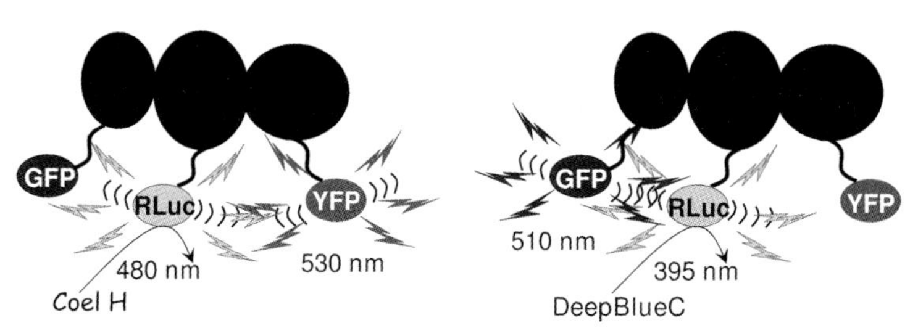

Fig. 13.4 Depending on the substrate that is degraded, the *R*luc emits a blue light at 480 nm (Coelenterazine H, BRET[1]), or at 395 nm (DeepBlueC, BRET[2]). The YFP (BRET[2]) is excited at 480 nm and emits at 530 nm, whereas the GFP (BRET[2]) absorbs at 395 nm to emit at 510 nm.

simultaneous interactions of three proteins can be relatively easily performed in BRET (see above). However, although BRET is perfectly compatible with high-throughput measurement in 96- or 384-well plates, to date BRET has not permitted the visualization of interactions in living cells under microscopic observation because of the very low intensity of the emitted bioluminescence, and thus GFP fluorescence. Today, this drawback may be overcome on the basis of recent inputs from physics and the development of the ultrasensitive cooled charge-coupled device (CCD) camera. Recent improvements in the experimental parameters have therefore allowed the exploration of protein interactions by BRET under the microscope (Ayoub et al., 2002; J. Perroy et al., unpublished results). There is no doubt that, in the near future, it will be possible to detect in real time BRET signals within a unique cell, and within specific cell compartments by using microscopy.

Finally, another interesting perspective of BRET is the imaging of a signal in living subjects, by using a CCD camera. Indeed, the results of a recent proof-of principle have suggested that interesting advances are about to be made in *in-vivo* imaging with BRET (De and Gambhir, 2005).

13.5
Time-Resolved FRET (TR-FRET) and Homogeneous Time-Resolved Fluorescence (HTRF)

Time-resolved FRET (TR-FRET) combines a FRET process with a time-resolved fluorescence detection to probe biomolecular interactions. Time-resolved fluorescence detection rejects most of the fluorescent background given both by biological media and by instrumentation components from the signal to be detected (Fig. 13.5).

TR-FRET uses a long-lifetime fluorescent FRET donor, a lanthanide complex, which can either be a lanthanide chelates or cryptates. The lanthanide chelates are based on the non-covalent association between a chelate and a lanthanide ion (mainly europium or terbium) (Hemmila et al., 1984), while lanthanide cryptates are formed by the inclusion of lanthanide ion (e.g., europium) into the tridimensional cavity of a ligand called "cryptand" (Alpha et al., 1987). Both, the cryptate and the chelate protect the lanthanide ion from a potential quenching of its environment, and act as antenna by collecting the energy from the excitation source before transferring it to the lanthanide ion (Alpha et al., 1990). However, the caged structure of the cryptate confers on it a kinetic stability that is dramatically higher than that of the lanthanide chelates. Rare-earth chelates can be dissociated in acidic media or in presence of divalent ions such as Mn^{2+}, while rare-earth cryptates could be used under drastic chemical conditions and are not affected by the presence of divalent ions in the media (Bazin et al., 2002).

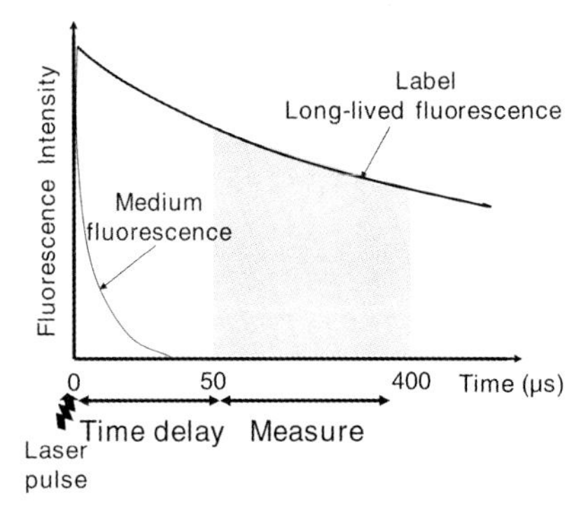

Fig. 13.5 The principle of time-resolved fluorescence detection. The fluorescent signal from the long-lived tracer (thick line) is integrated after a fixed delay time to remove from the measurement window (gray) the fluorescence background (thin line).

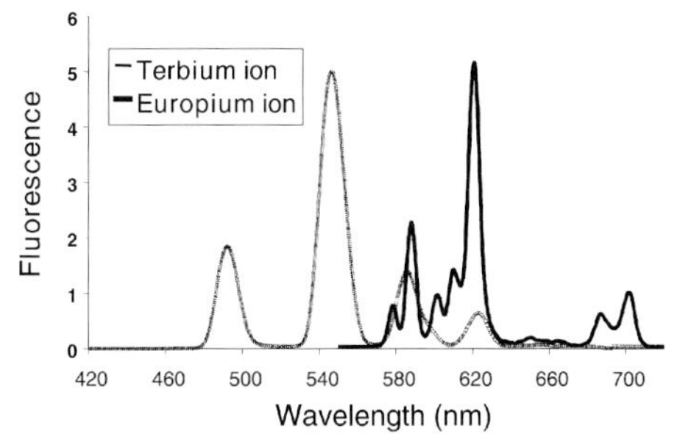

Fig. 13.6 Fluorescence emission spectra of the lanthanide complexes. The lanthanide ion included in the complex determines its fluorescence emission spectrum. The terbium ion (grey) induces a fluorescence emission between 470 and 620 nm, while the europium ion (black) induces a fluorescence emission between 580 and 710 nm.

Lanthanide complexes are excited in the ultraviolet (UV) wavelength range by pulsed light sources such as xenon flash-lamp or nitrogen lasers. Depending on the lanthanide ion used, their fluorescence occurs in a wavelength range between 450 nm and 710 nm, with typical narrow emission lines (Fig. 13.6). With the electronic transitions of the lanthanide ions being forbidden by quantum mechanistic rules, their fluorescence lifetime is exceptionally long, in the range of 100 to 1000 µs.

In FRET, the lifetime of the acceptor's emission contains a contribution equal to the donor's lifetime in presence of energy transfer (Schiller, 1976; Morrison, 1988; Mathis, 1993). In TR-FRET, the use of a long-lifetime donor leads to a long-lived emission of the acceptor. In addition, in order to isolate a FRET signal free from a short-lived fluorescent background, a time-resolved detection at the acceptor emission wavelength makes a clear distinction between the long-lived signal from the acceptor involved in the FRET process and the short-lived signal emitted by the freely diffusing acceptor (Mathis, 1993) (Fig. 13.7).

The HTRF technique, which pioneered the field of TR-FRET, is based on the use of europium cryptates as donor (Mathis, 1995). Both, a cross-linked allophycocyanin called XL665 and small fluorescent near-infrared dyes can be used as HTRF acceptors. The large spectral overlap of their absorption spectra with the europium cryptate emission leads to a high FRET efficiency (R_0 value >7 nm). Moreover, their high fluorescence quantum yields ensure an optimal release of the FRET signal in a spectral range (665 nm) where the cryptate signal is insignificant. This leads to an exceptionally large spectral selectivity which allows the establishment of an efficient ratiometric signal detection. The measurement of the cryptate emission signal at 620 nm reflects indeed the media absorption at the excitation wavelength. Therefore, the acceptor signal at 665 nm, which is also inversely proportional to the media absorbance, can be "ratioed" by the cryptate signal to yield a measurement which is then independent of the media optical

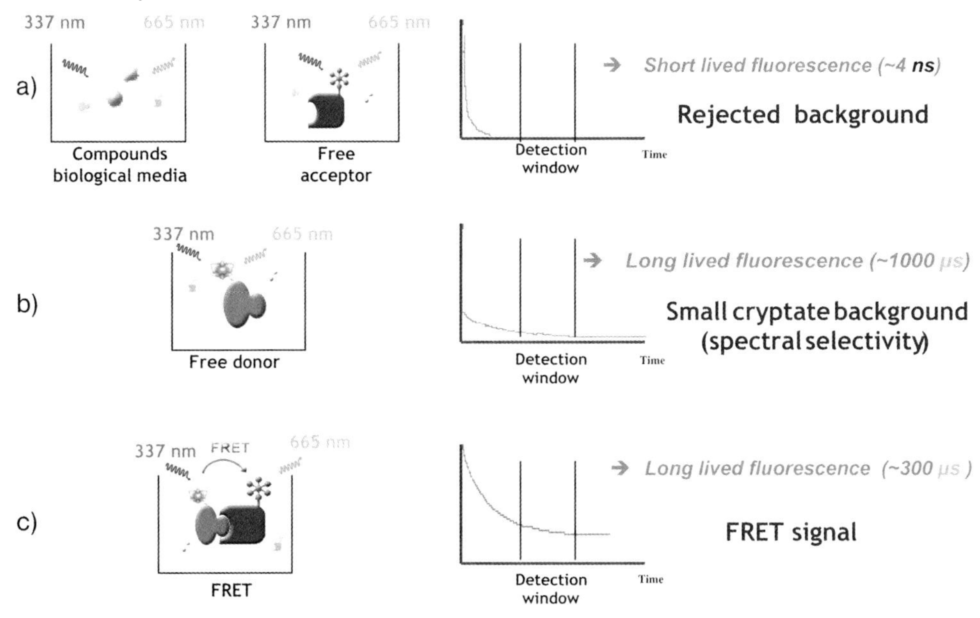

Fig. 13.7 Temporal selectivity of the TR-FRET signal. The Fig. shows as an example the temporal selectivity which can be obtained between europium cryptate and one of its acceptors. At the acceptor emission wavelength (665 nm), the FRET signal from the sensitized acceptor (C) and the small signal from the europium cryptate (B) could be easily discriminated from the free acceptor signal and from the background due to other compounds or the biological media (A).

properties at the excitation wavelength. The acceptor:europium cryptate signal ratio depends solely on the specific biological interactions under study (Mathis, 1993).

A large variety of biological events such as protein–protein interactions or enzymatic activities were monitored using HTRF with a very low detection limit (Bazin et al., 2002). HTRF is highly compatible with high-throughput measurements in 96-, 384-, or 1536-well plates (Ferrer et al., 2003). Until now, HTRF has not permitted the microscopic monitoring of biological interactions in living cells, but recent development in the design of a time-resolved microscope may overcome this drawback in the near future (Ghose et al., 2006).

13.6
New Developments in Fluorescent Labeling of Membrane Proteins

As mentioned above, one key limitation of the use of fusion proteins to label proteins in living cells is the size of the fluorescent (GFP variants) or luminescent (*R*luc) proteins. Experiments taking advantage of the long lifetime of europium as an energy donor are based on the use of labeled antibodies, making difficult the labeling of the intracellular side of the proteins. In addition, the large size of the antibodies, may also affect the normal functioning of the targeted proteins. To overcome these problems, new approaches have been developed for the specific labeling of membrane proteins in living cells.

An interesting approach first reported in 1998 by Tsien's group is based on the insertion in proteins of a tetra-cystein tag (Griffin et al., 1998) (the optimal sequences being HRWCCPGCCKFT or FLNCCPGCCMEP; Martin et al., 2005) that can be labeled with biarsenic derivatives such as 4′,5′-bis(1,2,3-dithioarsolan-2-yl)fluorescein (FlAsH) allowing the precise labeling of the protein with fluorescein (or rhodamine if ReAsH is used instead of FlAsH). One main advantage of this approach is to limit the size the fluorescent label, therefore limiting the effect on the functioning of the labeled protein. This is nicely illustrated with the labeling of the A2A-adenosine receptor in its intracellular loop 2 for example, as the insertion of a GFP variant prevented G-protein activation by the receptor, whereas insertion of a tetra-cystein tag did, not even after labeling with FlAsH (Hoffmann et al., 2005). However, this method can only be used on the intracellular side of the protein (Chen and Ting, 2005), and much care must be taken to maximally decrease the non-specific labeling of other cys-containing proteins using 1,2-ethanedithiol (Stroffekova et al., 2001).

Other approaches make use of specific binding sites for ligands that can be labeled with fluorophores, such as poly-his tags that bind Ni-nitrilotriacetate (NTA) fluorescent derivatives (Guignet et al., 2004; Ramanoudjame et al., 2006), or a minimal α-bungarotoxin binding sequence as short as 13 residues (Scherf et al., 1997), that can be labeled with fluorescent bungarotoxins (Sekine-Aizawa and Huganir, 2004; McCann et al., 2005).

Bacterial phosphopantetheine transferases known as Sfp (from *Bacillus subtilis*) have also been used recently to label cell-surface proteins in which a specific sequence (called the peptidyl carrier protein (PCP) tag, of 80 residues) was inserted. This enzyme transfers the 4′-phosphopantetheinyl group from coenzyme A (CoA) onto a conserved Ser of the tag, thereby allowing the insertion of any fluorophore fused to this group. A similar approach has also been developed taking advantage of the *Escherichia coli* Acyl carrier protein synthase (AcpS) that also transfers the same phosphopantetheinyl group of CoA on a minimal sequence (77 amino acids) of the Acyl carrier protein (ACP). These two methods can even be combined to insert two different fluorophores, taking advantage of the high selectivity of AcpS for the ACP tag, whereas Sfp can label both ACP and PCP tags. However, because the reaction requires the presence of an enzyme, this approach can only be used to label surface proteins on their extracellular side.

Enzymes that covalently react with their ligand have also been used as a method to label proteins with any type of fluorophore. One of such example is the Halo tag developed by Promega, that derives from a natural enzyme from *Rhodococcus rhodochrous*, and hydrolyzes haloalcanes into *n*-butanol. This enzyme is modified in its active site, such that the substrate can be labeled with various fluorophores to react and form a covalent bound with the enzyme. These substrates, in being able to pass the plasma membrane, can be used to label either extracellular or intracellular proteins, but the size of the Halo tag (33 kDa) is similar to that of GFPs. The second enzyme which has recently been developed to label protein is called "snap tag". This corresponds to the catalytic domain of the O_6-alkylguanine DNA alkyltransferase (AGT) that removes the alkyl groups from guanines in DNA. The reaction is such that the alkyl group is covalently linked to a cysteine from the active site. Of interest, this enzyme can also react with benzyl-guanine derivatives, even when labeled with any fluorophore on their benzyl moiety, allowing as such the covalent labeling of the enzyme. This enzyme is smaller than the GFPs (182 residues) and, depending on whether the benzyl-guanine derivatives used can penetrate the cells, can be employed either to label any cellular proteins or to label only those with an extracellular side.

13.7 Ligand–Receptor Interaction Monitored by FRET

A primary application of the energy transfer technology to study GPCRs is the characterization of ligand interaction with receptors. Galzi's group used various GPCRs fused to GFP at the extracellular N-terminus, and examined the binding of ligands labeled with Texas red or Bodipy. Such an approach allows rapid binding kinetic analysis and direct comparison of agonist binding and cellular responses (Vollmer et al., 1999; Palanche et al., 2001). These studies revealed the existence of at least two conformations of the neurokinin NK2 receptor. Indeed, the truncated neurokinin A (NKA)-(4-10) binds with a single phase and evokes only Ca-responses. In contrast, the full-length NKA exhibits both a rapid binding phase that correlates with Ca-responses, and a slower phase that correlates with cAMP formation (Palanche et al., 2001). A biphasic binding mode of PTH to its receptor was also observed using a similar FRET approach with a tetramethyl-rhodamine PTH and an N-terminally GFP-labeled receptor (Castro et al., 2005). In that case, the second phase in binding coincides with the conformational change in the PTH receptor, also monitored by FRET. This led the authors to propose a two-step binding mode for PTH on its receptor, the first step corresponding to the initial binding of the ligand, and the second to a different binding mode resulting from the conformational switch in the receptor. This is consistent with the conformational change that is associated with receptor activation and occurs sequentially after the ligand's first interaction with the receptor.

Because precise measurements of FRET efficiency allows the determination of distances between fluorophores, FRET between ligands and their receptor has also been used to obtain direct information on the conformation of a GPCR occupied with various ligands. For example, by using various cholecystokinin (CCK) receptor ligands labeled with Alexa488, and receptors labeled with Alexa568 at various positions at the level of a unique Cys residue, Harikumar and Miller (2005) provided evidence for a differential conformation between the agonist- and antagonist-occupied receptor.

13.8 Fast GPCR Activation Process Monitored in Living Cells

As indicated above, the efficiency of energy transfer depends on both the distance and the relative orientation of the fluorophore dipoles. Accordingly, conformational changes within membrane proteins can be monitored in real time when both the donor and the acceptor are fused within the same protein at specific positions. By inserting CFP and YFP at various position within the intracellular loops and C-terminal tail of GPCRs, Vilardaga et al. were the first to report real-time measurement of GPCR activation using FRET measurements in living cells (Vilardaga et al., 2003). When CFP was inserted in the second intracellular loop of the PTH receptor, and YFP at the extreme C-terminus, a significant decrease in FRET efficiency was observed upon PTH application, a phenomena that occurs with a fast kinetic ($\tau = 1\,s$), faster than the previously proposed kinetic of GPCR activation measured in purified, reconstituted system with the β_2 adrenergic receptor (Gether et al., 1995; Ghanouni et al., 2001). Even faster (25-fold) activation kinetic was measured with the α_2 adrenergic receptor using the same approach (Vilardaga et al., 2003), illustrating that this process may well be dependent upon the receptor type. By taking advantage of this FRET sensor directly inserted into the receptor, these authors also examined the mechanism of action of partial agonists. Their data are consistent with such ligand stabilizing a specific conformation of the receptor, and was different from that stabilized by a full agonist (Vilardaga et al., 2003) or an inverse agonist (Vilardaga et al., 2005).

However, criticism may be offered that these receptors have a decreased efficacy in G-protein coupling (as measured by the decrease in agonist potency at stimulating the intracellular cascade compared to the wild-type receptor), most likely due to the insertion of CFP in the intracellular loop of these receptors. Indeed, when the same approach was used with the adenosine A2A receptor (Hoffmann et al., 2005), or the metabotropic glutamate receptor (Tateyama et al., 2004), a complete loss of G-protein activation was observed. To overcome this problem, CFP in the third intracellular loop of the A2A receptor was replaced by a tetra-cysteine tag, and the receptor (also fused to CFP at its C-terminal end) was labeled with FlAsH in living cells (Hoffmann et al., 2005). The far smaller size of the tetra-cystein-FlAsH labeling compared to YFP, was then compatible with a normal coupling of the receptor to its G-protein partner. In addition, the use of FlAsH instead of YFP, confirmed the fast kinetic of receptor activation, and allowed a much better signal-to-noise ratio.

A CFP-YFP FRET approach was also used to monitor metabotropic glutamate (mGlu) receptor activation in living cells. The mGlu receptors are part of a specific class of GPCRs (called class C), characterized by a large extracellular region composed of a bilobate venus Flytrap (VFT) and a cysteine-rich domains, on top of the common 7TM core domain responsible for G-protein activation (Pin et al., 2003). These receptors are constitutive dimers linked by a disulfide bridge and, as illustrated by the crystal structure of the extracellular domain of the mGlu1 receptor (one of the eight subtypes of mGlu receptors), dimerization of these receptors is required for function (Kunishima et al., 2000; Tsuchiya et al., 2002). Indeed, agonist binding in the open form of the VFT results in the closure of the two lobes and, most importantly, to a relative movement of the two VFTs in the dimer. This relative movement was assumed to be associated with a concomitant movement of the two 7TM domains leading to G-protein activation (Kniazeff et al., 2004). To examine further this possibility, CFP and YFP were inserted in different parts of the mGlu receptors, either both in the same subunit, or each in two different subunits (Tateyama et al., 2004). Although no agonist-induced change in FRET was observed when both CFP and YFP were fused in the same subunit, a clear change in FRET was observed if YFP and CFP were introduced into each subunit. Of interest, depending on the relative position of YFP and CFP, either an increase or a decrease in FRET was observed, consistent with a relative movement of the two 7TM domains associated with receptor activation. However, this approach did not reveal conformational changes within each 7TM domain (but of course it did not exclude this likely possibility), and also the conclusions are limited by the inability of this fusion-receptor to activate G-proteins. Nonetheless, parallel experiments with wild-type receptors led these authors to propose recently that different signaling cascades (either Gs or Gq activation) may be activated by such dimeric receptors, depending on the relative position of the two subunits (Tateyama and Kubo, 2006).

13.9 FRET and BRET Validated the Constitutive Oligomerization of GPCR in Living Cells

For many years, GPCRs were assumed to function in monomeric forms, with one ligand activating one receptor that in turn activates one G-protein. However, such a view was challenged during the late 1980s with the first proposal that GPCRs may form dimers or oligomers, an idea then further supported by co-immunoprecipitation studies. In 2000, three studies brought direct evidence for a close proximity of GPCRs in living cells using different energy-transfer techniques. One study used a BRET approach with *R*luc and YFP-fused β_2 adrenergic receptor (Angers et al., 2000); the second study used a CFP-YFP FRET approach to demonstrate the dimeric nature of the yeast Ste2 receptor (Overton and Blumer, 2000); and the third used fluorophore labeled antibodies directed to N-terminal epitopes inserted in the somatostatin and dopamine receptors to show their possible heteromeric assembly (Rocheville et al., 2000). Since then, a number of studies have used these techniques to illustrate how wide this phenomenon is, with almost any GPCR examined so far being reported to oligomerize in living cells (Bouvier, 2001; Milligan and Bouvier, 2005; Pfleger and Eidne, 2005). Such oligomerization measured by FRET or BRET was always found specific, being observed between specific combination of receptors, but not with others.

One limitation of the use of FRET and BRET to monitor membrane protein interaction is that both surface receptor-fusion, as well as intracellular receptor oligomers (located either into endosomes, endoplasmic reticulum or in various types of vesicle during their trafficking to the surface) are detected. To firmly demonstrate that GPCRs dimers are indeed found at the cell surface, advantage may be taken of total internal reflection fluorescence, as used to monitor mGlu dimer at the cell surface (Tateyama et al., 2004). Alternatively, antibodies may be used to specifically label surface proteins. This was achieved using antibodies directed against epitopes fused to the N-terminal extracellular end of the receptors, labeled either with rhodamine and fluorescein (Rocheville et al., 2000; Patel et al., 2002a), or with europium cryptate or chelate as donor fluorophore, and XL665, alexa647 or d2 as acceptors allowing time-resolved FRET measurements (McVey et al., 2001; Kniazeff et al., 2004; Maurel et al., 2004; Urizar et al., 2005).

The second limitation is that these experiments are often conducted with recombinant fusion proteins over-expressed in heterologous expression system. However, when such experiments were conducted at receptor expression level similar to those found in native tissue, energy transfer was still observed, demonstrating that at least such signal is not only the consequence of a high receptor density (Mercier et al., 2002). Of interest, by measuring BRET signal in cells transfected with a fixed amount of β_2 adrenergic receptor-*R*luc, and increasing the amount of the same receptor fused to YFP ratio, saturation curves were obtained which allowed estimation of the apparent affinity of the two protomers (Mercier et al., 2002; Ayoub et al., 2004). This has then been widely used to examine the possible preferential association between various types of GPCR (Mercier et al., 2002; Wang et al., 2005). This revealed that β_1 and β_2 adrenergic receptors have the same ability to form homodimers or heterodimers. The same approach revealed that the melatonin MT1 receptor has the same ability to homodimerize or heterodimerize with MT2, whereas the latter displays a higher apparent affinity for MT1 than for itself (Ayoub et al., 2004). This supports the importance of the MT1/MT2 heterodimer in melatonin-mediated physiological effects.

Both, FRET and BRET assays were also used to examine whether dimer formation would be regulated by different types of ligand acting on each protomer. In some cases, changes in FRET or BRET signals were observed upon agonist application, and this was first proposed to reflect agonist-induced formation or stabilization of GPCR dimers (Patel et al., 2002a; Grant et al., 2004). However, such variation in energy transfer efficiency may also be the consequence of conformational changes within stable dimers, resulting in a change in either the distance or the relative orientation between the fluorophores. Indeed, a number of additional studies are more consistent with this second explanation. First, Western blot and co-immunoprecipitation commonly used to study GPCR dimerization did not show changes in monomer/dimer ratio after ligand application. Second, in most cases, ligand interaction with the receptor dimer did not result in a change in BRET efficiency (Angers et al., 2002; Pfleger and Eidne, 2005). Moreover, when changes are observed, these do not necessarily relate with the activity of the ligands, agonists and inverse agonists inducing similar changes, and depend on the receptor dimer being studied (Ayoub et al., 2002; Percherancier et al., 2005). This is also nicely illustrated with the covalently linked mGlu dimers in which an increase or a decrease in FRET efficiency is observed upon agonist binding, depending on the insertion sites of the CFP and YFP (Tateyama et al., 2004).

Taken together, and in agreement with many other approaches, energy transfer technology brings strong evidence for the existence of constitutive GPCR dimers/oligomers in living cells.

13.10 FRET and BRET Changed the Concept of G-Protein Activation

Since the discovery of heterotrimeric G-proteins as essential transducers of GPCRs, it has become well established that agonist activation of a GPCR allows its association to an heterotrimeric G-protein, and the release of GDP from the α subunit. Binding of GTP in the empty G-protein then induces a rapid dissociation of the receptor–G-protein complex, thereby allowing GαGTP on one side, and βγ on the other side to reach their own effectors. Although consistent with the rapid activation of a large number of G-proteins by a single activated receptor, this association–dissociation model was questioned more than ten years ago as it was difficult to reconcile with the high selectivity between GPCRs and certain G-proteins observed in native cells, and with the fast kinetic (Neubig, 1994).

Within the past five years a number of studies have examined the activation process of heterotrimeric G-proteins by various GPCRs, using either FRET or BRET. Because of the commonly accepted concept of G-protein dissociation resulting from GTP binding, energy transfer technologies soon appeared as a powerful way to monitor G-protein activation. This was first studied in *Dictyostelium discoideum* and, as expected, a decrease in FRET between α and βγ was interpreted as a direct evidence for G-protein dissociation in living cells (Janetopoulos et al., 2001). Such a decrease in FRET between Gα and βγ was also observed for the Go type of G-protein (Azpiazu and Gautam, 2004; Frank et al., 2005) and the yeast G-protein (Yi et al., 2003). However, other studies based on BRET or CFP-YFP FRET revealed a constant proximity between Gα and βγ, at least with Gαi and Gαs (Bunemann et al., 2003; Gales et al., 2005, 2006), with relative movement of βγ and α associated with activation. This is nicely illustrated by the observation that either an increase or a decrease in the energy transfer efficiency occurs during activation, depending on the insertion point of the CFP, YFP or *R*luc (Bunemann et al., 2003; Gales et al., 2005, 2006; Gibson and Gilman, 2006). It has therefore been proposed that the previously reported decrease in FRET may not necessarily reflect Gαβγ dissociation, but rather a relative movement between the two main components of the G-protein. Such a proposal is supported by the recent report that Gαq and βγ are still in proximity after binding of GRK2, a kinase involved in the desensitization of GPCRs, as observed in the crystal structure of the Gq-GRK2 complex (Tesmer et al., 2005).

G-protein activation was also examined by measuring FRET or BRET between the activating receptor an either Gα or $\beta\gamma$. One study examined the coupling of α2A adrenergic receptor and Gs, and the appearance of FRET after receptor activation, a signal that is largely increased when a non-dissociating mutant of Gαs is used, supporting the current model of the G-protein interacting transiently with the activated receptor only (the collision model). However, three other reports were not consistent with this model, and proposed a pre-association of the receptor and the G-protein, and the absence of complete dissociation of the G-protein from the receptor after activation. This was revealed by the large BRET signal measured under basal condition between the β_2 and α_2 adrenergic receptors and either $\beta\gamma$ and Gαi, respectively (Gales et al., 2006), or between the thrombin receptor PAR1 and Gαi (Ayoub et al., 2007). Such a basal BRET does not reflects some constitutive activity of the receptor, as it was not inhibited by pertussis toxin treatment, in contrast to the agonist-induced effect. Moreover, depending on the insertion point of *R*luc in the G-protein, either a rapid increase or decrease in BRET was observed after receptor activation. This later point is a clear indication that the change in BRET is not simply the consequence of G-protein recruitment by the activated receptor, but is rather due to a change in the relative position of these two partners. Nobles et al. (2005) examined the coupling between various GPCRs (M4 muscarinic, α_2A adrenergic, D_2 dopaminergic and adenosine A_1 receptors) and Gαo using FRET, and also brought evidence for a pre-association of these two partners. These data were therefore more consistent with the pre-assembly theory, the G-protein and the receptor being already part of the same protein complex, allowing a faster process, and a better control of the selectivity of the signaling cascades activated by the receptor. It has even been proposed that such receptor–G-protein complexes are formed in the endoplasmic reticulum (Dupre et al., 2006). Although in contradiction with the commonly accepted model of receptor–G-protein coupling, this proposal is consistent with the large number of reports showing that Gα and the receptor can be fused into a single polypeptide chain without affecting their normal functioning in cells (Hildebrandt, 2006).

13.11
GPCRs as Part of Large Signaling Complexes

Within the past few years, a number of biochemical and proteomic approaches have suggested that membrane receptors are likely assembled into large signaling complexes that include the receptor, the transducer, and the effector proteins. This possibility has been examined recently using energy-transfer technologies in living cells. It was mentioned earlier that G-proteins may be part of such signaling complexes (Dupre et al., 2006), but G-proteins have also been shown to be pre-assembled with their effectors. This was recently nicely illustrated for the G$\beta\gamma$ complex associated with the G-protein-regulated Inwardly Rectifying K-channels

(GIRK) (Riven et al., 2006). Indeed, as observed with the GPCR–G-protein interaction using FRET or BRET, a basal FRET is observed between Gβγ and GIRK and, upon activation with a GPCR, either an increase or a decrease in FRET efficiency is observed depending on the insertion points of the YFP and CFPs. This conclusion is supported by a different study based on BRET (Rebois et al., 2006). Taking into account that other studies report a co-assembly of GPCRs such as the $GABA_B$ receptor and GIRK channels (David et al., 2006), this is consistent with all components of this signaling cascade being co-assembled in living cells. Similarly, G-proteins and GPCRs have been shown the co-assemble with adenylyl cyclase (Rebois et al., 2006; Dupre et al., 2007), and Gαq was found recently to associate stably with its effector PLCβ (Dowal et al., 2006). Regulators of G-protein signaling (RGS proteins) that accelerate the GTPase activity, have also been found to be part of such putative pre-assembled complexes (Benians et al., 2005). If validated, such a proposal may provide an effective means of explaining the specificity of the signaling cascades, and will likely play an important role in determining the kinetic of the response.

However, it may prove to be surprising that, when such interactions were examined in living cells using FRET or BRET approaches, often a constitutive association of the partners is reported. Although in each of these studies negative controls are often presented, the demonstration of the β-arrestin recruitment by activated GPCRs is a good example that such pre-assembly is not always the case. Indeed, β-arrestin is cytosolic and reaches the cell membrane upon interaction with an activated and phosphorylated GPCR. This was visualized at an early stage in living cells using GFP-fused β-arrestin constructs (Barak et al., 1997), and the recruitment by GPCRs can be directly followed using either BRET (Bertrand et al., 2002) or FRET (Krasel et al., 2005), with no evidence of any pre-association between these two partners.

13.12
Conclusion and Future Prospects

The use of energy transfer technologies has allowed the direct "visualization" of GPCR functioning in living cells. The data generated have changed the commonly held views on the functioning of these receptors, the most important finding being the possibility of recording such GPCR-mediated responses in real time, and in living cells. These studies have also brought new evidence that these receptors are likely pre-assembled in specific signaling complexes, thereby allowing rapid and specific cellular responses. However, to date most studies have been performed using either recombinant fusion proteins, or using labeled antibodies directed against epitope tags. The use of such large domains carrying the requested chromophores represents a major limitation to the use of FRET studies. However, the development of alternative labeling techniques and of new fluorophores has begun to open new routes, including the possibility of measuring such signals in living animals (So et al., 2006).

Summary

Discovered during the late 1920s, resonance energy transfer technologies allow the proximity between two fluorophores to be monitored. These technologies are now widely used to monitor protein–protein interactions, as well as conformational changes of proteins in living cells. A key step in the development of these approaches was the discovery of fluorescent proteins in the mid 1990s, such as the green fluorescent protein (GFP) and its derivatives with different spectral properties. Since then, other protein-labeling techniques have been developed, that offer multiple alternatives for energy transfer measurements. In this chapter, the principle of resonance energy transfer will first be described, followed by some commonly used methods that allow such measurements in living cells. In addition, recent developments in protein-labeling techniques will be outlined. Finally, details of recent studies which utilize these methodologies and are changing current views of how G-protein-coupled receptors transmit signals inside the cells, are discussed.

References

Alpha B., Lehn J. M. and Mathis G. (1987) Energy transfer of Eu(III) and Tb(III) cryptates of macrobicyclic polypyridine ligands. *Angew. Chem. Int. Ed. Engl.*, **26**: 266–267.

Alpha B., Ballardini R., Balzani V., Lehn J. M., Perathoner S. and Sabbatini N. (1990) Antenna effect in luminescent Eu(III) and Tb(III) cryptates. A photophysical study. *Photochem. Photobiol.*, **52**: 299–306.

Angers S., Salahpour A., Joly E., Hilairet S., Chelsky D., Dennis M. and Bouvier M. (2000) Detection of beta 2-adrenergic receptor dimerization in living cells using bioluminescence resonance energy transfer (BRET). *Proc. Natl. Acad. Sci. USA*, **97**: 3684–3689.

Angers S., Salahpour A. and Bouvier M. (2002) Dimerization: an emerging concept for G protein-coupled receptor ontogeny and function. *Annu. Rev. Pharmacol. Toxicol.*, **42**: 409–435.

Ayoub M. A., Couturier C., Lucas-Meunier E., Angers S., Fossier P., Bouvier M. and Jockers R. (2002) Monitoring of ligand-independent dimerization and ligand-induced conformational changes of melatonin receptors in living cells by bioluminescence resonance energy transfer. *J. Biol. Chem.*, **277**: 21522–21528.

Ayoub M. A., Levoye A., Delagrange P. and Jockers R. (2004) Preferential formation of MT1/MT2 melatonin receptor heterodimers with distinct ligand interaction properties compared with MT2 homodimers. *Mol. Pharmacol.*, **66**: 312–321.

Ayoub M. A., Maurel D., Binet V., Prézeau L., Ansanay H. and Pin J.-P. (2007) Real-time analysis of agonist-induced activation of protease-activated receptor 1/Gαi protein complex measured by BRET in living cells. *Mol. Pharmacol.*, **71**: 1329–1340.

Azpiazu I. and Gautam N. (2004) A fluorescence resonance energy transfer-based sensor indicates that receptor access to a G protein is unrestricted in a living mammalian cell. *J. Biol. Chem.*, **279**: 27709–27718.

Barak L. S., Ferguson S. S., Zhang J. and Caron M. G. (1997) A beta-arrestin/green fluorescent protein biosensor for detecting G protein-coupled receptor activation. *J. Biol. Chem.*, **272**: 27497–27500.

Bazin H., Trinquet E. and Mathis G. (2002) Time resolved amplification of cryptate emission: a versatile technology to trace biomolecular interactions. *J. Biotechnol.*, **82**: 233–250.

Benians A., Nobles M., Hosny S. and Tinker A. (2005) Regulators of G-protein signaling form a quaternary complex with the agonist, receptor, and G-protein. A novel explanation for the acceleration of signaling activation kinetics. *J. Biol. Chem.*, **280**: 13383–13394.

Bertrand L., Parent S., Caron M., Legault M., Joly E., Angers S., Bouvier M., Brown M., Houle B. and Menard L. (2002) The BRET2/arrestin assay in stable recombinant cells: a platform to screen for compounds that interact with G protein-coupled receptors (GPCRS). *J. Recept. Signal Transduct. Res.*, **22**: 533–541.

Bockaert J. and Pin J.-P. (1999) Molecular tinkering of G-protein coupled receptors: an evolutionary success. *EMBO J.*, **18**: 1723–1729.

Bouvier M. (2001) Oligomerization of G-protein-coupled transmitter receptors. *Nat. Rev. Neurosci.*, **2**: 274–286.

Bunemann M., Frank M. and Lohse M. J. (2003) Gi protein activation in intact cells involves subunit rearrangement rather than dissociation. *Proc. Natl. Acad. Sci. USA*, **100**: 16077–16082.

Castro M., Nikolaev V. O., Palm D., Lohse M. J. and Vilardaga J. P. (2005) Turn-on switch in parathyroid hormone receptor by a two-step parathyroid hormone binding mechanism. *Proc. Natl. Acad. Sci. USA*, **102**. 16084–16089.

Chan S. S., Arndt-Jovin D. J. and Jovin T. M. (1979) Proximity of lectin receptors on the cell surface measured by fluorescence energy transfer in a flow system. *J. Histochem. Cytochem.*, **27**: 56–64.

Chen I. and Ting A. Y. (2005) Site-specific labeling of proteins with small molecules in live cells. *Curr. Opin. Biotechnol.*, **16**: 35–40.

Dale R. E., Eisinger J. and Blumberg W. E. (1979) The orientational freedom of molecular probes. The orientation factor in intramolecular energy transfer. *Biophys. J.*, **26**: 161–193.

David M., Richer M., Mamarbachi A. M., Villeneuve L. R., Dupre D. J. and Hebert T. E. (2006) Interactions between GABA-B(1) receptors and Kir 3 inwardly rectifying potassium channels. *Cell Signal.* **18**: 2172–2181.

De A. and Gambhir S. S. (2005) Noninvasive imaging of protein-protein interactions from live cells and living subjects using bioluminescence resonance energy transfer. *FASEB J.*, **19**: 2017–2019.

Dowal L., Provitera P. and Scarlata S. (2006) Stable association between Galpha (q) and phospholipase Cbeta 1 in living cells. *J. Biol. Chem.*, **5**: 5.

Drews J. (2000) Drug discovery: a historical perspective. *Science*, **287**: 1960–1964.

Dupre D. J., Baragli A., Rebois R. V., Ethier N. and Hebert T. E. (2007) Signalling complexes associated with adenylyl cyclase II are assembled during their biosynthesis. *Cell Signal.* **19**: 481–489.

Dupre D. J., Robitaille M., Ethier N., Villeneuve L. R., Mamarbachi A. M. and Hebert T. E. (2006) Seven transmembrane receptor core signalling complexes are assembled prior to plasma membrane trafficking. *J. Biol. Chem.* **281**: 34561–34573.

Ferrer M., Zuck P., Kolodin G., Mao S. S., Peltier R. R., Bailey C., Gardell S. J., Strulovici B. and Inglese J. (2003) Miniaturizable homogenous time-resolved fluorescence assay for carboxypeptidase B activity. *Anal. Biochem.*, **317**: 94–98.

Förster T. (1948) Intermolecular energy migration and fluorescence. *Ann. Phys.*, **2**: 55–75.

Frank M., Thumer L., Lohse M. J. and Bunemann M. (2005) G Protein Activation without Subunit Dissociation Depends on a G{alpha}i-specific Region. *J. Biol. Chem.*, **280**: 24584–24590.

Gales C., Rebois R. V., Hogue M., Trieu P., Breit A., Hebert T. E. and Bouvier M. (2005) Real-time monitoring of receptor and G-protein interactions in living cells. *Nat. Methods*, **2**: 177–184.

Gales C., Van Durm J. J., Schaak S., Pontier S., Percherancier Y., Audet M., Paris H. and Bouvier M. (2006) Probing the activation-promoted structural rearrangements in preassembled receptor-G protein complexes. *Nat. Struct. Mol. Biol.*, **13**: 778–786.

Gether U., Lin S. and Kobilka B. K. (1995) Fluorescent labeling of purified beta 2 adrenergic receptor. Evidence for ligand-specific conformational changes. *J. Biol. Chem.*, **270**: 28268–28275.

Ghanouni P., Steenhuis J. J., Farrens D. L. and Kobilka B. K. (2001) Agonist-induced conformational changes in the G-protein-coupling domain of the beta 2 adrenergic receptor. *Proc. Natl. Acad. Sci. USA*, **98**: 5997–6002.

Ghose S., Trinquet E., Laget M., Bazin H. and Mathis G. (2006) Rare earth cryptates for the investigation of molecular interactions in vitro and in living cells. *J. Alloys and Compounds* (in press).

Gibson S. K. and Gilman A. G. (2006) Gialpha and Gbeta subunits both define selectivity of G protein activation by alpha2-adrenergic receptors. *Proc. Natl. Acad. Sci. USA*, **103**: 212–217.

Grant M., Collier B. and Kumar U. (2004) Agonist-dependent dissociation of human somatostatin receptor 2 dimers: a role in receptor trafficking. *J. Biol. Chem.*, **279**: 36179–36183.

Griffin B. A., Adams S. R. and Tsien R. Y. (1998) Specific covalent labeling of recombinant protein molecules inside live cells. *Science*, **281**: 269–272.

Guignet E. G., Hovius R. and Vogel H. (2004) Reversible site-selective labeling of membrane proteins in live cells. *Nat. Biotechnol.*, **22**: 440–444.

Harikumar K. G. and Miller L. J. (2005) Fluorescence resonance energy transfer analysis of the antagonist- and partial agonist-occupied states of the cholecystokinin receptor. *J. Biol. Chem.*, **280**: 18631–18635.

He L., Bradrick T. D., Karpova T. S., Wu X., Fox M. H., Fischer R., McNally J. G., Knutson J. R., Grammer A. C. and Lipsky P. E. (2003) Flow cytometric measurement of fluorescence (Forster) resonance energy transfer from cyan fluorescent protein to yellow fluorescent protein using single-laser excitation at 458 nm. *Cytometry A*, **53**: 39–54.

Hemmila I., Dakubu S., Mukkala V. M., Siitari H. and Lovgren T. (1984) Europium as a label in time-resolved immunofluorometric assays. *Anal. Biochem.*, **137**: 335–343.

Hildebrandt J. D. (2006) Bring your own G protein. *Mol. Pharmacol.*, **69**: 1079–1082.

Hoffmann C., Gaietta G., Bunemann M., Adams S. R., Oberdorff-Maass S., Behr B., Vilardaga J. P., Tsien R. Y., Ellisman M. H. and Lohse M. J. (2005) A FlAsH-based FRET approach to determine G protein-coupled receptor activation in living cells. *Nat. Methods*, **2**: 171–176.

Janetopoulos C., Jin T. and Devreotes P. (2001) Receptor-mediated activation of heterotrimeric G-proteins in living cells. *Science*, **291**: 2408–2411.

Kniazeff J., Bessis A.-S., Maurel D., Ansanay H., Prezeau L. and Pin J.-P. (2004) Closed state of both binding domains of homodimeric mGlu receptors is required for full activity. *Nat. Struct. Mol. Biol.*, **11**: 706–713.

Krasel C., Bunemann M., Lorenz K. and Lohse M. J. (2005) Beta-arrestin binding to the beta2-adrenergic receptor requires both receptor phosphorylation and receptor activation. *J. Biol. Chem.*, **280**: 9528–9535.

Kunishima N., Shimada Y., Tsuji Y., Sato T., Yamamoto M., Kumasaka T., Nakanishi S., Jingami H. and Morikawa K. (2000) Structural basis of glutamate recognition by a dimeric metabotropic glutamate receptor. *Nature*, **407**: 971–977.

Lorenz W. W., McCann R. O., Longiaru M. and Cormier M. J. (1991) Isolation and expression of a cDNA encoding *Renilla reniformis* luciferase. *Proc. Natl. Acad. Sci. USA*, **88**: 4438–4442.

Marshall J., Molloy R., Moss G. W., Howe J. R. and Hughes T. E. (1995) The jellyfish green fluorescent protein: a new tool for studying ion channel expression and function. *Neuron*, **14**: 211–215.

Martin B. R., Giepmans B. N., Adams S. R. and Tsien R. Y. (2005) Mammalian cell-based optimization of the biarsenical-binding tetracysteine motif for improved fluorescence and affinity. *Nat. Biotechnol.*, **23**: 1308–1314.

Mathis G. (1993) Rare earth cryptates and homogeneous fluoroimmunoassays with human sera. *Clin. Chem.*, **39**: 1953–1959.

Mathis G. (1995) Probing molecular interactions with homogeneous techniques based on rare earth cryptates and fluorescence energy transfer. *Clin. Chem.*, **41**: 1391–1397.

Maurel D., Kniazeff J., Mathis G., Trinquet E., Pin J.-P. and Ansanay H. (2004) Cell surface detection of membrane protein interaction with homogeneous time-resolved fluorescence resonance energy transfer technology. *Anal. Biochem.*, **329**: 253–262.

McCann C. M., Bareyre F. M., Lichtman J. W. and Sanes J. R. (2005) Peptide tags for labeling membrane proteins in live cells with multiple fluorophores. *Biotechniques*, **38**: 945–952.

McVey M., Ramsay D., Kellett E., Rees S., Wilson S., Pope A. J. and Milligan G. (2001) Monitoring receptor oligomerization using time-resolved fluorescence resonance energy transfer and bioluminescence resonance energy transfer. The human delta-opioid receptor displays constitutive oligomerization at the cell surface, which is not regulated by receptor occupancy. *J. Biol. Chem.*, **276**: 14092–14099.

Mercier J. F., Salahpour A., Angers S., Breit A. and Bouvier M. (2002) Quantitative assessment of beta 1- and beta 2-adrenergic receptor homo- and heterodimerization by bioluminescence resonance energy transfer. *J. Biol. Chem.*, **277**: 44925–44931.

Milligan G. and Bouvier M. (2005) Methods to monitor the quaternary structure of G protein-coupled receptors. *FEBS J.*, **272**: 2914–2925.

Miyawaki A., Llopis J., Heim R., McCaffery J. M., Adams J. A., Ikura M. and Tsien R. Y. (1997) Fluorescent indicators for Ca2+ based on green fluorescent proteins and calmodulin. *Nature*, **388**: 882–887.

Morin J. G. and Hastings J. W. (1971) Energy transfer in a bioluminescent system. *J. Cell Physiol.*, **77**: 313–318.

Morrison L. E. (1988) Time-resolved detection of energy transfer: theory and application to immunoassays. *Anal. Biochem.*, **174**: 101–120.

Nagai T., Ibata K., Park E. S., Kubota M., Mikoshiba K. and Miyawaki A. (2002) A variant of yellow fluorescent protein with fast and efficient maturation for cell-biological applications. *Nat. Biotechnol.*, **20**: 87–90.

Neubig R. R. (1994) Membrane organization in G-protein mechanisms. *FASEB J.*, **8**: 939–946.

Nobles M., Benians A. and Tinker A. (2005) Heterotrimeric G proteins precouple with G protein-coupled receptors in living cells. *Proc. Natl. Acad. Sci. USA*, **102**: 18706–18711.

Overton M. C. and Blumer K. J. (2000) G-protein-coupled receptors function as oligomers in vivo. *Curr. Biol.*, **10**: 341–344.

Palanche T., Ilien B., Zoffmann S., Reck M. P., Bucher B., Edelstein S. J. and Galzi J. L. (2001) The neurokinin A receptor activates calcium and cAMP responses through distinct conformational states. *J. Biol. Chem.*, **276**: 34853–34861.

Patel R. C., Kumar U., Lamb D. C., Eid J. S., Rocheville M., Grant M., Rani A., Hazlett T., Patel S. C., Gratton E. and Patel Y. C. (2002a) Ligand binding to somatostatin receptors induces receptor-specific oligomer formation in live cells. *Proc. Natl. Acad. Sci. USA*, **99**: 3294–3299.

Patel R. C., Lange D. C. and Patel Y. C. (2002b) Photobleaching fluorescence resonance energy transfer reveals ligand-induced oligomer formation of human somatostatin receptor subtypes. *Methods*, **27**: 340–348.

Percherancier Y., Berchiche Y. A., Slight I., Volkmer-Engert R., Tamamura H., Fujii N., Bouvier M. and Heveker N. (2005) Bioluminescence resonance energy transfer reveals ligand-induced conformational changes in CXCR4 homo- and heterodimers. *J. Biol. Chem.*, **280**: 9895–9903.

Perroy J., Pontier S., Charest P. G., Aubry M. and Bouvier M. (2004) Real-time monitoring of ubiquitination in living cells by BRET. *Nat. Methods*, **1**: 203–208.

Pfleger K. D. and Eidne K. A. (2005) Monitoring the formation of dynamic G-protein-coupled receptor-protein complexes in living cells. *Biochem. J.*, **385**: 625–637.

Pfleger K. D. and Eidne K. A. (2006) Illuminating insights into protein-protein interactions using bioluminescence resonance energy transfer (BRET). *Nat. Methods*, **3**: 165–174.

Pin J.-P., Galvez T. and Prézeau L. (2003) Evolution, structure and activation mechanism of family 3/C G-protein coupled receptors. *Pharmacol. Ther.*, **98**: 325–354.

Prasher D. C., Eckenrode V. K., Ward W. W., Prendergast F. G. and Cormier M. J. (1992) Primary structure of the *Aequorea victoria* green-fluorescent protein. *Gene,* **111**: 229–233.

Prendergast F. G. (2000) Bioluminescence illuminated. *Nature,* **405**: 291–293.

Prinz A., Diskar M. and Herberg F. W. (2006) Application of bioluminescence resonance energy transfer (BRET) for biomolecular interaction studies. *Chembiochem,* **7**: 1007–1012.

Ramanoudjame G., Du M., Mankiewicz K. A. and Jayaraman V. (2006) Allosteric mechanism in AMPA receptors: a FRET-based investigation of conformational changes. *Proc. Natl. Acad. Sci. USA,* **103**: 10473–10478.

Ramsay D., Kellett E., McVey M., Rees S. and Milligan G. (2002) Homo- and hetero-oligomeric interactions between G protein-coupled receptors in living cells monitored by two variants of bioluminescence resonance energy transfer. Hetero-oligomers between receptor subtypes form more efficiency than between less closely related sequences. *Biochem. J.*, **365**: 429–440.

Rebois R. V., Robitaille M., Gales C., Dupre D. J., Baragli A., Trieu P., Ethier N., Bouvier M. and Hebert T. E. (2006) Heterotrimeric G proteins form stable complexes with adenylyl cyclase and Kir3.1 channels in living cells. *J. Cell Sci.*, **119**: 2807–2818.

Riven I., Iwanir S. and Reuveny E. (2006) GIRK channel activation involves a local rearrangement of a preformed G protein channel complex. *Neuron,* **51**: 561–573.

Rizzo M. A., Springer G. H., Granada B. and Piston D. W. (2004) An improved cyan fluorescent protein variant useful for FRET. *Nat. Biotechnol.*, **22**: 445–449.

Rocheville M., Lange D. C., Kumar U., Patel S. C., Patel R. C. and Patel Y. C. (2000) Receptors for dopamine and somatostatin: formation of hetero-oligomers with enhanced functional activity. *Science,* **288**: 154–157.

Scherf T., Balass M., Fuchs S., Katchalski-Katzir E. and Anglister J. (1997) Three-dimensional solution structure of the complex of alpha-bungarotoxin with a library-derived peptide. *Proc. Natl. Acad. Sci. USA,* **94**: 6059–6064.

Schiller P. W. (1976) The measurement of intramolecular distances by energy transfer. In: Chem F. C. and Edeldoch H. (Eds.), *Biochemical Fluorescence Concepts.* M. Dekker, New York, Vol. 1, pp. 285–303.

Schlyer S. and Horuk R. (2006) I want a new drug: G-protein-coupled receptors in drug development. *Drug Discov. Today,* **11**: 481–493.

Sekine-Aizawa Y. and Huganir R. L. (2004) Imaging of receptor trafficking by using alpha-bungarotoxin-binding-site-tagged receptors. *Proc. Natl. Acad. Sci. USA,* **101**: 17114–17119.

Selvin P. R. (2000) The renaissance of fluorescence resonance energy transfer. *Nat. Struct. Biol.*, **7**: 730–734.

Shimomura O., Johnson F. H. and Saiga Y. (1962) Extraction, purification and properties of aequorin, a bioluminescent protein from the luminous hydromedusan, Aequorea. *J. Cell. Comp. Physiol.*, **59**: 223–239.

So M. K., Xu C., Loening A. M., Gambhir S. S. and Rao J. (2006) Self-illuminating quantum dot conjugates for in vivo imaging. *Nat. Biotechnol.*, **24**: 339–343.

Sokol D. L., Zhang X., Lu P. and Gewirtz A. M. (1998) Real time detection of DNA : RNA hybridization in living cells. *Proc. Natl. Acad. Sci. USA,* **95**: 11538–11543.

Stroffekova K., Proenza C. and Beam K. G. (2001) The protein-labeling reagent FLASH-EDT2 binds not only to CCXXCC motifs but also non-specifically to endogenous cysteine-rich proteins. *Pflugers Arch.*, **442**: 859–866.

Stryer L. (1978) Fluorescence energy transfer as a spectroscopic ruler. *Annu. Rev. Biochem.*, **47**: 819–846.

Stryer L. and Haugland R. P. (1967) Energy transfer: a spectroscopic ruler. *Proc. Natl. Acad. Sci. USA,* **58**: 719–726.

Tateyama M. and Kubo Y. (2006) Dual signaling is differentially activated by different active states of the metabotropic glutamate receptor 1alpha. *Proc. Natl. Acad. Sci. USA*, **103**: 1124–1128.

Tateyama M., Abe H., Nakata H., Saito O. and Kubo Y. (2004) Ligand-induced rearrangement of the dimeric metabotropic glutamate receptor 1alpha. *Nat. Struct. Mol. Biol.*, **11**: 637–642.

Tesmer V. M., Kawano T., Shankaranarayanan A., Kozasa T. and Tesmer J. J. (2005) Snapshot of activated G proteins at the membrane: the Galphaq-GRK2-Gbetagamma complex. *Science*, **310**: 1686–1690.

Tsien R. Y. (1998) The green fluorescent protein. *Annu. Rev. Biochem.*, **67**: 509–544.

Tsuchiya D., Kunishima N., Kamiya N., Jingami H. and Morikawa K. (2002) Structural views of the ligand-binding cores of a metabotropic glutamate receptor complexed with an antagonist and both glutamate and Gd^{3+}. *Proc. Natl. Acad. Sci. USA*, **99**: 2660–2665.

Urizar E., Montanelli L., Loy T., Bonomi M., Swillens S., Gales C., Bouvier M., Smits G., Vassart G. and Costagliola S. (2005) Glycoprotein hormone receptors: link between receptor homodimerization and negative cooperativity. *EMBO J.*, **24**: 1954–1964.

Vilardaga J. P., Bunemann M., Krasel C., Castro M. and Lohse M. J. (2003) Measurement of the millisecond activation switch of G protein-coupled receptors in living cells. *Nat. Biotechnol.*, **21**: 807–812.

Vilardaga J. P., Steinmeyer R., Harms G. S. and Lohse M. J. (2005) Molecular basis of inverse agonism in a G protein-coupled receptor. *Nat. Chem. Biol.*, **1**: 25–28.

Vogel S. S., Thaler C. and Koushik S. V. (2006) Fanciful FRET. *Sci. STKE*, **2006**: re2.

Vollmer J. Y., Alix P., Chollet A., Takeda K. and Galzi J. L. (1999) Subcellular compartmentalization of activation and desensitization of responses mediated by NK2 neurokinin receptors. *J. Biol. Chem.*, **274**: 37915–37922.

Wallrabe H. and Periasamy A. (2005) Imaging protein molecules using FRET and FLIM microscopy. *Curr. Opin. Biotechnol.*, **16**: 19–27.

Wang D., Sun X., Bohn L. M. and Sadee W. (2005) Opioid receptor homo- and heterodimerization in living cells by quantitative bioluminescence resonance energy transfer. *Mol. Pharmacol.*, **67**: 2173–2184.

Xu Y., Piston D. W. and Johnson C. H. (1999) A bioluminescence resonance energy transfer (BRET) system: application to interacting circadian clock proteins. *Proc. Natl. Acad. Sci. USA*, **96**: 151–156.

Yi T. M., Kitano H. and Simon M. I. (2003) A quantitative characterization of the yeast heterotrimeric G protein cycle. *Proc. Natl. Acad. Sci. USA*, **100**: 10764–10769.

Index

Biophysical Analysis of Membrane Proteins. Investigating Structure and Function. Edited by Eva Pebay-Peyroula

ISBN: 978-3-527-31677-9

n